Birkhäuser

Applied and Numerical Harmonic Analysis

More information about this series at http://www.springer.com/series/4968

Elena Prestini

The Evolution of Applied Harmonic Analysis

Models of the Real World

Second Edition

Elena Prestini
Dipartimento di Matematica
Università di Roma—Tor Vergata
Rome
Italy

ISSN 2296-5009 ISSN 2296-5017 (electronic)
Applied and Numerical Harmonic Analysis
ISBN 978-1-4939-7961-5 ISBN 978-1-4899-7989-6 (eBook)
DOI 10.1007/978-1-4899-7989-6

Mathematics Subject Classification (2010): 42-03, 42A16, 42A24, 42A38, 42B05, 42B10, 01A50, 01A55, 01A60, 92C55, 65T50

Softcover reprint of the hardcover 2nd edition 2016

Printed on acid-free paper

This book is published under the trade name Birkhäuser
The registered company is Springer Science+Business Media New York
(www.birkhäuser-science.com)

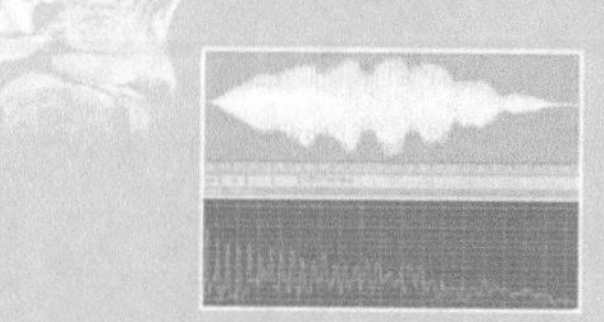

Cover of the first edition

Foreword to the Second Edition

The major contribution to this second edition is the addition of chapter ten which emphasizes applications to atmospheric science including measurements and probing instruments. As in the previous chapters the common thread is the Fourier method. The pinnacle of this chapter is that the most definite quantitative achievement in turbulence theory advanced by Kolmogorov is rooted in the statistical Fourier decomposition. The physical explanations and mathematics have enough details to be self-sufficient. The chapter will be of particular interest to atmospheric scientists and engineers.

Readers with experience in remote sensing will appreciate the discussion about the speed of sound in air and its wave properties of propagation, refraction, and reflection. The birth of atmospheric acoustics is attributed to men of peace (clergyman, and scientists) utilizing the works of men of war; we learn from the past that the firing of guns was used to make the first definitive measurements of the speed of sound in the atmosphere. To this day warfare technology provides unexpected benefits to society. Professor Prestini tells us that the field of radar meteorology owes its spectacular successes to the radar developments during World War II and thereafter. This can't help but leave us wondering how much good would come to humanity if its engagement in military compared to civilian sciences were reciprocal to what it is today. And what would ensue if we could eliminate conflicts altogether? The freed resources could tackle the numerous problems facing humanity including the enormous one of global warming. Speaking of which, the chapter lucidly articulates the evolution of understanding the physical principles that drive variations in climate. Milankovitch theory and its prediction of the three principal cycles of maximum insolation is a vivid example of the simplicity and predictive power rooted in Fourier analysis. Discussion of the discovery and subsequent effect of greenhouse gasses reminds us that too much of a perceived good (i.e., warming by CO_2 to prevent the earth from being "held fast in the iron grip of frost") can produce a bad outcome. The scientific method evoked so many times by the author points to the threat global warming poses to society.

As in many examples throughout the first nine chapters, details about the principal protagonists and their works are exposed and related to later developments and present understandings of the subjects. We are informed that Galileo was the first to recognize the wave nature of sound. Particularly striking is Leonardo da Vinci's brilliant conclusion that "the motion of the water, which resembles that of hair, which has two motions, of which one is caused by the weight of the hair, the other by the direction of the curls; thus the water has eddying motions." This partitioning of fluid motion into mean and turbulent flow predates Reynolds's idea by four centuries.

Sodars, radars, and lidars are the prime remote sensors that among other parameters can measure winds via the Doppler effect. Considerable space is devoted to these instruments and readers' understanding is enhanced by contrasting and/or comparing the nature of resulting measurements and the underlying properties of wind tracers.

The chapter provides well-articulated physical insights and explanations of atmospheric phenomena with sufficient mathematical exposure to be understandable but not overbearing. Particularly cute is the example of turbulence creation due to flow around a person and its capture followed by transport of odor, in this case perfume. The material is not meant to be a substitute for a working knowledge about the many subjects it touches upon, but informs about the essence and the diverse paths science takes in its quest for truth. Those with previous experience will be reminded about the complexities of atmospheric science and its interdisciplinary character. Those with less experience will be enriched with the newly acquired knowledge. To some, the chapter will serve as an excellent introduction into various modern aspects of atmospheric science. Reading this chapter was a pleasure to me and I hope you will enjoy it as much as I did.

Dr. Dušan Zrnić
Senior Scientist at the NOAA/National Severe Storms Laboratory
Affiliate Professor of Electrical Engineering and Meteorology
University of Oklahoma, Norman, OK, USA

Foreword to the First Edition

Two hundred years ago Baron Jean Baptiste Joseph Fourier (1768–1830) championed a mathematical idea that was to have a profound and enduring influence far beyond what could be imagined at the time; understanding the significance of his work requires some historical perspective.

Arithmetic and trigonometry were already well polished. For example, the trigonometrical table published by Claudius Ptolemy (2nd c. A.D.) in the days of the emperors Hadrian and Antoninus was more precise and more finely tabulated than the four-figure table of sines used today in schools. Euclidean geometry was thoroughly familiar. Calculus, starting with Archimedes (287–212 B.C.), who had determined the volume of a sphere by the limiting process now familiar as integration, had been given its modern formulation by Gottfried Wilhelm Leibniz (1646–1716) and Isaac Newton (1642–1727). Calculus became a mighty tool of astronomy in explaining the movement of the Earth and planets and had an impact on philosophy by its success in showing that numerous puzzling features of the Universe were explicable by deduction from a few physical premises: the three laws of motion and the law of gravitation. But integral and differential calculus had not been brought to perfection; they suffered from limitations when confronted with entities that were deemed nonintegrable or nondifferentiable and from difficulties in handling certain notions involving infinity and infinitesimals.

Even today, in physics, as distinct from mathematics, one recognizes that the infinitely small has not been observed, nor has the infinitely large. Would it not therefore seem logical that finite entities, such as the integers, should suffice for physical theory?

Nevertheless, as is well known, theoretical physicists expertly wield the power of modern mathematical analysis, confidently canceling infinities and adding up infinitesimals. Modern analysis, their indispensable and versatile tool, sprang from Fourier's "Analytical Theory of Heat," published in 1822 (Fourier [1978]). It took half a century and works by several mathematicians from Dirichlet to Cantor to interpret Fourier's method of representing a function as a sum of infinitely many harmonics, laying the basis for modern analysis.

Fourier's idea was conceived while he was investigating the conduction of heat, in particular the heat flow resulting from a spatially dependent distribution of temperature along a metal bar. When the starting distribution is physically measurable, functions that are not integrable, not differentiable, not continuous, or not finite do not arise. But Fourier's theorem seemed to be capable of wider interpretation, and such proved to be the case as mathematicians, some of whom strongly criticized Fourier for failing to supply a rigorous proof, refined the very concept of a real function and lifted restriction after restriction on functions that could be analyzed by Fourier's method.

The idea of adding together a finite number of components of simple form to represent some previously observed phenomenon has a long history. It was used by Ptolemy when he accounted for the observed motion of the planets around the Earth. His system of deferents and epicycles was a geometrical device that added a periodic (annual) correction to the position of a point in circular orbit around the Earth to arrive at the longitude of a planet in the sky; the equant was another geometrical device that incorporated a trigonometrically calculable periodic correction for orbital eccentricity. Where further adjustment to improve agreement with celestial appearances was warranted, the addition of another epicycle would extend the trigonometric series to three terms. This procedure, in use for over a thousand years, was a precursor to the Fourier series which, though nominally infinite, are evaluated to a finite number of terms for practical purposes such as the annual publication of planetary ephemerides.

Even earlier, Babylonian planetary positions were predicted by summing periodic corrections that were triangular rather than sinusoidal (and therefore easier to calculate). Fourier's contribution called for successive sinusoids that were in a harmonic relationship; for example, synthesizing the mean annual temperature variation for a particular location would require the first sinusoid to have a period of one year, the second to have a period of half a year, the third one-third of a year, and so on. If only a few terms are to be summed this proposal has the same pragmatic ring as the practices of the Babylonians and Greeks. But Fourier provocatively implied that if the summing were continued forever, the final sum would agree exactly with the original data. This is certainly not obvious. In spite of that, another astronomical precedent was set, some years before Fourier, by Carl Friedrich Gauss (1777–1855), who computed a cometary orbit as a sum of sinusoids. Not only did he observe the harmonic relationship of the periods chosen, but he fixed the amplitudes and phases of the components by the numerical algorithm now known in computing circles as the *fast Fourier transform.* In 1965 J. W. Cooley and J. W. Tukey brought the fast algorithm, now universally referred to as the FFT, to the attention of a much wider audience than the few to whom it was already known.

Harmonic analysis is a mathematical tool, but Professor Prestini interprets it in its modern mature form in a variety of scientific and technological aspects so diverse that any connection among them would hardly be suspected. Certainly it is well known that radio and television broadcasting, and wireless telephony, are conducted in frequency bands that are allocated by government, while music produced by a piano is clearly composed of the approximately sinusoidal air vibrations launched by the strings when hit by the hammers. It is by appeal to the familiar vibrating string that Professor Prestini introduces the mathematical theory.

To understand the analysis of music, speech, television, and cell-phone transmission into sinusoids and, conversely, the synthesis of the corresponding time-varying signals from sinusoids is to possess the key to understanding how astronomical images, medical images, and images of protein molecules are formed. In these cases we are dealing with sinusoids in space instead of in time and with two (or three) dimensions instead of one. However, once the temporal analysis and synthesis are grasped, the transfer of the concepts to the spatial domain is facilitated. And in each such application this book explains how the traditional vocabulary of each field is related to the terminology of vibrating strings. We learn that what we called the resolution of a television image is analogous to the pitch of a musical sound: Pitch is connected with the number of vibrations per second, while spatial resolution has to do with the number of lines per centimeter. If you know some property of a beep of a certain pitch and temporal duration, you know a corresponding property of a patch of image of a certain resolution and spatial extent.

Learning the analogies between the various fields treated confers powerful knowledge, but each field also has its own distinctive and fascinating history of development which the author incorporates into each chapter.

The final result is an impressive overview of a range of modern technologies, an implicit lesson in the interplay between technology and one segment of mathematics, and a stimulus to thinking about possible futures of the fields discussed and about inventions in other fields that have not yet been made. Why exact mathematics, a product of the human mind, should have proved so useful in science and technology, remains a mystery. The book can be thoroughly recommended to any reader who is curious about the physical world and the intellectual underpinnings that have led to our expanding understanding of our physical environment and to our halting steps to control it; and everyone who uses instruments that are based on harmonic analysis will benefit from the clear verbal descriptions that are supplied.

September 2003

Ronald Bracewell
Stanford University

Preface to the Second Edition

A unifying view of many fields of science is a main feature of this book. This is done via Fourier analysis, a branch of mathematics that—with its dual stance of time and frequency (or space and frequency)—happens to permeate most of modern science.

Following this first edition thread—besides some change and an update of Section 3.12—a new chapter has been added. It deals with five intertwining topics: turbulence, sodar, radar, lidar, and climate. They are described through their historical developments, followed by some more technical considerations. Then modern results, also providing images for the figures, are discussed.

From the point of view of harmonic analysis, wind is an interesting topic. Among sections concerning remote sensing (Sections 10.6–10.18), a few are dedicated to wind in order to describe different methods to determine wind velocity and direction (Sections 10.11, 10.14, 10.15, 10.18). The important role played by harmonic analysis in the theory of turbulence is mentioned in Section 10.5 and shown in Section 10.9.

In a few instances this new chapter briefly outlines the life of scientists that, with their intuition and dedication, made fundamental advances.

In the last section that deals with Earth temperature and climate, the time scale changes. The values of astronomical parameters, such as Earth eccentricity and tilt, are reconstructed hundred of thousand years back and predicted hundred of thousand years hence. Moreover in two hundred years the temperature of the Earth, a problem dear to Fourier, has become a matter of worldwide attention. This problem, from which sprang the theory of harmonic analysis, opens the book and the problem of the evolution of the temperature of the Earth closes the book.

I hope readers will find this book useful and enjoyable.

Acknowledgments

When my editor contacted me to suggest a second edition, my reply was not really enthusiastic: the first edition has been so much work and took so long. Then a friend Alberto Mugnai, a scientist at the Institute of Atmospheric Sciences and Climate

(ISAC) of the National Council of Research (CNR) of Italy, mentioned a colleague of him, Giuseppe Mastrantonio, very much interested in harmonic analysis and went ahead to arrange a meeting. Mastrantonio is a historical scientist, having collaborated with Giorgio Fiocco to the realization of the first sodar in Italy. I talked to him on the matter, and my curiosity lit up again. I am grateful for his help during all stages of this work, for providing much literature on sodar, as well as many pictures. He also introduced me to his colleagues Luca Baldini, Gian Paolo Gobbi, Davide Dionisi, and Fernando Congeduti.

Baldini is a radar expert; Gobbi, Dionisi and Congeduti are lidar experts. I like to thank them all for generously devoting their time to answer my questions and providing beautiful pictures as well. As for the lidar I was also in contact with Alcide Di Sarra of the Italian Ente Nazionale Energie Alternative (ENEA). And I like to thank Mariella Morbidoni, the head librarian of the Tor Vergata Area of Research (CNR-ARTOV), for her help and kindness.

Regarding turbulence, combustion, and aerospace issues I wish to thank Claudio Bruno at the time at United Technology Research Center (UTRC), in Connecticut.

Regarding climate I ought many thanks to André Berger of the Catholic University of Louvain in Belgium, who never failed to kindly answer my several e-mails. He also provided the image of a figure. I thank Eric A. Smith, Retired Professor of Atmospheric Science, and Christos Efthymiopoulos of the Athens Academy of Sciences for bibliographical information.

I thank Dušan Zrnić of the National Oceanic and Atmospheric Administration for suggesting improvements in the radar Sections 10.13–10.15. Also I thank him together with his colleague Richard Doviak, as well as Peyman Givi of the University of Pittsburgh, and James M. McDonough of the University of Kentucky for images that appear in the figures.

Finally I wish to thank Editors Allen Mann and Mitch Moulton for suggesting this second edition, and Ben Levitt with whom I worked in close contact during the latest stage of this publication.

Rome, Italy
June 2016

Elena Prestini

Preface to the First Edition

The suggestion to write a book on mathematics came from my Italian editor Ulrico Hoepli Jr. during a Christmas party years ago, about the time I returned to the University of Milano as an associate professor after being an instructor at Princeton University.

The suggestion appealed to me but I could not work on it right away. Years later, with a former colleague of mine Michele Sce, whom I wish to remember here, we proposed the contents and set a general plan. To my sorrow we could not go much further due to his sudden illness.

At that juncture the strongest input, which eventually shifted the topic of the book, came from my students. Over the years many, majoring in physics, chemistry, biology, and computer science, had attended my courses on mathematics. I wanted to write a book that would be easy for them to read and close to their scientific interests so that their studies would be better motivated.

I also wished to write a book for a broad audience. On more than a few occasions, after being introduced as a mathematician, I was told something like: "At school I liked math but never understood what it was used for." I wanted to meet this understandable curiosity and show how my field of interest is put to good use. Fortunately Fourier analysis has an extremely wide range of applications—reaching the everyday lives of everyone—in line with the goal Fourier assigned to mathematics: "the public good and the explanation of natural phenomena."

From that background the first book *Applicazioni dell'Analisi Armonica*, which appeared in 1996 in Italian, and the present book, double the size, were born. My Italian editor generously released his rights to this English version. I wish to thank him for that, for having accepted the change of topics, and for his original suggestion as well.

I also wish to thank John Benedetto, series editor in chief of Birkhäuser's Applied and Numerical Harmonic Analysis book series, for encouragement and support of this project.

An important contribution came from Ronald N. Bracewell of Stanford University, himself the author of a standard text on Fourier analysis and applications and whose range of interests covers a good part of the topics dealt with here.

He read my original book since, to my surprise, he knows Italian and kindly offered to assist me with English. Actually he did much more. I thank him for his steady support of the project, for his scientific advice, and for consultation on my English. The result is a presentation of much of modern science, from the billions of years of cosmic evolution to chemical reactions followed to the nanosecond, from the cosmic world populated by gigantic galaxies to the atomic world, and up to the ideal world of mathematics. Scientists, who are engaged in pushing ahead the frontier of knowledge, appear to share the feelings expressed in the poetic words of the ancient astronomer:

> When I search into the multitudinous revolving spirals of the stars my feet no longer rest on the earth, but, standing by Zeus himself, I take my fill of ambrosia, the food of gods. (Claudius Ptolemy, 2nd century A.D.)

Overview

The opening chapter of this book is devoted to Fourier's life, which was full of adventures and misadventures and equally filled with scientific and public work.

The second chapter introduces the mathematics that is common to the different applications. The historical introduction is followed by a presentation of basic concepts such as the Fourier series and Fourier transform, fundamental theorems such as the Dirichlet theorem and the inversion formula, and basic properties having to do with dilations and translations. The mathematical introduction given in Chapter 2 is as elementary as possible and illustrated by examples.

The enormous influence exerted upon science by Fourier's idea of decomposing a function into a sum of sinusoids or "waves"—influence that kept gathering power throughout the century that followed Fourier's work—is made apparent in the remaining seven chapters. They are devoted to different applications, each chosen for its significance and diversity. In all chapters the mathematical core is preceded by a historical account and followed by a description of recent advances.

Chapter 3, which begins with the history of the great electrical revolution—from Alessandro Volta to Alexander Bell and Guglielmo Marconi and up to satellite communications—has as a main topic signal processing, probably the best-known application of Fourier analysis. Filters, noise, quantization, sampling, and the sampling theorem are dealt with. The discrete Fourier transform (DFT) and the well-known algorithm for the fast Fourier transform (FFT) follow. The origin of this algorithm is traced back to Gauss in his successful attempt to compute the orbit of an asteroid. At this point the topic of space exploration appears as a natural link and ends the chapter. Here the historical part is dominated by the figures of the well-known German-born scientist Wernher von Braun and his Russian counterpart Sergei Korolev, while the technical part deals with aspects of propulsion and combustion.

In Chapter 4, after recalling how sound came to be understood and measured, the attention shifts to music, computer music, and its history. The analysis of musical sounds is dealt with, beginning with a single note and ending with the dynamic spectra of the performances of three tenors, which allows a quantitative and comparative analysis of their styles. Finally musical synthesis and some of the special effects it is used for are illustrated.

The topics in Chapter 5 are all interleaved with the long struggle to understand the nature of light, from Newton to Einstein. The Fourier transform in two dimensions, needed for the mathematical model of the so-called optical transforms, is introduced. Diffraction at infinity is the underlying physical phenomenon, which also lies at the heart both of radioastronomy, as illustrated in Chapter 9, and of x-ray crystallography, as illustrated in Chapter 6. Also relevant to x-ray crystallography is synchrotron radiation. With its unique features and the huge storage rings needed to produce it, it is the final topic of the chapter.

Chapter 6 is dedicated to x-ray crystallography, to its history from Nicolaus Steno to Max von Laue and Lawrence Bragg, and to the demanding task, so essential to the understanding of life, of determining the spatial atomic structure of human proteins, expected to number over a million. The first result, due to molecular biologist Max Perutz, dates from the 1960s. It could take work extending over a great deal of this century to see the end of it. The chapter would not be complete without mentioning the revolutionary work of biochemist James Watson and physicist Francis Crick on DNA, which was based on the experimental data of the crystallographer Rosalind Franklin.

Chapter 7 introduces the Radon transform, which is closely connected with the Fourier transform. Advanced in the early 1900s by the mathematician from whom it takes its name, it found its most important application in the CT scanners first built by the engineer Godfrey Hounsfield following the theoretical work by physicist Allan Cormack.

Chapter 8 begins with the history of nuclear magnetic resonance (NMR). This phenomenon, originally detected in the late 1920s by the physicist Isidore Rabin, aroused the interest of scientists from different fields, such as physicists Felix Bloch and Edward Purcell, chemists like Richard Ernst and Paul Lauterbur, and medical doctors such as Raymond Damadian. Their work lies at the foundations of the well-known magnetic resonance imaging (MRI). Also NMR spectroscopy, a technique that, like crystallography, is employed to determine the spatial atomic structure of proteins, is illustrated. The case of the prion protein, involved in the dreaded BSE, provides an example.

Chapter 9 takes us back to space. Introduced by a brief history of one of the oldest human activities, astronomical observations, this chapter focuses on radioastronomy. From its beginning in the 1930s, radioastronomy has revealed new celestial objects such as quasars and pulsars, led to the theory of the Big Bang, and following very recent measurements, pointed to an endless expansion of our universe.

Aims and Scope

In times like ours in which scientists tend to have a restricted and focused point of view, which is reflected in the scientific literature, I believe a book adopting a generalistic approach may have some value.

The book offers a brief introduction to several fields that are active and central in contemporary science together with the required Fourier analysis and landmarks in their fascinating historic development. Numerous illustrations make the reading easier, while the extensive bibliography may satisfy those in search of deeper knowledge, both scientific and historical.

Who might be interested in this book? To my mind, among the primary readers should be working scientists, particularly mathematicians and especially Fourier analysts wishing to know more about the applications of their own field of expertise, as well as scientists in one of the fields of application who want to know what workers in other fields do. In Ronald Bracewell's words: "To specialists already equipped with basic knowledge it could represent a way of efficiently building on their existing background, expanding their areas of expertise, and perhaps qualifying them for other jobs. One has to bear in mind that mathematicians are heavily outnumbered by people engaged in image engineering and in biomedicine. The rapid expansion of medical imaging will result in crystallographers, radioastronomers, and others being drawn from these areas in order to fill the immediate needs in medicine. Elena Prestini's book could facilitate the transfer into medical imaging of people whose primary exposure at graduate level was to one of the less populated specialities."

This book may also interest those concerned with history of science. I describe the ancient origins of every topic, move quickly to modern times—the primary focus—and finally reach the present.

The general reader might be interested in this work as well. I am thinking for instance of young people in the process of deciding the course of their studies. They might find useful information herein and derive inspiration from reading about the work of celebrated scientists. I have paid a lot of attention to the organization of the material and worked toward reaching clarity of exposition. This together with the selection of topics, I hope, will keep the readers interested until the very end.

While significant parts of every chapter are accessible to the general reader, some knowledge of calculus is required to understand the technical part. The book can be used as a complement in teaching at both the undergraduate and graduate levels, whenever the teacher feels comfortable with it. For instance I require my students majoring in chemistry—the corresponding level in the U.S. would be advanced undergraduate—to read it all (the original Italian version), and in my lectures I go over the technical parts of the topics relevant to chemistry. Even though it requires some extra effort from the students, I have found that it pays off for them to know some of the outstanding accomplishments in which the mathematical techniques that they are required to master are involved.

Acknowledgments

A book of such a wide scope cannot be reasonably written by a researcher working in isolation. It has been a pleasure of mine to discuss matters concerning the several fields presented here sometimes by e-mail but more often in personal meetings with the following scientists listed according to the order of the chapters:

Claudio Bruno, Scuola di Ingegneria Aerospaziale, Università di Roma "La Sapienza," Rome, Italy

Eugenio Gorgucci, Istituto di Fisica dell'Atmosfera, Consiglio Nazionale delle Ricerche, Frascati, Italy

Claudio Zanotti, Centro Nazionale Propulsione e Materiali, Consiglio Nazionale delle Ricerche, Milan, Italy

Giuseppe di Giugno, Scientific Director, Istituto Ricerca per Industria e Spettacolo, Colleferro, Italy

Giorgio Adamo, Discoteca di Stato, Rome, Italy

Massimo Altarelli and Renzo Rosei, Elettra Sincrotrone, Trieste, Italy

Zahid Hussain and Keith H. Jackson, Advanced Light Source, Berkeley, California, USA

Stephen G. Lipson, Department of Physics, Technion University, Haifa, Israel

Gianfranco Chiarotti and Sergio Tazzari, Dipartimento di Fisica, Università di Roma "Tor Vergata," Rome, Italy

Silvio Cerrini, Director of Istituto di Strutturistica Chimica, Consiglio Nazionale delle Ricerche, Monterotondo Scalo, Italy

Adriana Zagari, Dipartimento di Chimica, Università di Napoli, Naples, Italy

Vukika Srajer, Department of Biochemistry and Molecular Biology, University of Chicago, Chicago, Illinois, USA

Erik L. Ritman, Department of Biophysiology and Biophysics, Mayo Graduate School of Medicine, Rochester, Minnesota, USA

Maurizio Paci, Dipartimento di Scienze e Tecnologie Chimiche, Università di Roma "Tor Vergata," Rome, Italy

Francesco De Luca, Dipartimento di Fisica, Università di Roma "La Sapienza," Rome, Italy

Johannes Zuegg, School of Medical Research, Australian National University, Canberra, Australia

Ferenc A. Jolesz, Director MRI Division, Harvard Medical School, Cambridge, Massachusetts, USA

Gary H. Glover and Laura Logan, Department of Radiology, Stanford University, Stanford, California, USA

Alfonso Cavaliere, Dipartimento di Fisica, Università di Roma "Tor Vergata," Rome, Italy

Paolo De Benardis, Dipartimento di Fisica, Università di Roma "La Sapienza," Rome, Italy

Franco Scaramuzzi, Ente Nazionale Energie Alternative, Frascati, Italy.

To all of them go my thanks for their kind and generous collaboration in the course of writing this book.

I also thank my colleague Francesco De Blasi for his comments, Renato Guidetti of the University of Milan for his help with the bibliography, Paul Lyman for his help with British history, David E. Bruno for some of the drawings, and Giancarlo Baglioni for his help with the electronic pictures.

I wish to thank Ann Kostant, Elizabeth Loew, and Tom Grasso of Birkhäuser for the care put into editing my original text.

Finally my thanks go to the Mathematical Sciences Research Institute in Berkeley where I spent the month of July 2000, working hard while enjoying its warm hospitality and the beautiful landscape.

Rome, Italy
September 2003

Elena Prestini

Contents

Chapter 1
Joseph Fourier: The Man and the Mathematician

1.1 An adventurous life

Orphaned when still a child, Joseph Fourier went through the French Revolution, the Napoleonic era, and the following Restoration as a high-ranking public servant. He learned on his own "how hard the way up and down another man's stairs is"[1] by experiencing directly how risky it is to please changing lords such as Robespierre, Napoleon, and King Louis XVIII. He started as a convinced Jacobin, maybe impulsive at the time, and ended up as a cautious liberal.

He was a distinguished administrator and diplomat, but more than that he was a scientist whose achievements were truly revolutionary in character. The most relevant aspect of his life, the one for which he is still remembered, is the contribution he made to the advancement of mathematics and physics. It is somewhat miraculous that his most important and creative work took place when he was prefect of Isère, an onerous administrative position he held with first class results. It suffices to mention the construction of the French part of a spectacular road through the Alps from Grenoble to Turin and the draining of twenty million acres of marshes around the village of Bourgoin midway between Lyon and Grenoble, a century-old proposal that nobody previously had been able even to start.

He ranks among the most important scientists of the nineteenth century for his studies on the propagation of heat, the consequences of which reach down to the present day. The question of terrestrial temperature was principally in his mind in establishing the mathematical theory of heat, and the paternity of the expression "greenhouse effect"— *effet de serre*—is attributed to Fourier. He was the first on record to hint at it, writing back in 1827: "The problem of global temperatures, one of the most important and difficult of the whole of natural philosophy, is composed by rather different elements that ought to be considered from a unique general

---

[1] "come è duro calle lo scendere e 'l salir per l'altrui scale" *The Divine Comedy*, Paradise, XVII, 59–60, by Dante Alighieri (1265–1321).

E. Prestini, *The Evolution of Applied Harmonic Analysis*,
Applied and Numerical Harmonic Analysis,
DOI 10.1007/978-1-4899-7989-6_1

point of view." The novelty of his method, at least initially, perplexed outstanding mathematicians of his time, from Lagrange to Laplace and Poisson. The publication of Fourier's work was consequently delayed as many as 15 years during which he tenaciously defended, explained, and extended it.

Moreover, it ought to be mentioned that Fourier is among the founders of modern Egyptology: He followed Napoleon in his Egyptian campaign, spent 3 years there, and organized an expedition to Upper Egypt that made important discoveries. They form the basis of a monumental work *Description of Egypt*—21 volumes—for which he supplied a historical introduction. Much under Fourier's influence, Jacques Joseph Champollion-Figeac and his brother Jean François became important archeologists and Egyptologists.

1.2 The beginnings

Jean Baptiste Joseph Fourier was born on March 21, 1768 in the ancient and beautiful town of Auxerre, 150 kilometers south of Paris in a position dominating the river Yonne, with beautiful churches, above all the ancient fifth century abbey St. Germain and the Gothic cathedral of St. Étienne (Fig. 1.1). Joseph was the name of his father, a master taylor of Auxerre, who had been born to shopkeepers in a small town in Lorraine. Joseph might have been attracted to Auxerre by the rich and powerful

Figure 1.1 The Gothic cathedral of St. Étienne from the east bank of the river Yonne. (Photograph by Jasette Laliaux)

ecclesiastical establishment of the town, which had had its own bishop since Gallo-Roman times. He probably expected some special consideration in memory of his paternal great uncle Pierre Fourier, a leading figure of the Counter Reformation in Lorraine in the sixteenth and early seventeenth centuries. Joseph Fourier married twice. With his first wife he had three children and with the second one 12. Jean Joseph was the ninth of these. His mother Edmie died in 1777 and his father the following year.

Not yet ten years old, Jean Joseph was left an orphan. By all means the young Fourier could have then been considered "lost." It did not go that way thanks to a certain Madame Moitton who recommended him to the bishop of Auxerre, thus sparing him a life of apprenticeship and servitude. At 12 he entered the local École Royale Militaire, run by the Benedictines.

The various military schools of the country, 11 in all, were placing special importance on the teaching of science and mathematics, due to the requirements of those pupils who were going to enter the specialist corps of artillery and engineers. They were periodically visited by a panel of inspectors, among them the *académicien* and mathematician Adrien Marie Legendre (1752–1833). Fourier, an exceptionally gifted student of happy nature and quick mind, soon showed a great passion for mathematics. He was only 13 when he got into the habit of collecting candle ends to use at night during the long hours he spent studying mathematics in some sort of store room. In so doing he succeeded in unintentionally scaring the deputy principal who, one night while making his rounds of the school, saw a light through the keyhole and rushed in, fearing a fire. Fourier's overall excellence is confirmed by the prizes he won during 1782 and 1783: in rhetoric, mathematics, mechanics, and even singing. Then a year-long period of illness followed, perhaps due to the excessive intensity of his studying.

His main ambition at the time was a military career. At the age of seventeen, having completed his studies, he wished to enter the artillery or the engineers. In spite of the support of the inspectors of the school, the minister of war rejected his application. The fact that he was not a noble appears to have played a role.

As a second choice he decided to enter the Benedictines. To St. Benoît-sur-Loire, seat of an ancient and splendid basilica—hosting the body and relics of St. Benedict, which were transported there from Monte Cassino in the seventh century—Fourier arrived in 1787. He stayed two years preparing for his vows while teaching other novices. During this period of spreading riots and chaos, preparatory to the revolution, Fourier remained indifferent to political news and obtained results in the theory of equations. In 1789 he sent a paper to the Académie des Sciences in Paris. That was the year of the French Revolution and Fourier's paper was one of the casualties.

The monastic orders, combining great wealth and a steadily shrinking number of inmates, had long been a strong temptation to a government continually on the verge of bankruptcy. During October 1789, by decree of the Constituent Assembly, it was forbidden to take any further religious vows. A few months later all religious orders were suppressed and subsequently all belongings confiscated.

Fourier himself, who in the meantime had manifested some uncertainty about the choice he had made, did not take his vows and returned to Auxerre to teach mathematics, rhetoric, history, and philosophy in his old school, now having as a second title that of collège national. A commissioner of the local directory who visited the college in October 1792 reported favorably on the health of the inmates and the liberal atmosphere. He deplored only the tendency to drive out Latin to make way for mathematics, so much in demand by the parents of the pupils. Fourier carried out his duties there with dedication and remarkable success that earned him a fine reputation.

1.3 The revolutionary

The town of Auxerre fortunately saw little or no bloodletting during the Revolution, but the local Société Populaire, associated with the revolutionary Jacobin party, was one of the most militant provincial clubs in the country. Fourier's involvement in politics did not occur officially until February 1793. The occasion seems to have been a speech before the local assembly following a decree of the Convention for the draft of 300,000 men. While the quotas of the various departments—the administrative regions into which France was divided in 1790—were fixed, it was left to each department through the vote of the citizens to decide how to meet its quota: by lot, by volunteering, or by other means. This hot issue was debated in Auxerre by a general assembly. Fourier intervened and proposed a plan that was then adopted. In a letter written later, in June 1795 from prison, he makes clear that he fully shared the ideals of the Revolution (Hérivel [975]): "As the natural ideas of equality developed it was possible to conceive the sublime hope of establishing among us a free government exempt from kings and priests, and to free from this double yoke the long usurped soil of Europe. I readily became enamored of this cause, in my opinion the greatest and most beautiful which any nation has ever undertaken."

In March 1793 he was invited to join the local Comité de Surveillance. It is not known whether Fourier wished the invitation or, on the contrary, would have preferred to decline it, after receiving it. Certainly, a refusal would not have been without risks, since by and large the town of Auxerre shared the Republican ideals and a refusal could have identified Fourier as an opponent of the patriot party. From what is known of his later involvement in politics, it is more likely that he eagerly embraced the chance to take part in the defense of the Republic, which was threatened by military reverses in Belgium and by the rebellion in Vendée.

Events were moving fast due to the mounting of internal opposition, which prompted the formation of the Tribunal Révolutionnaire, and the near famine conditions of the population. Already by September of the same year the innocuous committees of surveillance (for strangers and travelers) had been entrusted with universal surveillance and soon were to become an integral part of the Terror, having to proceed by the Law of Suspects of September 17, 1793 to arrest "those who by their conduct, relations or language spoken or written, have shown themselves partisan of tyranny or federalism and enemies of liberty." (During 1793–94 over 200,000 citizens were detained under this law and about 17,000 death sentences were handed down by the revolutionary tribunals and military commissions.) At this point Fourier, feeling "less suited than many others to execute this law," attempted to withdraw and submitted his written resignation. It was not well received. As Fourier relates (Hérivel [1975]):

> This move produced an opposite effect to what I had intended. In a reply sent to me I was reminded of a law which forbade any official to abandon his post and my resignation was rejected. At the same time other persons openly accused me of abandoning my colleagues at a moment when my help was about to become the most useful to them. I was reproached with the feebleness of my conduct and some even doubted the purity of my intentions.

Fourier was not a fanatic; he firmly believed in the ideals of the Revolution and to that he devoted his intelligence, eloquence, and zeal but always retained an independent judgment. This together with his juvenile impulsiveness led him into a dangerous situation. He was sent to the neighboring department of Loiret with the mission of collecting horses for the war effort. He completed that "with every possible success."

Unfortunately on his way through Orléans, in the course of his mission, he became involved in a local dispute. To Orléans, plagued by near starvation and declared in a "state of rebellion," a representative of the people was sent by the Convention. Immediately upon arrival he "purged" the administrative corps of the city, made numerous arrests, and threatened to bring in a movable guillotine like the one in Paris. Thereafter he turned against members of the local Société Populaire, members of his own party. It was at this point that Fourier, "behaving in conformity with the principles of Revolution" took the defense "perhaps imprudent but at least disinterested" of the heads of three local families. Fourier was immediately denounced to the Convention, his commission revoked, and himself declared "incapable of receiving such commissions [in the future]." This took place on October 29, thirteen days after the execution of the queen.

Fourier, in fear, returned to Auxerre where he would have faced the greatest possible danger if the Société Populaire and the Comité de Surveillance had not successfully intervened on behalf of their "young and learned compatriot." He remained a member of the local revolutionary party and kept on teaching in his old school, which was going to be closed down in August 1794. April of that year saw the execution of Danton and of his associates amid the mounting Terror.

Nothing is known of Fourier's feeling during this period except for what he wrote afterward, when he claimed he had spoken out in Auxerre against the excesses of the Revolution. By June 1794 he had become president of the Revolutionary Committee in Auxerre and, therefore, the foremost representative of the Terror in town. This high position is known from an entry in the local archives reporting his arrest: Fourier, always feeling injustice at the decree of the Convention that declared him unfit for "similar commissions" in the future, had gone in person to Robespierre in Paris to plead his case. Perhaps he made a bad impression on him. Certainly upon his return to Auxerre on July 4 he was arrested. Because of the high reputation he enjoyed in town, many interceded in his favor and he was released, only to be rearrested a few days later. Then an official delegation was sent to Saint Just in Paris to demand his release, but salvation came from a totally different route: on July 27 (9 Thermidor) the Convention ordered the arrest of Robespierre and Saint Just. They were to be promptly guillotined the next day without trial. The Terror was over. Fourier's life was saved and he regained his freedom.

1.4 At the École Normale

Fourier, back in Auxerre, was soon going to return to Paris for an event that would have a great impact on his life: the opening of the École Normale.

The Revolution had rid itself of the existing educational system but failed to replace it with anything new. That was not a good prospect for its civilian and military functions. To help repair the damage and remedy an acute shortage of elementary school teachers, by a decree dated October 30, 1794, the Convention set up a national college in Paris, the École Normale.[2] There were to be 1500 students chosen and financed by the districts of the Republic. Fourier was nominated by the neighboring district of St. Florentin, since Auxerre had already made its choice while he was in prison. He accepted after having requested and been granted authorization from "the constituted bodies of the commune of Auxerre." The view of one of his fellow students is interesting (Hérivel [1975]):

> When the pupils [of the École Normale] gathered together, France had only just emerged from beneath the axe of Robespierre. The agents of this tyranny were everywhere regarded with abhorrence: but the fear which they had inspired, joined to a fear of their return to power, retained for them some vestige of credit. They profited from this, by seizing the opportunity of quitting the scene of their vexatious act. Several had themselves named pupils of the École Normale. They carried there with the ignorance proper to them the hate, distrust and contempt which followed them everywhere. Beside them were men full of wisdom, talent and enlightenment, men whose names were celebrated in all Europe.

Fourier was likely to have few regrets at the prospect of leaving Auxerre where, as former president of the Revolutionary Committee, he was a marked man. Compared to the violent whirlpools of the Revolution, at the École Normale must have felt like paradise. The professors were chosen from among the foremost men in the country: Joseph Louis Lagrange (1736–1813), Pierre Simon Laplace (1749–1827) and Gaspard Monge (1746–1816) taught mathematics and for chemistry there was Claude Louis Berthollet (1748–1822). All of them were going to play a role in Fourier's life in the years to come.

The school opened with impressive dispatch on January 20, 1795, amid great enthusiasm. In a letter to J. A. R. Bonard, his former teacher of mathematics in Auxerre, Fourier conveyed a vivid description of the early sessions of the school (Hérivel [1975]):

> The École Normale holds its sessions at the Jardin des Plantes, in a middle-sized place of circular shape; the pupils who are very numerous are seated in rows on the tiers of a very

[2] The present day École Normale descends from the homonymous school established 16 years later during the Napoleonic era, dedicated to training professors of secondary and higher education.

> high amphitheater; there is no room for everyone and every day there are a fair number who find the door closed; if one is obliged to leave during the sessions, one cannot enter again. At the back of the room, within an enclosure separated by a railing, are seated several Parisian scientists and the professors. In front, on a slightly higher platform are three armchairs for the professors who are to speak and their assistants. Behind them, and on a second platform, are the representatives of the people in the uniform of deputies on detached service. The session opens at 11 o'clock when one of the deputies arrives; there is much applause at this moment and when the professor takes his place. The lessons are almost always interrupted and terminated by applause. The pupils keep their hats on, the professor who is speaking is uncovered; three quarters of an hour or an hour later, a second professor takes his place, then a third, and the usher announces that the session is ended.

Fourier, an experienced teacher himself, gives an account of the lecturing habits and idiosyncrasies of the professors. Lagrange, whom Fourier followed with eyes full of admiration, as deserved by "the first among European men of science," had a "rather poor reception" from the students. Incapable of preserving order, he showed his Italian origin by "a very pronounced accent" (he was born in Turin and spent his youth there). He could make some rather comic sentences, such as "There are still on this matter many important things to say, but I shall not say them." Nevertheless to Fourier "the hesitation and simplicity of a child," that sometimes could be seen in Lagrange, made only more apparent "the extraordinary man he is."

Of Laplace, "among the first rank men of science," Fourier writes that "he speaks with precision, but not without a certain difficulty" and that "the mathematical teaching he gives has nothing extraordinary about it and is very rapid." Berthollet, acknowledged as "the greatest chemist we have either in France or abroad," Fourier continues, "only speaks with extreme difficulty, hesitates and repeats himself ten times in one sentence" to conclude that "his course is only understood by those who study much or understand already." Monge instead "speaks with a loud voice" and "the science about which he lectures [descriptive geometry] is presented with infinite care and he expounds it with all possible clarity."

He reports on others including a certain Sicard, "well known as a teacher of deaf-mutes, full of enthusiasm and patience" but "mad." "His theory of grammar, which is brilliant in certain respects, is one of the craziest I know of" (Hérivel [1975]).

On May 1795, a few months after inauguration, the school was closed due to objective difficulties. In particular the seminars, which had been intended as the backbone of the system, were a failure because so few students had learned enough to be able to contribute. For the majority of the pupils the École Normale had been a waste of time, but for Fourier it was a turning point of his career: At those seminars he made his mark. Meanwhile disturbing rumors were coming from Auxerre.

1.5 From imprisonment to the École Polytechnique

The fall of Robespierre marked the beginning of the settling of scores with those associated with him. Many had used their power to commit all sorts of injustices and atrocities: It had been the "Terror." In Auxerre, Fourier's opponents wasted no time.

On March 20, 1795 in an address to the National Convention, arguing that the pupils at the École Normale were chosen under the reign of Robespierre and his *protégés* and that Fourier "had long professed the atrocious principles and infernal maxims of the tyrants," they protested his preparations to become a teacher of their children. They wanted to prevent him from securing a teaching position and the indemnities, due to him as a pupil at the École Normale, suspended. On May 12, an order was issued in Auxerre regarding the disarmament of a number of terrorists including Fourier, followed on May 30 by a second-order commanding the detention of all those terrorists who had failed to comply with the original order.

Fourier, in an attempt to disarm his enemies, wrote to the municipality asking to which duly constituted authority he should present himself to be disarmed; he even resigned a new position given to him at the École Centrale des Travaux Publiques (Central School of Public Works). All in vain. On the night of June 7 he was awakened by armed guards and taken to prison without notification and having scarcely been given the time to dress himself. The pretext was failure to present himself to be disarmed; the true charge, that of having inspired terrorism in the years 1793–1794.

It is not known whether Fourier ever underwent trial or was even interrogated, but we know his line of defense from letters, written from prison, to the representative of the people and to the chairman of the Committee of General Security (Hérivel [1975]). To the denunciation of terrorism and to the reproach of having been a member of the Committee of Surveillance, he replied with eloquence, "I was entrusted by their own votes with a surveillance determined by the law. I received this position without soliciting it, I continued in it without the power of withdrawing from it and I exercised it without passion." As to the charge of terrorism, he continued, "I am unable here to advance all the reasons that make these charges unfounded. I shall only insist on the incontestable facts that no one in the commune of Auxerre was condemned to death or judged by the Revolutionary Tribunal at Paris, that no revolutionary tax was established of any kind whatsoever, that the property of those detained was never confiscated, that no cultivator, artisan or merchant was arrested." Moderation had been the dominant feature of his doings. There remained something to be said about those who underwent arrest and detention. Fourier does not avoid the issue:

> There remain therefore those citizens who being nobles or priests or relations of *émigrés* found themselves included under the law of September 17 and who experienced a temporary detention when they showed themselves declared enemies of the Revolution. They accuse me of not having opposed their arrest and never pardon me for having signed the warrants for their arrests. They pretend to believe I could have released them and wanted me to make use of the trust which had been placed in me. Being unable to accuse us of abusing powers they reproach us with excessive rigor, but far from having merited this insult I believe that I have accorded to humanity, friendship, generosity even, all that was allowed by the letter of the law and the rigor of the times.

As an example of this he pointed to his early imprisonment under Robespierre and added, "I have experienced terror more than I inspired it" and "I owed to 9 Thermidor both life and liberty." Finally, he wrote with anguish, "To exclude me from a school of mathematics is to take away from me an entirely legitimate possession which I have acquired by my work and which I retain by cultivating it daily."

By August Fourier was released, the reasons unknown. Perhaps it was an intervention of Lagrange or Monge, or more likely the change in the Convention of the political climate with the end of repression against the Jacobins in face of a mounting royalist threat. Fourier resumed his teaching position at the École Centrale, soon to be renamed École Polytechnique by a decree of September 1, 1795. The École Centrale des Travaux Publiques had been instituted by a decree of the Convention of March 11, 1794. It did not have the general aims of the École Normale since it was a military academy: Its graduates, with 3 years training in science, engineering, and applied arts, were intended to provide the military elite. It took only about 400 students annually. Opened in November 1794 with a trial run of fifty students, it was converted into a full organization when the École Normale closed.

The director was Monge and with his support Fourier was appointed to an assistant teaching post, from which he had resigned in vain to avoid arrest. Now back, he helped run the course in descriptive geometry—by Monge, mathematician and military engineer — that dealt largely with the use of science and mathematics in military contexts, the art of attack and defense, and the organization of simulated battle situations. Also he taught courses in Lagrange's curricula in analysis and was involved with the selection of the entrants to the École. Fourier believed they should have "outstanding talents regardless of how much they have actually been taught," an opinion similar to the one expressed by Monge at the time of the setting up of the school. Among those entrants was Siméon-Denis Poisson (1781–1840) who was to be Fourier's pupil at the École, his deputy as professor of analysis during his absence in Egypt, and finally Fourier's opponent over some mathematical questions concerning the analytical theory of heat. During this period at the École, Fourier once again built up a fine reputation for his teaching and in 1797 succeeded Lagrange in the chair of analysis and mechanics. While continuing his investigation into the theory of equations he also began work on problems in applied mathematics and in 1798 published his first paper in the *Journal de l'École Polytechnique*.

1.6 Fourier, Napoleon, and the Egyptian campaign

The quiet days at the École Polytechnique ended abruptly and unexpectedly as a result of a letter from the minister of the interior dated March 27, 1798:

> The Minister of Interior to Citizen Fourier Professor at the École Polytechnique:
>
> Citizen, the Executive Directory having in the present circumstances a particular need of your talent and of your zeal, has just disposed of you for the sake of public service. You should prepare yourself and be ready to depart at the first order. If you are actually charged with any employment or if you occupy any place at the expense of the Republic you will conserve it during your mission and the salary attached to it will be paid to your family.

The minister of the interior was obeying the orders of the Directory (a five-man executive committee that gave the name to the new regime begun in October 1795) instructing him to "put at the disposition of General Bonaparte engineers, artists and other subordinates of your ministry together with the different things he will demand of you for the purpose of the expedition to which he has been assigned."

On May 19 Fourier sailed from Toulon, following Bonaparte, together with generals, officers, members of the scientific and literary commission, and about 30,000 soldiers and sailors, all stowed in some 180 ships. The destination, unknown to most, was Egypt. (Once more it was Monge, in charge together with Berthollet of the selection of the scientific commission, to choose Fourier.) Successfully avoiding an encounter with the English fleet that was scouring the Mediterranean at the orders of Admiral Nelson, the French armada captured Malta where 7,000,000 gold francs were acquired from the suppression of the ancient order of the Knights of Malta. On July 1, Alexandria was sighted and captured the following day. Fourier temporarily settled down in the nearby town of Rosetta, north of Alexandria (Fig. 1.2), and took up a position in the provincial purchasing commission while the army headed toward Cairo with a march through the desert. On July 24 the Mameluke forces under Murâd Bey were defeated at the Battle of the Pyramids. The next day Napoleon entered Cairo. This sequence of successes was halted and de facto canceled by the destruction of the French fleet in Abukir Bay. The feelings of the soldiers are well captured by the physicist and engineer Étienne Louis Malus [1892]: "From then on we realized that all our communications with Europe were broken. We began to lose hope of ever seeing our native land again." On learning of the disaster the Directory ceased to assist the expedition.

As nothing happened, Bonaparte kept on going and set up a council with the native leaders in the hope of bringing their affairs under the ultimate control of the French. Fourier was later to sit at the meetings of the council as French commissioner. Of the many tasks that Bonaparte was engaged in, the foundation of the Institut d'Égypte, dated August 20, 1798, is what concerns us most. Organized as the Institut de France, of which Bonaparte was a member, it consisted of four classes: mathematics, physics, political economy, and literature and fine arts. The class of mathematics was the most distinguished, including Monge, Bonaparte, Fourier, and Malus. In the physics class was Berthollet, in literature and fine arts the artist Dominique Vivant Denon (1747–1825). This latter, who would initiate in the Napoleonic era the policy of enriching the Louver with works of art from conquered lands, was to provide a pictorial record of many aspects of the Egyptian expedition with his large collection of drawings. The Institute was located in the former palace of the beys, the great room of the harem serving for the meetings. The first one took place on 25 August and had Monge elected president, Bonaparte vice president, and Fourier permanent secretary. Napoleon had assigned to the Institute as tasks the progress and propagation of the sciences in Egypt, the collection and publication of natural, historical, and other data on Egypt. Last, but not least, he was expecting assistance in the civil and military administration of the land. Right away at the first meeting he raised several issues such as improving the army's baking ovens, ways of brewing beer without hops, methods in use to purify the Nile water, the choice between windmills and watermills, and the location of resources for manufacturing gunpowder. Fourier kept busy writing notes on the questions posed by Napoleon but also found time for an old love of his, the general

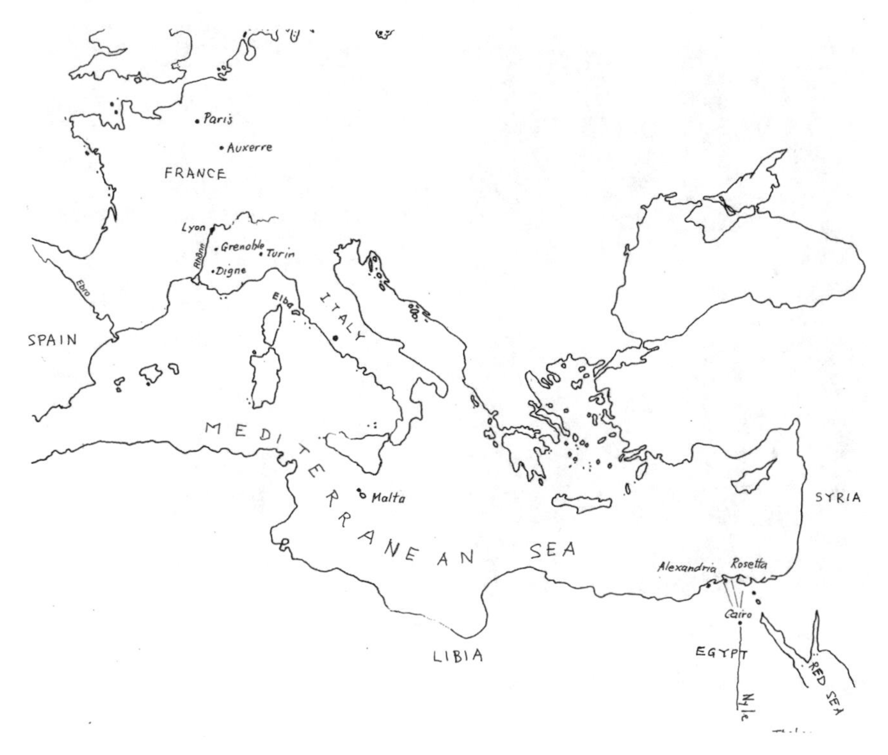

Figure 1.2 Map of the Mediterranean sea and surrounding regions.

resolution of algebraic equations, and wrote four mathematical memoirs to be read before the Institute.

In February 1799 Bonaparte launched an ill-fated campaign in Syria, in July he stopped a Turkish invasion at the battle of Abukir, and in August he left for Paris upon learning of a troubled situation there. He brought with him the trusted Monge and Berthollet. Fourier was kept out of the plan until the very last moments. When he finally learned of it, he became so agitated at the thought of being left behind that he followed them into the street and could hardly be persuaded to let them go.

Before departing, Bonaparte found time to plan a mixed scientific and literary expedition to Upper Egypt under the joint leadership of Fourier and the assistant secretary of the Institute. The discoveries of the expedition, which spent about two months in the fall of 1799 investigating the monuments and inscriptions in Upper Egypt, would years later form the basis of the *Description of Egypt* (Fig. 1.3).

Bonaparte left behind a difficult situation, challenged by the English forces and civilian unrest. General Jean Baptiste Kléber, the new commander in chief, was assassinated in Cairo in June 1800 and was succeeded by General Jacques François Menou. Under both commanders Fourier was given many important administrative and judicial positions. By Menou he was put virtually in control of all nonmilitary affairs and entrusted with a most relevant assignment, the conduct of the French side

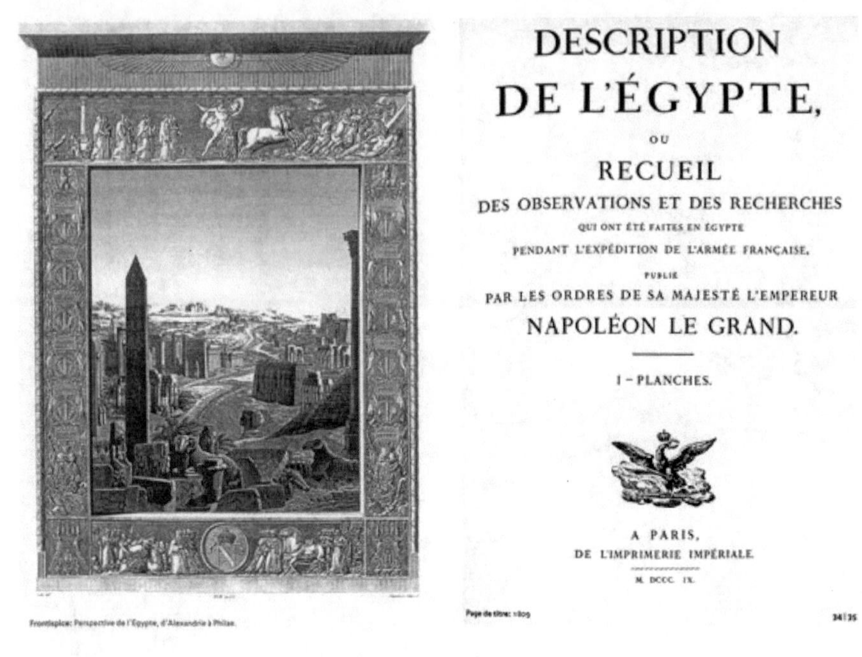
DESCRIPTION
DE L'ÉGYPTE,
OU
RECUEIL
DES OBSERVATIONS ET DES RECHERCHES
QUI ONT ÉTÉ FAITES EN ÉGYPTE
PENDANT L'EXPÉDITION DE L'ARMÉE FRANÇAISE,
PUBLIÉ
PAR LES ORDRES DE SA MAJESTÉ L'EMPEREUR
NAPOLÉON LE GRAND.

I - PLANCHES.

A PARIS,
DE L'IMPRIMERIE IMPÉRIALE.
M. DCCC. IX.

Figure 1.3 The opening pages of the *Description of Egypt*.

of the negotiations with the Egyptian beys. He was successful in persuading them to sign an alliance with the French at a time when they were none too strong militarily. At the negotiations the chief bey Mourâd was represented by his wife, the beautiful and intelligent Sitty-Nefiçah with whom Fourier negotiated also the freedom of some of her slaves who were of interest to the French generals.

The English forces that had never ceased to create difficulties for the occupation were about to see the end of their efforts. On March 1801 they landed at Abukir Bay. At this point the members of the Institute thought it was time to leave for France and Fourier with them, in spite of General Menou having expressed himself as follows: "Your departure in the actual circumstances appeared to me, and still appears to me, and always will appear to me immoderate and ill-conceived." They embarked at Alexandria, were immediately arrested by the English, and returned to Alexandria. The deterioration of the French position was soon going to force Menou to surrender. He signed the terms of capitulation on August 30, 1801. The French were allowed to keep part of their scientific collections and findings they had put together with care during the occupation but not the important and highly symbolic Rosetta stone. The stone, which marks the beginning of modern Egyptology, is a perpetual reminder of the scientific importance of the French expedition, while its collocation at the British Museum is a memento of its military failure. The French forces, such as remained, withdrew in the autumn of 1801. Fourier organized much of the departure and returned to France in the middle of November 1801. They were transported on English ships at the expense of the Sultan and his allies. Fourier resumed his teaching

Figure 1.4 Fourier dressed as prefect. (Portrait by unknown artist, possibly Claude Gautherot, Museum of Auxerre.)

post at the École Polytechnique and was able to give a few lectures before Bonaparte, meanwhile established as First Consul, appointed him prefect of the Department of Isère on February 1802 (Fig. 1.4). The adventures, abruptly terminated, made room for the years of maturity, for high rank administrative duties, and for the *Analytical Theory of Heat*.

1.7 The prefect of Isère and the *Analytical Theory of Heat*

The Department of Isère, crossed by the river of the same name and bordering Italian territory, had its center in Grenoble. It was one of the 83 regions into which France had been divided by a decree of the Constituent Assembly of February 3, 1790. In 1799 the departmental directories were abolished and replaced by a single person, the prefect, sole representative of the executive power. Fourier's feelings about the position, which Bonaparte presented as "an earnest of my confidence in citizen Fourier," are not known. He accepted it and departed for the remote town of Grenoble. His tasks as prefect were varied and demanding, even though he could count on administrative officers and in particular on three private secretaries. He had to oversee the enforcement of the laws and the implementation of the directives constantly issued from Paris, mainly pertaining to taxes and recruitment, first for the consular and later the imperial armies. At the same time he had to put forward to the central government the needs and requests of the area, backward in development but independent in spirit. Also he was supposed to keep the government in Paris informed about the preservation of order. Soon Bonaparte's regime began to develop into a form of police state and Fourier had to execute unpleasant orders concerning the opening of letters, the suppression of antigovernmental pamphlets, and the supervision of the local official news sheet *Annals of the department of Isère, administrative, political and literary journal* to keep away scandals and revolution.

Much more pleasant it must have been for him to get to know the prominent local people, even though he had to compile notebooks on them for the Ministry of the Interior. Indeed the policy of Napoleon was to unite the greatest number of Frenchmen, regardless of their original sympathies and status, in support of his policy and person. From the beginning Fourier was on excellent terms with the nobility who praised his old world manners, the charm of his conversation, and appreciated the many services he did them like helping *émigrés*, returning from exile, in gaining back their original properties and allowing immunity from the recruitment of the guards of honor in exchange for money, to be used in raising a body of paid volunteers. He was also popular among the wealthy middle class, the so-called *bourgeoisie*, for his excellent administration and was on good terms with the local clergy. At a ceremony in the cathedral of Grenoble in which Fourier received the oaths of the curés nominated in the diocese, the bishop celebrated the mass while incense was given to the prefect. On the other hand Fourier's relations with the members of the local Jacobin party are unknown.

The year 1804 was eventful. In February Bonaparte visited Grenoble and in the following May crowned himself emperor. Fourier wrote to Napoleon to tell him that 82,084 of the electorate supported the plebiscite on hereditary descent, with twelve dissenters, and he spent more than three months in the capital for the celebrations. Then he became involved with the visit in November of Pope Pius VII to Paris, since the route passed through Isère.

Besides political affairs Fourier attended to the development and welfare of the department, which was much in need. He reopened schools and colleges and revitalized the mining and craft industries. His major achievement was the draining of the marshes around the village of Bourgoin midway between Lyon and Grenoble. The huge area, covering some twenty million acres, was rather useless and at that time responsible for annual epidemics of fever. There had been projects planned and attempted many times over more than a century but agreement among the 37 communes bordering the swamps was never achieved. Fourier succeeded where his predecessors had failed, thanks to his persistence, patience, and diplomatic ability. The negotiations stretched over a period of some four years during which he visited all the communes involved, one by one, and met individually most of the inhabitants to convince them to temporarily give up their rights of pasture for the sake of a better future. In 1807 a treaty was finally signed and in 1812 work was completed. The value of the land greatly increased as well as the health of the inhabitants. During the drainage archeological remains were found and Fourier instructed Jean Champollion-Figeac, then librarian of Grenoble, to preserve them in the library. Fourier had Champollion as a cultural friend in Grenoble, introduced him to Egyptology, and used the power of his position to spare him conscription. In 1822 Champollion would do his celebrated work on the decipherment of Egyptian hieroglyphics.

Another large scale project initiated by Fourier was the opening on the French side of an important and spectacular road through the Alps, the Grenoble–Turin, via Briançon and Pinerolo. Fourier overcame the opposition of the minister of the interior, himself a native of the countryside in question, by appealing directly to Napoleon. Knowing the man, Fourier had him presented with a map and a one page memoir setting forth the advantages, not least the military ones. Two days later the request was granted. In Fourier's times, nevertheless, the road's construction did not go beyond Briançon, partly due to the fall of Napoleon in 1815, but when completed it was the quickest and shortest route from Lyon to Turin. Later Fourier recorded that of all his prefectorial assignments this one gave him the greatest satisfaction.

Another time-consuming accomplishment was his contribution to the *Description of Egypt*. Fourier had been chosen to unite and publish the collection of works—which had to be done by another former member of the Institute, due to his many prefectorial duties—and to write the general introduction. This was essentially a survey of the history of the ancient civilization to the time of the French expedition and renaissance under French patronage. Fourier took great care and for the final polishing even isolated himself in a country residence, having to deny rumors of illegal holidays to the minister of the interior. He knew how Napoleon would scrutinize it. Finally in the autumn of 1809 the preliminary discourse was submitted for Napoleon's approval. The emperor kept it for a long time and made corrections in his own hand, to conform the description of the Egyptian campaign to his personal view. The work was published in 1810. Fourier received the title of baron and an annual pension.

Fourier must have enjoyed his achievements and the demonstrations of loyalty and affection he received, but he always hoped to be allowed to resign from the job and to return to an intellectual life. Monge and Berthollet in vain tried to persuade Napoleon, and Fourier even thought of going into exile. In the end he stayed at his post and tried to make life as interesting as possible for himself as well as for others.

During those very same years in which he was intensively solicited by various engagements, imposed on him by external factors, Fourier found the time to take up a problem totally new with respect to his preceding research and was able to pursue it in the silence of the mind to complete solution. This process peaked in the memoir titled *Analytical Theory of Heat*, his masterpiece dated 1807. The reason for this new scientific interest is not known, since Fourier never described how or when he came to be motivated to address this problem. It is known nevertheless that he always regretted the Egyptian heat and never quite got used to the chilling climate of the Alps. Soon he suffered bad attacks of rheumatism. Thus he never went out, even in the hottest weather, without his overcoat and was often accompanied by a servant with another coat in reserve. In Grenoble, where the winters are more severe than in Paris, his concern for adequate heating must have been even greater. From this to the question of loss and propagation of heat in solids and radiation in space, to the problem of conserving heat, the step can be short.

He started working on it already in the spring of 1802, when he just had arrived in Grenoble, and had difficulties until 1804. Seemingly only in the second half of 1804 did he find a new approach. During the next 3 years Fourier obtained, in the brief intervals of research time available, the main body of his contribution to mathematical physics. Presumably he used some of the time of his retreat to work on the introductory paper on the Egypt volumes to write a memoir *Theory of the Propagation of Heat in Solid Bodies* (Fig. 1.5). He presented it to the Institut de France, reading an abstract in front of the First Class. (The full text is reproduced in Grattan-Guinness [1972].)

Fourier applied his method to two kinds of problems. The first was to find the eventual steady distribution of temperature at all points of a body given a steady supply of heat at some points of that body; an example of that kind of problem is provided by a thin bar, heated by a furnace at one end and immersed in air held at a given temperature at its surface. The second kind was to find the temperature of a body, initially heated throughout to a given temperature distribution and then allowed to cool in an environment of a given temperature at every point at all subsequent times. An example of the second kind of problem, far more complex than the first, is provided by the Earth, and apparently it was this very problem that stimulated Fourier's search for a general theory of the propagation of heat in solid bodies. Thus in his *Mémoire sur les Températures du Globe Terrestre* (Oeuvres, 2, p. 114) he states: "The question of terrestrial temperature always seemed to me one of the most important objects of cosmological studies and I had it principally in my mind in establishing the mathematical theory of heat."

The permanent secretary for the mathematical and physical sciences of the Institute was the astronomer Jean Baptiste Joseph Delambre (1749–1822). He asked Lagrange, Laplace, Silvestre François Lacroix (1756–1843), and Monge to examine the paper. The members of the commission, if anything, had already formed a high opinion of Fourier, but the memoir was not well received. Lagrange rejected the very foundation of its mathematical side, namely the use of the trigonometric series (now called Fourier series). He thought it was not rigorous and never changed his mind. Also some criticism was directed against Fourier's derivation of the equation of the

-1

Lû le 21 Xbre 1807
Commissaires Mrs Lagrange,
LaPlace, Monge et Lacroix.

Delambre
Secr. perp.

Théorie de la propagation de la
Chaleur dans les Solides

Objet
de cet ouvrage.

Lorsque la chaleur est inégalement distribuée entre les différens points d'un corps solide, elle tend à se mettre en équilibre et passe successivement des parties qui sont plus échauffées dans celles qui le sont moins. En même temps la chaleur se dissipe d'elle-même par la surface et se perd dans le milieu ou dans le vuide. Cette tendance à une distribution uniforme et ce refroidissement spontanée qui a lieu à la surface des corps, sont les deux causes qui changent à chaque instant la température des différens points. La question de la propagation de la chaleur consiste à déterminer quelle est la température de chaque point d'un corps à un instant donné. On suppose que les températures initiales soient connues, et on demande suivant quelle loi varient les températures. Les exemples suivans feront connaître plus clairement la nature de ces questions.

Supposons qu'un cercle ou anneau métallique, dont le plan est horizontal, soit exposé à la flamme d'un foyer dans une partie de sa circonférence. L'action de ce foyer étant durable et uniforme, les autres parties s'échaufferont continuellement, et, après un certain temps, chaque point de l'anneau aura acquis presqu'entièrement la plus haute température à laquelle il puisse parvenir. Cette limite, ou maximum de température, n'est pas la même pour tous les points. Elle est d'autant moindre qu'ils sont plus éloignés de ceux où le foyer est immédiatement appliqué.

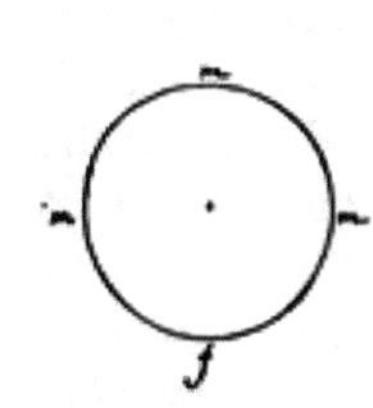

La ligne fmmmf représente la circonférence moyenne de l'anneau, c'est-à-dire, celle qui joint tous les centres des sections circulaires, faites par des plans perpendiculaires au plan de l'anneau, et qui passent par le centre de cet anneau. Le diamètre de cette section est assez petit pour qu'on puisse sans erreur sensible, regarder comme égales les températures des différens points d'une même section. Le point f est celui au dessous duquel le foyer se trouve placé.

Lorsque les températures sont devenues permanentes, le foyer transmet à chaque instant une quantité de chaleur qui compense exactement celle qui se dissipe par la surface dans tous les points de l'anneau.

Si maintenant on supprime le foyer, la chaleur continuera de se propager dans l'intérieur du solide, mais la quantité de cette chaleur qui se dissipe par la surface ne sera plus remplacée par

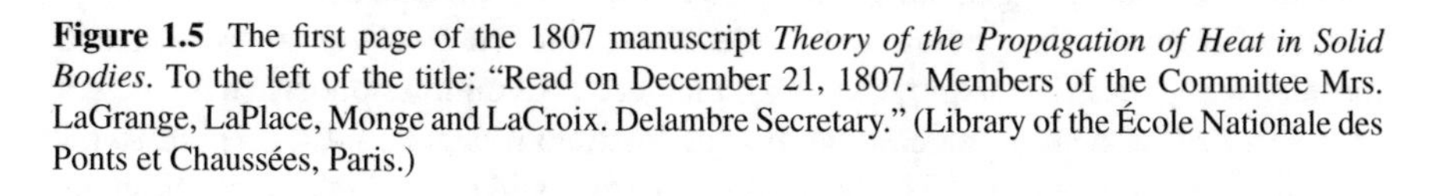

Figure 1.5 The first page of the 1807 manuscript *Theory of the Propagation of Heat in Solid Bodies*. To the left of the title: "Read on December 21, 1807. Members of the Committee Mrs. LaGrange, LaPlace, Monge and LaCroix. Delambre Secretary." (Library of the École Nationale des Ponts et Chaussées, Paris.)

motion of heat in a continuous solid. Laplace, who had carried out experiments on the subject, seems to have taken that position. But the chief opponent here was Jean Baptiste Biot (1774–1862), a protégé of Laplace, who had published in 1804 a short paper containing not only experimental results but some efforts at a mathematical formulation of heat diffusion. Fourier made a reference to it but confined himself to its experimental results. While Biot did not derive the right equation and did not attempt to solve the one he obtained, it appears that Fourier got some hint out of Biot's mathematical model. That must have been all the more evident to Biot who, on top of everything, was aware that Fourier himself had written an equation in an earlier draft of his memoir, possibly sent to him or to Poisson. Thus Biot became transformed into an opponent.

Nothing has survived of the commission's criticisms, except for Fourier's replies preserved in two letters to Laplace and Lagrange concerning the convergence of trigonometric series, the most critical issue. Fourier also sent to the Institute additional notes and wrote to Delambre asking for the date of publication of his manuscript. Nothing happened for the simple reason that the commission failed to report. There only appeared in the Bulletin of the Philomathic Society an unenthusiastic summary and review by Poisson who was interested in heat diffusion himself. A few months after Fourier had presented his manuscript, in a report to Napoleon on the development of science since the Revolution, Delambre wrote that in mathematics the chances of progress were very slight, with insurmountable difficulties preventing major breakthroughs and leaving only minor points of detail to be cleared up.

A new turn took place at the beginning of 1810 when the Institute announced the subject of its grand prize in mathematics for the year 1811: The mathematical theory of heat and its experimental verification. There appears to be no trace of the decision process that led to setting the subject of the prize. Certainly the members of the First Class of the Institute were divided on the ground that no report had yet been made on the 1807 memoir, but in the end that was the decision. Fourier submitted the old memoir, revised and extended by new sections on the cooling of infinite solids and on terrestrial and radiant heat; in all it was 215 pages. The commission, made up of Lagrange, Laplace, Lacroix, Malus, and René-Just Haüy (1743–1822), had to examine another candidate also. The prize was awarded to Fourier, but the motivation must have left him disappointed. There were no doubts about the importance of his work but ambivalent feelings remained, which were clearly stated in the report: "This theory contains the true differential equations of the transmission of heat both in the interior of bodies and at their surface. The novelty of the subject combined with its importance has made the class determined to crown this work, while observing however that the manner in which the author arrives at these equations is not exempt from difficulties and that his analysis to integrate them still leaves something to be desired on the score of generality and even rigor." (Hérivel [1975]).

Out of Fourier's disappointment a letter to the Institute likely grew, for although the letter disappeared, Delambre's reply remained: The commissioners had full power in such a matter. Fourier understood that the Institute was in no hurry to publish his memoir (indeed only in 1815 on his return to Paris could he get Delambre started on it). Thus he began to prepare a third version of his research in the form of a book and at the same time he obtained new results on other aspects of the problem. In spite of his administrative duties he worked on this in the following three years and was about to complete it when the political situation turned dramatic once more.

1.8 Fourier and the "Hundred Days"

After the disastrous Russian campaign of 1812 and the debacle of the battle of Leipzig of 1813, in January 1814 France was being attacked at its frontiers for the first time since 1795. Fourier, among other imperial officials, must have wondered about Napoleon's fate and his own. The extraordinary successes Napoleon achieved with his army of young conscripts during the first 3 months of 1814 proved insufficient to reverse the trend. The French people, tired of his interminable wars, were not behind him any longer and on March 31 Paris surrendered to allied troops while Napoleon had moved east to attack their rearguard. In the meantime Grenoble was besieged by Austrian forces.

When the news of the surrender of Paris and the abdication of Napoleon were learned, the Austrian forces occupied Grenoble on behalf of Louis XVIII, brother of the executed Louis XVI, who had since become king of France. The return of the king was unexpected and Fourier must have been concerned about his position and even personal condition, but he was allowed to continue as prefect, certainly due to the integrity of his conduct and the support he had among members of the old nobility. The situation therefore became the more embarrassing when it was learned that, on his way to exile in Elba, Napoleon was going to pass through Grenoble. Fourier let him know, through the prefect of Lyon, that it would be dangerous to pass through Grenoble owing to the excited condition of the population; at the same time, however, he gave instructions to prepare for Napoleon's stay in the prefecture. Then in a state of extreme upset he waited. Suddenly, the day Napoleon was supposed to come, a messenger arrived announcing that Napoleon would not pass through Grenoble after all. Fourier retired for the rest of the day to recover.

Some time later he was confirmed in his position of prefect and even received a visit by the king's brother, Count d'Artois, later to become King Charles X. His position secured, he just had to deal with some unrest on the issue of returning the land, sold during the revolution, to their rightful owners, nobles, and clergy. Then on March 1, 1815 a letter from the prefect of the neighboring department of Var reached Fourier. It read: "Bonaparte at the head of 1,700 men disembarked yesterday at the Gulf Juan, reached Grasse this morning and according to those soldiers who have been questioned is heading for Lyon by Saint-Vallier, Digne, and Grenoble." The letter ended with assurances that the news, no matter how extraordinary it might seem, was entirely true. This time there was no room for diplomacy. Fourier had to take a clear stance. It was 4 p.m. By 7 p.m. he had worked out various contingency plans for the defense of the department in collaboration with the major of Grenoble and the commanding officer of the garrison. He also had started a letter to the minister of the interior, which was completed by 7 a.m. the next day after a troubled night. He assured the minister that the inhabitants of Grenoble were firmly behind the king and that no motive of fear would have turned him from his duty toward king and country. He asked for instructions and added, "I know personally the audacious enemy who threatens us and I do not doubt that before very long he will send us emissaries." That did not happen.

By the afternoon Fourier had a proclamation put up in town in which he reminded the citizens of their duty to the king and those who might be inclined to forget the severe punishments they would incur, in conformity with the law.

In spite of the sincere efforts of Fourier and other prefects, Napoleon's advance in the south of France proved to be irresistible. On March 7 the gates were forced open and Napoleon entered Grenoble amidst the enthusiasm of the populace. Meanwhile Fourier was leaving the town from another gate, with him the commanding officer of the garrison. Before leaving he had a room prepared for Napoleon in the prefect's residence. Also he left a letter in which he expressed his feelings of obligation toward the king and his wishes not to offend his old master.

It was the night of March 7 and Fourier was headed toward Lyon to join the Count d'Artois. Napoleon, irritated by Fourier's failure to greet him in Grenoble, suspended him from office and required him to evacuate the territory of the 7th military division within five days. Failure to do so meant arrest as an enemy of the nation. On the evening of March 8 Napoleon was already moving toward Lyon with an ever-increasing army. The Count d'Artois quickly retreated. Maybe it became clear to Fourier that the king's cause was lost in southern France or maybe he was told that the worst was over after Napoleon had been presented with his letter, for on March 9 he turned back to Grenoble and on the morning of the 10th met with Napoleon in Bourgoin. Helpless, he could only accept Napoleon's wishes. The meeting did not go badly: The next day he learned he had been made prefect of the Rhône and was installed at Lyon on March 12. He carried out all reasonable administrative duties requested by his position, such as recruitment for the imperial army and surveillance of political suspects; at the same time he tried to reduce as much as possible the injustice and suffering associated with the change of regime. But his new post did not last long. Before the end of May he resigned, in disagreement with the harsh orders of the new minister of the interior, the mathematician and French Revolutionary figure Lazare Carnot. Apparently they had to do with a purge of administrators, some in his own prefecture, suspected of royalist sympathies. Whatever the reason, his credits were not entirely destroyed. On June 10 Napoleon granted him a retirement pension of 6,000 francs.

1.9 Return to Paris

Fourier went to Paris where he could enjoy freedom from administrative duties and opportunities to meet again with his old acquaintances such as Laplace, Monge, and Berthollet. But on 18 June Napoleon was defeated at Waterloo. The Napleonic era was over, the Bourbon monarchy back.

Soon Fourier found himself in a desperate financial position. The retirement pension granted to him by Napoleon was totally lost, the first installment being due on July 1. Also the 4,000 francs that went with his Napoleonic barony of 1809 was annulled. He had little money when he came to Paris, due to his generosity and the habit of living up to the top of his income. It seems that he thought of emigrating to England, to be far from political persecution, when a friend came to help. It was the prefect of the Seine, count of Chabrol, who had entered the École Polytechnique in 1794 when Fourier taught there and also joined the Egyptian campaign. In disregard of reactions from the extreme right, he offered Fourier the directorship of the Statistical Bureau. Fourier accepted and held the position for the rest of his life. The salary was modest, but Fourier must have considered himself fortunate. It went quite differently for Monge, faithful and prominent Bonapartist: In spite of his merits for having resurrected the French system of higher education, he was forbidden to enter his beloved École Polytechnique, was forced out of the Académie des Sciences, and died in poverty.

Now what Fourier cared about most was to firmly establish himself in the scientific community and also to obtain a pension for his many services to the state in teaching, in the Egyptian campaign, and as prefect of Isère. He applied to the minister of the interior on November 20, 1815, recalling all the above, hoping that his Jacobin activities of 1793–94 had been forgotten, justifying his action during the Hundred Days as having preserved Lyon "from greatest disasters" and attributing his dismissal to "unjust and arbitrary" measures required of him. Fourier's application was acknowledged and after considerable time refused. He did not give up and wrote again suggesting that "no political motive should efface the memory of so many services from which the State and many generations will receive real and lasting advantages." The reply addressed to "Baron Fourier" stated that "the King had recently adjourned his decision on this matter." Later in response to his insistence, the technicality that he had retired before reaching the age of sixty backed up a final refusal. Only years later was he granted a pension, curiously from the minister of police, for "important services of information" that we may suppose to be of statistical nature.

In the meantime on May 27, 1816 Fourier was elected "free *académicien*"—a position different from the ordinary one that required the vacancy of a "chair"—to the scientific class of the Institut de France. In vain, since King Louis XVIII did not confirm the election. Shortly afterward the old academies were restored in place of the classes of the Institut de France—with both Monge and Carnot excluded—and already by May 1817 a vacancy had arisen for physics in the Académie des Sciences. Fourier was elected by an overwhelming majority of 47 votes out of 50. This time he was better prepared and obtained the king's approval, being supported by an old friend of his from Grenoble, the viscount Dubouchage, now minister of marine. In a letter of recommendation Dubouchage had stated, referring to Fourier's activities as prefect in Isère, that "his conduct has won him the especial gratitude of the families most devoted to the royal cause who found themselves most exposed to oppressive measures." Fourier was now firmly established, had another source of income, and could throw himself into work. In the next 5 years he submitted to the Académie eight memoirs, in mathematics and statistics, and sat on a large number of commissions.

In August 1822 Delambre, the permanent secretary of the Académie for the mathematical sciences, died and Fourier was elected in his place with 38 votes while his opponent Biot had only 10. In this new position he was responsible for all the official correspondence on the mathematical side, for composing *éloges* as he did for Delambre and Laplace, and for producing the final reports on the state of mathematics. He also found time to publish a number of papers, but although he extended his studies to thermoelectric effects he did not show any interest in a pamphlet published by Sadi Carnot (1796–1832). Lazare Carnot's son was putting forth the idea that change of heat could be a source of motive power: His theory of cyclic processes was a great contribution to science indeed.

Life in Paris was not without scientific controversies, partly due to the fact that Fourier's work had not been published. Only in 1822 did his book *Analytical Theory of Heat* appear, soon followed by his 1811 prize paper in the *Mémoires de l' Académie Royale des Sciences*. By that time much more had been achieved on the subject by Fourier and by others, so on the opening page he added a footnote pointing out that it was an unaltered printing of his manuscript. It had happened that Poisson in a paper of 1815, mentioning Fourier's prize essay, echoed the criticisms of the commission's report and attempted to give an alternative treatment of the propagation of heat in solids. Also Biot in his *Traité de Physique* of 1816 claimed to have been the first, back in 1804, to enunciate and apply the equation for the steady state distribution of the temperature in an iron bar heated at one end. Claims like these always cause great resentment because of the strong emotional bonding that ties every researcher to his own scientific contributions. Fourier reaffirmed the validity and priority of his solution. Here is Fourier's punch line: "One does not extend the bounds of science by presenting in a form said to be different results which one has not found oneself and, above all, by forestalling the true author in publication," and with a double-sided compliment to Poisson "M. Poisson has too much talent to exercise it on the work of others. Science waits and will obtain from him discoveries of great superior order."

With the great Lagrange dead and in 1827 Laplace, there was still much talent left on the Parisian mathematical scene. Besides Poisson, their natural heir, and besides the outsider Fourier there was another outsider of value Augustin Louis Cauchy (1789–1857). His initial work appeared in the early 1810s and continued ceaselessly for almost half a century. In Cauchy extraordinary scientific talent lived alongside smallness of character: He was always ready to put down the work of others to prove the superiority of his own, which in the great majority of cases was clear anyhow. In 1814 Cauchy wrote an important essay on using complex variables in the evaluation of integrals that laid the foundation of his theory of complex variables. The paper met with opposition from Legendre, remained with the secretariat of the Institut de France, and received a poor summary from Poisson, who was interested in the problem too. Then, very much like Fourier, Cauchy won a prize with a paper on the motion of water waves that also remained with the secretariat. In this paper he began to develop integral solutions to partial differential equations and two years later, in 1817, in another paper he found what is now called "Fourier's integral theorem," which was in Fourier's unpublished prize manuscript. Cauchy did have access to Fourier's memoir, having been nominated the preceding year to one of the vacancies of the Académie arising from the expulsion of Monge and Lazare Carnot. Fourier did not wait to acquaint him with his priority and Cauchy acknowledged it.

Two young and talented mathematicians then briefly appeared on the mathematical scene: Galois and Abel. Évariste Galois (1809–1832), known for that part of algebra named "Galois theory" after him, in 1829 submitted to the Académie a paper solving what was a long-standing open problem on the solutions of algebraic equations. Fourier was then permanent secretary. The paper was sent to Cauchy to examine but got lost. Twice again Galois submitted it to receive three years later, and only six months before his death in a duel, the assessment "unintelligible" signed by Poisson. Things went only slightly better for the Norwegian Niels Henrik Abel (1802–1829). In 1826 he submitted to the Académie his masterpiece on transcendental functions. The paper, passed to Legendre and Cauchy, who never looked at it, got published after various misadventures 35 years later, in 1841. Abel could not rejoice, having died some ten years earlier.

In Paris Fourier added new acquaintances to the old ones. Devoted friends of his were the engineer and mathematician Claude Louis Marie Henri Navier (1785–1836) to whom Fourier's papers passed on his death, Peter Gustav Lejeune Dirichlet (1805–1859) who made important contributions to some of Fourier's ideas in pure mathematics, Joseph Liouville (1809–1882) who was interested in heat diffusion—his first paper dealt with the case of the inhomogeneous bars—and Jacques Charles François Sturm (1803–1855), later to work with Liouville on partial differential equations. Fourier who never married, but was said to be fond of the company of intelligent women, particularly cared about the friendship of Sophie Germain (1776–1831), a self-taught mathematician of considerable talent who corresponded with Gauss and Lagrange. Their friendship lasted from at least 1820 until Fourier's death and a number of his letters to her have survived. Maybe Sophie Germain, who died of cancer, was the suffering woman that Fourier recommended to a certain Doctor Herminier in an undated letter. He wrote "She is worthy of all your interest by reason of the rarest and most beautiful qualities. For myself, who love her tenderly, I would be most grateful for anything you could do for her."

Fourier at last received many honors. In 1823 he was made a foreign member of the Royal Society, in 1827 was elected to the Académie Française and the Académie de Médecine, and on the death of Laplace, was made president of the Conseil de Perfectionnement of the École Polytechnique. But in his last five years he suffered increasingly, mainly from chronic rheumatism, which pushed him to wear heavy woolen clothes and keep the stove in his apartment lit at all times. In addition he developed breathing problems. To be able to work, in his last months, he had to be placed in a boxlike chair from which only his head and arms protruded. He worked relentlessly, producing almost illegible manuscripts: Now he finally had the opportunity to devote himself to mathematics and he wanted to secure his fame by working as long as he could. Things looked different from some forty years earlier when, young and unknown, he had written in a letter to Bonard: "Yesterday was my 21st birthday, at that age Newton and Pascal had [already] acquired many claims to immortality." But time was running out. On May 4, 1830 he was stricken by some attack while descending stairs and died on the 16th in his apartment opposite the Jardin du Luxembourg, now 73 Boulevard St. Michel.

It seems nice and truthful what Grattan-Guinness [1972] writes of Fourier: "He preserved his honor in difficult times and when he died he left behind him a memory of gratitude among those who had been under his care as well as important problems for his scientific colleagues."

Chapter 2
Introduction to Harmonic Analysis

2.1 How trigonometric series came about: the vibrating string

"The eighteenth century stands out in mathematical history as an era of great genius. Through the work of an astonishing array of masters the science was extended and broadened by the opening of many new fields. Technical skill attained to extraordinary high levels and new ideas were crowded one upon the other" (Langer [1947]). The subject of trigonometric series received much attention: it was stated that they could represent "any" function, and formulas were found, but agreement among the masters of the time was not there. It must be said that the issues were far from simple. The tricky one of computations dealing with infinitely many terms, around which much of calculus revolves, was not being properly investigated and clearly regulated. The notion of the sum of infinitely many terms was imprecise; as if no different from the finite case, it was a common practice to rearrange terms infinitely many times as well as to integrate and differentiate term by term infinitely many times.

(All these operations, as is well-known nowadays, might not produce the expected results, unless specific conditions are met. We mention a couple of examples for clarification. In most calculus books, it is proved that

$$1 - \frac{1}{2} + \frac{1}{3} - \frac{1}{4} + \frac{1}{5} - \frac{1}{6} + \frac{1}{7} - \frac{1}{8} + \frac{1}{9} - \cdots = \lg 2$$

and also that

$$1 + \frac{1}{3} - \frac{1}{2} + \frac{1}{5} + \frac{1}{7} - \frac{1}{4} + \frac{1}{9} + \frac{1}{11} - \frac{1}{8} + \cdots = \frac{3}{2} \lg 2.$$

The second series is obtained by rearranging the terms of the first one: Every two positive terms are followed by a negative one. This seemingly harmless modification changes the sum of the series. As a second example, we mention the telescopic series

E. Prestini, *The Evolution of Applied Harmonic Analysis*,
Applied and Numerical Harmonic Analysis,
DOI 10.1007/978-1-4899-7989-6_2

$$\sum_{k=0}^{\infty}(x^{k+1}-x^k).$$

All terms $(x^{k+1}-x^k)$ are continuous on $[0, 1]$. However, the partial sum $\sum_{k=0}^{n}(x^{k+1}-x^k)$, being equal to $-1+x^n$, as n tends to infinity, converges to -1 for $0 \le x < 1$ and to 0 for $x = 1$. This is manifestly a discontinuous function at $x = 1$. Nevertheless, under specific conditions, rearranging terms does not change the sum of the series and similarly a series of continuous functions has a continuous sum.)

The context was that of power series, widely used to represent functions that manifestly have to be differentiable infinitely many times. An example is

$$e^x = 1 + x + \frac{x^2}{2!} + \frac{x^3}{3!} + \cdots .$$

Then, functions like sines and cosines of multiples of an angle were introduced in place of powers of x for the same purpose (Fig. 2.1). That a finite sum of such terms represents a function is obvious (Fig. 2.2). The opposite — namely which class of functions can be represented by a sum, possibly infinite, of such terms with coefficients in the real numbers — proved to be a difficult issue. Clearly the function must be periodic, but conditions more strict than that can be expected since, conceivably, the trigonometric representability may attach only to functions of a special class. Which one was it?

The issue was complicated by the very definition of function, not clearly stated. This concept, which lies at the very heart of mathematical analysis, was imprecise to the point that it had different meanings to different mathematicians. For instance, a function was thought of as a graph by Leonard Euler (1707–1783); to Jean le Rond d'Alembert (1717–1783) function and analytic formula were one. On top of that, a

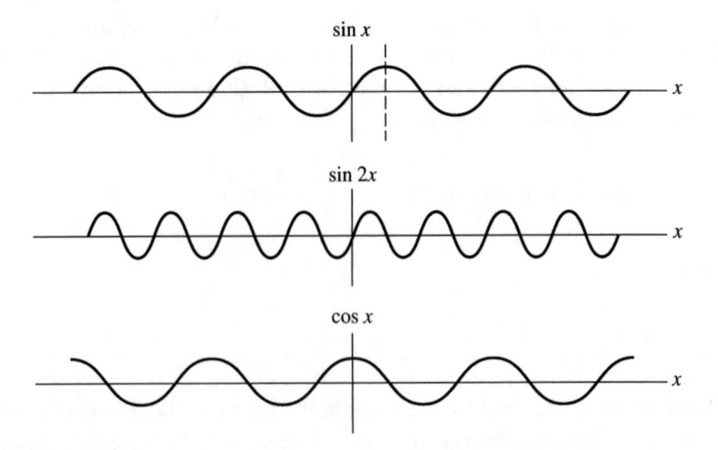

Figure 2.1 The graphs of $\sin x$, $\sin 2x$, and $\cos x$. The last one is obtained by translating the graph of $\sin x$ by $\frac{\pi}{2}$ to the left.

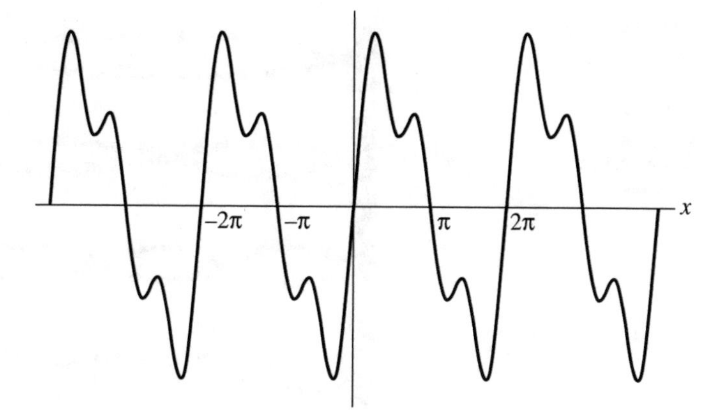

Figure 2.2 Periodic graph of the function $10 \sin x + 3 \sin 2x + 4 \sin 3x$.

function defined by different laws on different portions of an interval, such as the one graphed in Figure 2.11, was regarded as a conglomerate of different functions.

No wonder that the question of representability in trigonometric series remained controversially open for more than half a century, from the middle of the 1700s until Dirichlet's work of 1829. The restrictions on functions turned out to be so subtle that in their pursuit a final clarification imposed itself.

At the beginning of the eighteenth century, near the close of Isaac Newton's life (1643–1727) and with Gottfried Wilhelm von Leibniz (1646–1716) just dead, the effectiveness of calculus — their creation — as an instrument for the treatment of problems in mechanics was generally recognized. Mechanics with its abundance and variety of problems exerted a strong fascination on those interested in attaining analytic mastery over the manifestations of nature.

Problems pertaining to the motions of single mass particles were solved to a reasonable extent. Beyond them the forefront of advance dealt with matters of greater complexity: motions of bodies with many degrees of freedom, reactions of flexible continuous mass distributions, vibrations of elastic bodies.

In particular, the motions of tautly stretched elastic strings of length l, held fixed at its extremes, in response to a displacement from the state of equilibrium, received a great deal of attention and these investigations are among the most important in the eighteenth century development of the rational mechanics of deformable media. In many cases, the response can be acoustically perceived, ranging from the hum of a heavy structural wire to the notes of musical string instruments. It can also be visually noticeable being sometimes marked by features such as the presence of nodal points that maintain a state of rest while the string between them is in agitation (Fig. 2.3).

It was at this juncture that many divergent conceptions, which were to play a crucial role in the development of mathematics and of mechanics as well, came in to confrontation.

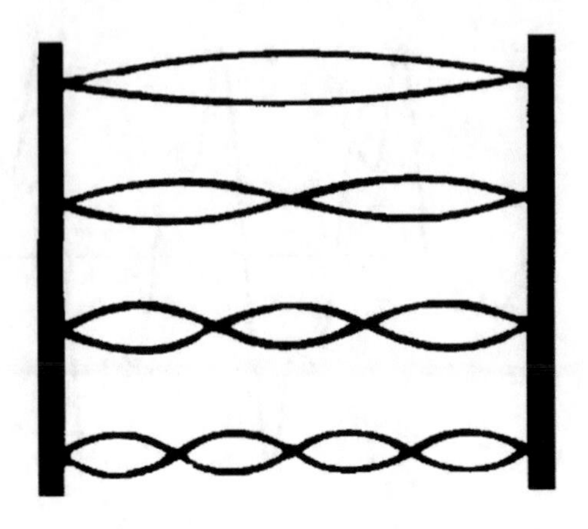

Figure 2.3 The first four modes of vibration of a string held at the two extremes: In the first one the fundamental tone is emitted, in the remaining ones the first three overtones.

It was believed that continuous material bodies could be approximated by a system of discrete mass particles of finite number n. These discrete systems could be made to merge into the continuous case by letting the size of the individual particles diminish indefinitely and by letting their number go to infinity.

In the case of the homogeneous string, the natural approximation that presents itself is that of a string of n beads of equal mass mounted at equally spaced points along the string, which itself is assumed to be weightless (though strong), perfectly flexible, and elastic. Friction is ignored, and if M is the total mass of the particles and T is the tension, it is furthermore assumed that the ratio M/T is negligible for the purpose of taking gravity out of the picture and concentrating attention on tension alone. That is the case for musical strings. For example each string of a grand piano, evidently of small mass, undergoes a tension that corresponds to a force from 75 to 90 kilos.

Clearly, the original problem has been replaced by an idealized abstract one. To do so always requires deep understanding and sound judgment, for the subsequent theory stands or falls according to results that at strategic points have to agree with the data of observations.

As early as 1728, Johann Bernoulli (1667–1748) considered the case of loaded strings in which at most eight particles were involved (Burkhardt [1908]). He derived the equation of motion and focused his attention on special analytic solutions, those giving rise to "normal" vibrations. In the simplest of such "normal" vibrations, the string emits its fundamental tone; its motion has no nodes.

In 1747, by considering the general discrete case of n beads and by letting n tend to infinity, d'Alembert deduced the partial differential equation of the motion $y(x, t)$ of the continuous string in the (x, y)-plane[1]

$$\frac{\partial^2 y}{\partial t^2} = c^2 \frac{\partial^2 y}{\partial x^2}$$

[1] J. le R. d'Alembert, "Recherches sur la courbe que forme une corde tendue mise en vibration," *Mémoires de l'Académie Royale de Berlin*, 3 (1747: publ. 1749), 214–219.

where $c = \sqrt{T/M}$ has the dimensions of velocity and the value of l is assumed to be 1. The equation, states that the transverse acceleration at position x is proportional to the curvature at time t. This result is still a standard. Then by an ingenious use of familiar formulas from calculus he solved the equation, finding its functional solution

$$y(x, t) = \Phi(x + ct) + \Psi(x - ct)$$

where Φ and Ψ are arbitrary functions with continuous second-order derivatives. The above formula has an intuitive interpretation: The term $\Phi(x+ct)$ can be interpreted as a uniform motion of the initial state $\Phi(x)$ to the left with speed c; likewise $\Psi(x-ct)$ can be interpreted as a similar motion of $\Psi(x)$ to the right.

Then the conditions for the fixed ends at $x = 0$ and $x = 1$, the initial displacement of the string $y = f(x), 0 \leq x \leq 1$, at time $t = 0$, and the zero initial velocity imply

$$y(x, t) = \frac{f(x + ct) + f(x - ct)}{2} \tag{2.1}$$

provided $f(x)$ is extended by the condition that it be odd and periodic of period 2 (Rogosinski [1959]).

Even though (2.1) makes sense under no regularity assumptions whatsoever, d'Alembert's critical mind saw no reason that all possible motions should conform to his formula. He was unwilling to admit the applicability of his analysis to cases in which the initial position $f(x)$ was not twice differentiable, as in Figure 2.4. The solution $y(x, t)$, having to satisfy a second-order differential equation, must be twice differentiable, he argued. Therefore, the same must be true of its position $y(x, 0)$ at time $t = 0$, that is $f(x)$. (The issue was resolved much later by admitting as solutions displacement functions $y(x, t)$ satisfying the equation everywhere except at isolated points (Riemann–Weber [1927]) as in Figs. 2.4, 2.5, 2.6.) The case of a string set into motion by plucking, as in "pizzicato" (Fig. 2.4), seems to have been Euler's motivation to the problem. Euler, guided by his physical intuition, thought that every motion originating from a state of rest would necessarily stem from some initial shape $f(x)$ of the string. In a paper of 1748[2] he derived d'Alembert's results by a method of his own. Then, he stated that the case of "discontinuous" functions (at that time "discontinuous" meant nondifferentiable, that is functions with corners) must be encompassed to allow for the above situation. D'Alembert was not convinced and kept on going his own way in a subsequent paper.[3]

There the disagreement rested until 1755 when Daniel Bernoulli resumed it. Daniel — together with his father Johann and paternal uncle Jakob (1655–1705), the most distinguished of the eight mathematicians that the Swiss family produced in three generations — having interested himself in acoustics, recognized the relation

[2]L. Euler, "De vibratione chordarum exercitatio," *Nova Acta Eruditorum*, (1749), 512–527. The paper also appeared in Euler's translation from Latin into French as "Sur la vibration des cordes," *Mémoires de l'Académie Royale de Berlin*, 4 (1748: publ. 1750), 69–85.

[3]J. le R. d'Alembert, "Addition au mémoire sur la courbe que forme une corde tendue mise en vibration," *Mémoires de l'Académie Royale de Berlin*, 6 (1750: publ. 1752), 355–360.

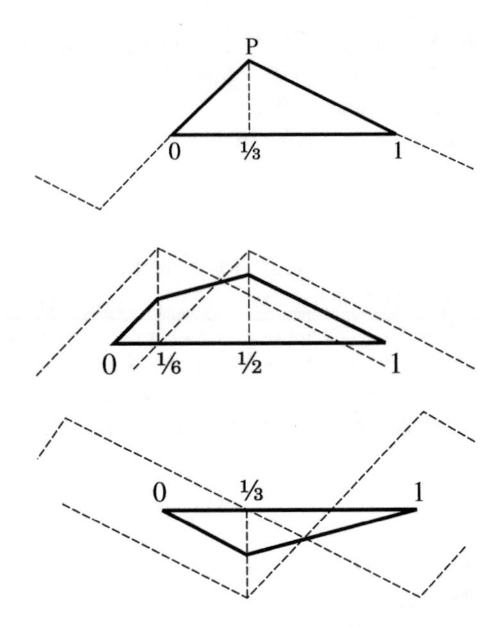

Figure 2.4 The "plucked" string at time $t = 0$, for $c = 1$ and $P = (1/3,\ 1/3)$.

Figure 2.5 D'Alembert's solution at time $t = 1/6$.

Figure 2.6 D'Alembert's solution at time $t = 2/3$.

between the "normal" vibrations and the overtones that the string could emit. Since it was generally accepted at the time that a musical string ordinarily responds with a combination of its fundamental and overtones, Bernoulli made a step, moderate only in this framework, and published a memoir[4] bluntly stating that every motion of the string is expressible in the form

$$y(x,\ t) = \sum_{n=1}^{\infty} A_n \sin \frac{n\pi x}{l} \cos \frac{n\pi ct}{l}$$

with appropriate coefficients A_n. This meant a representation in sine series for the initial $f(x)$. There it was

$$f(x) = \sum_{n=1}^{\infty} A_n \sin \frac{n\pi x}{l}. \tag{2.2}$$

Euler, the most important mathematician of the century, placed the weight of his authority against that formulation.[5] The infinity of unknown constants in (2.2) might seem to allow sufficient generality for representation in trigonometric series, he argued, but in fact overriding reasons showed this could not be hoped for. "Since

[4]D. Bernoulli, "Réflexions et éclaircissemens sur les nouvelles vibrations des cordes exposées dans les mémoires de l'Académie de 1747 et 1748," *Mémoires de l'Académie Royale de Berlin*, 9 (1753: publ. 1755), 147–172.

[5]L. Euler, "Remarques sur les mémoires précédents de M. Bernoulli," *Mémoires de l'Académie Royale de Berlin*, 9 (1753: publ. 1755), 196–222.

the initial position of the string is absolutely arbitrary, it may happen, and it will happen very often, that this initial position is not expressible by any equation be algebraic or transcendental and that it does not satisfy any continuity law." Such initial curve will not be included in Bernoulli's formula, "nice as it is." "Bernoulli's solutions shall be regarded as very special ones," Euler concluded. D'Alembert sided with that opinion and moreover insisted that (2.2) implied differentiability properties for $f(x)$.

Bernoulli kept his stance. In his mind, trigonometric series were engraved into the subject (Fig. 2.1, Fig. 2.3).

Later, in 1759, the then twenty-four-year-old Lagrange, in a long paper[6] that established his early reputation in the mathematical world, examined afresh the behavior of the weightless loaded string with an unspecified number of particles and obtained a proof for the continuous string that placed no restrictions upon the shape of the curve marking the initial position. In that, following Euler, he supported the functional solution and rejected trigonometric series. Nevertheless he came close to Bernoulli's formula, without realizing it.

Also it has to be mentioned that in a paper[7] written in 1777 and published only in 1798 after his death, assuming functions known upon some ground or other to be representable in terms of cosine series of the type (2.2), Euler found the now standard formula for the coefficients (see (2.3) in Section 2.3).

2.2 Heat diffusion

Then it was Fourier's turn. He derived from physical fundamentals the partial differential equation diffusion of heat diffusion in continuous bodies such as a lamina, a thin bar, an annulus, a sphere. In searching for solutions (see Section 2.6 for a modern treatment), he started with the lamina and right away trigonometric series popped up again. The particular lamina on which he was working was of finite width and semi-infinite length. He centered it on the x-axis, starting at the origin. The width ran from -1 to 1 along the y-axis (Fig. 2.7).

The choice of the boundary conditions was a separate issue. Fourier arranged for himself very simple ones with the y-edge held at temperature 1 in a chosen unit of measure (by supplying heat) and the x-sides held at 0 temperature (by a flow of cold air). He was concerned with the steady state temperature at all points of the lamina. In this case, characterized by temperature $u(x, y)$, which is independent of time (that is, $\partial u/\partial t = 0$), the equation turned out to be

$$\frac{\partial^2 u}{\partial x^2} + \frac{\partial^2 u}{\partial y^2} = 0.$$

[6] J. L. Lagrange, "Recherches sur la nature et la propagation du son," *Miscellanea Taurinensia*, 1 (1759), classe mathématique i-x and 1–112.

[7] L. Euler, "Disquisitio ulterior super seriebus secundum multipla cuiusdam anguli progredientibus," *Nova Acta Academiae Scientiarum Petropolitanae*, 11 (1793: publ. 1798), 114–132.

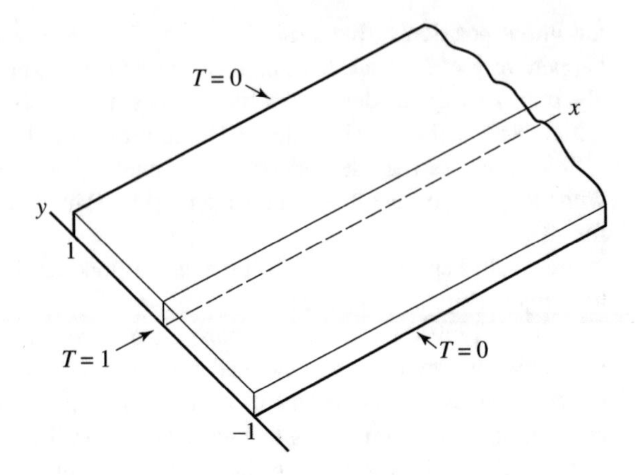

Figure 2.7 The thin lamina of Fourier's memoir.

To solve it mathematically, Fourier looked for solutions $u(x,\ y)$ that could be written as $\varphi(x)\psi(y)$. The method, now called "separation of variables," had already appeared in d'Alembert's paper of 1750 mentioned before. Upon substitution into the above equation, it is found that φ and ψ satisfy ordinary differential equations

$$\varphi(x)/\varphi''(x) = -\psi(y)/\psi''(y) = A$$

that are easily solved. With $A = 1/n^2$ he found the solutions $u(x,\ y) = e^{-nx} \cos ny$. Then he claimed that the general solution was given by a combination of these with arbitrary coefficients.

As for the vibrating string, when Fourier came to impose the boundary condition $u(0,\ y) = 1$, for y in between -1 and 1, he found

$$1 = a_1 \cos \frac{1}{2}\pi y + a_2 \cos \frac{3}{2}\pi y + a_3 \cos \frac{5}{2}\pi y + \cdots .$$

The next question was to determine the infinitely many coefficients that appeared in it. Before doing that Fourier observed that one might doubt the existence of such values since the right-hand side is evidently zero for $y = -1$ and "right afterwards" (*subitement*) it has to take the value 1; but he promised that such difficulty was going to be completely clarified in the sequel. He proceeded by differentiating the above formula term by term infinitely many times and, by substituting $y = 0$, he obtained a linear system of infinitely many equations in the infinitely many unknowns a_n with $n = 1, 2, 3, \ldots$. Then by a long and complicated method, which shows his iron determination, he found the correct value $4/\pi$ of the first coefficient and subsequently of all the others.

Next, Fourier explained what kind of function his series represented: It is periodic of period 4; for y between -1 and 1 it takes the constant value 1; at $y = \pm 1$ it is zero; and for y between 1 and 3 it is equal to -1. Speaking in modern terms, Fourier found the "Fourier series" of the *square wave* (Fig. 2.8). He computed the Fourier series for

Figure 2.8 The square wave, first to be developed in Fourier series.

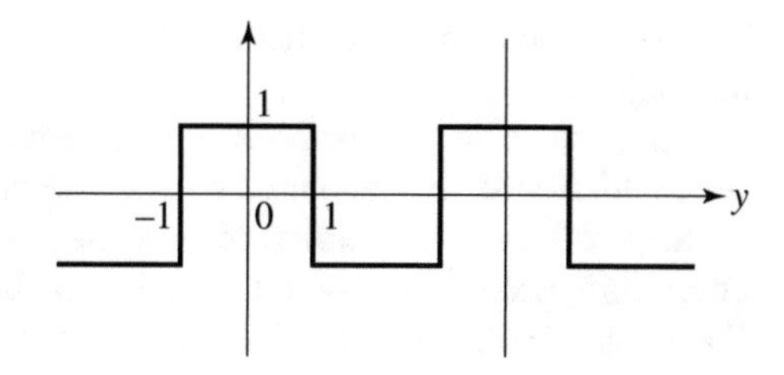

other specific periodic functions, such as the *sawtooth wave* and the *triangular wave* (Fig. 2.9), as well as the sine series of a "wave" that is constant, say 1, on the interval $(0,\ \alpha)$ and zero on the interval $(\alpha,\ \pi)$. Euler had stated[8] that functions of this kind "no doubt will be very difficult, not to say impossible" to represent in trigonometric series. Fourier did not mention Euler at this juncture. Rather he commented "these results fully confirm Daniel Bernoulli's opinion."

This already shows how much Fourier had advanced in the understanding of the nature of infinite sums; it is also apparent that Fourier had acquired a geometric concept of "periodicity" — that of a graph that repeats itself forever, infinitely many times — much wider than the old "algebraic" one, based on trigonometric functions.

Then Fourier went ahead with a first extension of the above, looking for the representation of a general function $f(x)$ by sine series, as in (2.2) in the case $l = 1$. First he expressed an infinitely differentiable $f(x)$, as well as the sine terms of (2.2), by Taylor's series. Then he equated the coefficients of the like powers of x. Ingeniously he derived an integral expression for A_n, after establishing it as a solution of a suitable ordinary differential equation of second order.

Fourier's examination of the foundational aspects of his new result concerned the generality of the function to which it applied. From physical considerations, he presumed that the result was true for "any" function, even those not differentiable. Indeed his formula involved the computation of an integral which, having the geometrical meaning of an area, does not require any differentiability at all.

Then, Fourier showed a much simpler method of obtaining the same formula, which is the now standard one of multiplying (2.2) through by $\sin \pi n x$ and integrating term by term. The formula and even the method was that used by Euler in the mentioned paper of 1777, published posthumously in 1798. (Fourier, who was reported to have learned of it later as indicated by Lacroix, did not let go without comment the charge of his having failed to refer to earlier works on the subject. For, in a letter, he wrote: "I am sorry not to have known the mathematician who first made use of this method because I would have cited him. Regarding the researches of d'Alembert and Euler could one not add that if they knew this expansion they made but a very imperfect use of it.")

About his last method he remarked that "it is just a useful abbreviation," but it is totally insufficient to solve all of the difficulties that his theory of heat presents: One

---

[8]L. Euler, "Sur le mouvement d'une corde, qui au commencement n'a été ébranlée que dans une partie," *Mémoires de l'Académie Royale de Berlin*, (1765: publ. 1767), 307–334.

has to be directed by other methods too, "needed by the novelty and the difficulty of the subject."

Really none of his methods is conclusive; rather it appears that, just as with Bernoulli, Fourier had acquired an intimate understanding of the physical meaning of the problem. Over a period of two years before the writing of his 1807 memoir, Fourier repeated all important experiments that had been carried out in England, France, and Germany and added experiments of his own. They gave him a number of striking experimental confirmations of his new theory, which went together with his overcoming of the difficulties advanced by the old masters. Fourier mentioned the motion of fluids, the propagation of sound and the vibrations of elastic bodies as other applications. He was fully aware of having opened up a new era for the solution of partial differential equations, having showed how discontinuous functions could be introduced in them: It was the era of linearization that would dominate mathematical physics for the first half of the nineteenth century and which has remained important ever since. The diffusion equation is a linear equation: Linear combinations of solutions are still solutions. It was not the first such equation to appear in history by any means (the wave equation is another), but the method Fourier used to solve it opened up enormous new possibilities. Before Fourier a differential equation was linear or nonlinear according to circumstances, but after him an effort was made to render a nonlinear physical problem into a linear model in order to exploit the power and generality of the method Fourier developed in his 1807 memoir.

2.3 Fourier coefficients and series

Under the broad assumption of absolute integrability, usually satisfied in practice, a function $f(x)$ defined on $(-\pi, \pi)$ can be developed in *Fourier series*

$$\begin{aligned} f(x) &= \frac{a_0}{2} + (a_1 \cos x + b_1 \sin x) \\ &+ (a_2 \cos 2x + b_2 \sin 2x) + \cdots + (a_n \cos nx + b_n \sin nx) + \cdots . \end{aligned} \tag{2.3}$$

The coefficients, named *Fourier coefficients*, are defined by

$$\begin{aligned} a_0 &= \frac{1}{\pi} \int_{-\pi}^{\pi} f(x)\, dx \\ a_1 &= \frac{1}{\pi} \int_{-\pi}^{\pi} f(x) \cos x\, dx \\ b_1 &= \frac{1}{\pi} \int_{-\pi}^{\pi} f(x) \sin x\, dx \end{aligned}$$

$$a_2 = \frac{1}{\pi}\int_{-\pi}^{\pi} f(x)\cos 2x\,dx$$
$$b_2 = \frac{1}{\pi}\int_{-\pi}^{\pi} f(x)\sin 2x\,dx \ldots,$$

with $a_0/2$ having the physical meaning of the average value of f over one period. A heuristic way to obtain the above formulas, for instance in the case of a_2, consists in multiplying both sides of (2.3) by cos $2x$ and integrating term by term, making use of the orthogonality relations $\int_{-\pi}^{\pi} \cos 2x \sin mx\,dx = 0$ for all integers m and $\int_{-\pi}^{\pi} \cos 2x \cos mx\,dx = 0$, unless $m = 2$, in which case π is obtained. That the right-hand side of (2.3), a periodic function of period 2π, coincides with $f(x)$ over $(-\pi, \pi)$ is a theorem due to P. J. L. Dirichlet that holds under a mild hypothesis (see next Section 2.4). Typical examples are the *sawtooth* and *triangular wave* in Figure 2.9 and the *square wave* in Figure 2.8.

A result that goes under the name of the *Riemann–Lebesgue theorem* proves that the coefficients a_n and b_n tend to zero as n tends to infinity. Thus an approximation of $f(x)$ can be obtained by summing finitely many terms. For a given function, how good such an approximation is might vary from point to point (Fig. 2.14 and Fig. 2.15).

If f is either an odd or an even function, then only the sines or the cosines appear in (2.3). Maybe just for fun, in his 1807 memoir, Fourier wrote the absolute value of $\sin x$ — an even function indeed — as a cosine series, namely $|\sin x| = 4\pi^{-1}(\frac{1}{2} - \frac{1}{1\cdot 3}\cos 2x - \frac{1}{3\cdot 5}\cos 4x - \cdots)$.

Fourier series can be written in two other equivalent ways. The first one makes use of the exponential form of complex numbers

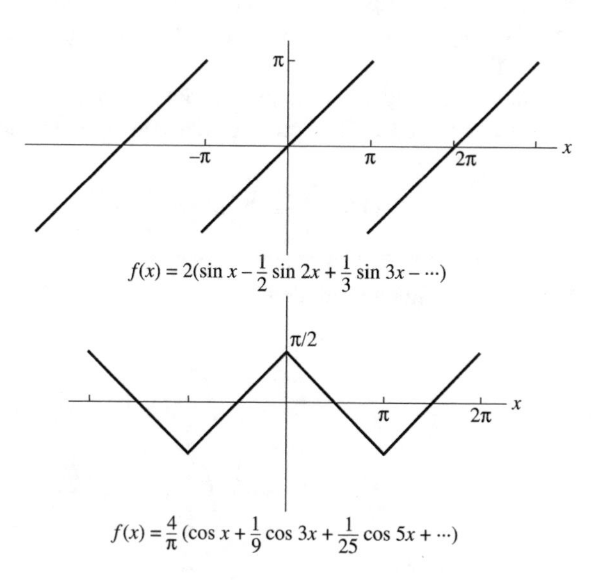

Figure 2.9 The sawtooth and the triangular wave with corresponding Fourier series.

$$f(x) = \sum_{n=-\infty}^{\infty} c_n\, e^{inx} \tag{2.4}$$

where the *complex Fourier coefficients* are defined as

$$c_n = \frac{1}{2\pi} \int_{-\pi}^{\pi} f(x)\, e^{-inx}\, dt.$$

The totality of Fourier coefficients is called the *spectrum* and the indices n that label them *frequencies*. The second one remains among real numbers and it accounts for the name, *harmonic analysis*, given to this theory

$$f(x) = \frac{A_0}{2} + \sum_{n=1}^{\infty} A_n \cos(nx + \phi_n). \tag{2.5}$$

The equivalence between (2.3) and (2.4) with $c_n = \frac{1}{2}(a_n - ib_n)$ and $c_{-n} = \frac{1}{2}(a_n + ib_n)$, n a nonnegative integer, follows from Euler's formula $e^{inx} = \cos nx + i \sin nx$. For the equivalence between (2.3) and (2.5) it suffices to switch to polar coordinates A_n and $-\phi_n$ by writing $a_n = A_n \cos \phi_n$ and $b_n = -A_n \sin \phi_n$ and to use the addition formula of the cosine.

The terms of the series in (2.5) are called *harmonics*. To $n = 1$ there corresponds the *fundamental harmonic* having *frequency* $1/2\pi$, the inverse of the period. To greater values of n there correspond the *higher harmonics* having $n/2\pi$ as frequency. Over one period the fundamental makes one complete oscillation (one *cycle*) and the higher harmonics n complete oscillations (n *cycles*). The numbers A_n and ϕ_n are called, respectively, *amplitudes* and *phases* of the harmonics. Geometrically, to a phase ϕ there corresponds a translation: The formula $\sin x = \cos(x - \frac{1}{2}\pi)$ means that the graph of $\sin x$ can be obtained by translating the cosine's graph by $\pi/2$ to the right.

More generally for functions f defined over $(-\frac{1}{2}T,\ \frac{1}{2}T)$, for instance those periodic of period T, the Fourier series corresponding to (2.3) reads

$$f(x) = \frac{a_0}{2} + \sum_{n=1}^{\infty} a_n \cos\left(\frac{2\pi nx}{T}\right) + b_n \sin\left(\frac{2\pi nx}{T}\right) \tag{2.6}$$

where the coefficients are defined by

$$a_n = \frac{2}{T} \int_{-\frac{T}{2}}^{\frac{T}{2}} f(x) \cos \frac{2\pi nx}{T} dx,$$

$$b_n = \frac{2}{T} \int_{-\frac{T}{2}}^{\frac{T}{2}} f(x) \sin \frac{2\pi nx}{T} dx.$$

Here, a_n stands for $a_{2\pi n/T}$ and similarly for b_n. So it is seen that if f is defined on a large interval the spectrum is denser; on the contrary if T is small. In any case it remains a *discrete spectrum*. Similar considerations hold for (2.4) and (2.5).

2.4 Dirichlet function and theorem

In a short time the disputes raised by Fourier's work reached a final clarification. It was the dawn of modern mathematics when, in 1837, Dirichlet proposed the by now standard concept of function $y = f(x)$ that associates a unique y to every x. Dirichlet gave the following striking example, named *Dirichlet function* after him:

$$D(x) = \begin{cases} 0 & \text{if } x \text{ is irrational,} \\ 1 & \text{if } x \text{ is rational,} \end{cases}$$

for x in the interval $(-\pi, \pi)$. That $D(x)$ it is not an "ordinary" function can be easily perceived: Between any two rational numbers (where D takes the value one), as close as they might be, there are always infinitely many irrational numbers (where D takes the value zero). Similarly, between any two irrationals there are always infinitely many rationals. This seeming word game has an implication: Due to the infinitely many "jumps," from zero to 1, the graph of D cannot be drawn, not even qualitatively. In spite of that the Dirichlet function is well defined: At $t = 0.5$ it takes the value 1, at $t = \sqrt{2}$ the value zero, just to give a couple of examples. More generally, for every x in $(-\pi, \pi)$ it suffices to establish whether x is rational or irrational to know the value of $D(x)$.

The example proposed by Dirichlet makes clear that the "graph" has no role to play in the concept of function; rather it is an incidental feature.

It is interesting to check the applicability of the Fourier theory to $D(x)$. Likely we are bound to meet with some "weird" phenomenon. First of all, to write the Fourier series all Fourier coefficients have to be computed; then the equality between the series and original function has to be checked. With Riemann integration theory, not even the first step goes through: Dirichlet's function is not integrable so that even a_0 cannot be calculated. Much later, in 1901, a more powerful theory of integration was proposed by Henri-Léon Lebesgue (1875–1941). (A modern presentation can be found in Rudin [1966]; an old and more elementary one is given in de la Vallée-Poussin [1950].) The Lebesgue integral "takes" the Dirichlet function and allows its Fourier coefficients to be computed: They are all zero. (Indeed, in the theory of the Lebesgue integral, the set of all rational numbers has measure zero and so it is disregarded. Only the value D takes on the irrational numbers matters.) It remains to note that the corresponding Fourier series is zero for all x's and so it does not sum to $D(x)$.

Even less eccentric functions have to be ruled out, such as those that oscillate infinitely many times. An example is in Figure 2.10 (note that the graph is only

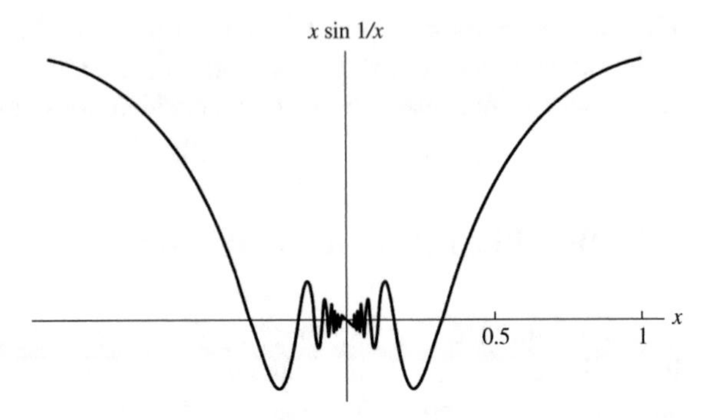

Figure 2.10 Graph of a function that oscillates infinitely many times in proximity of the origin.

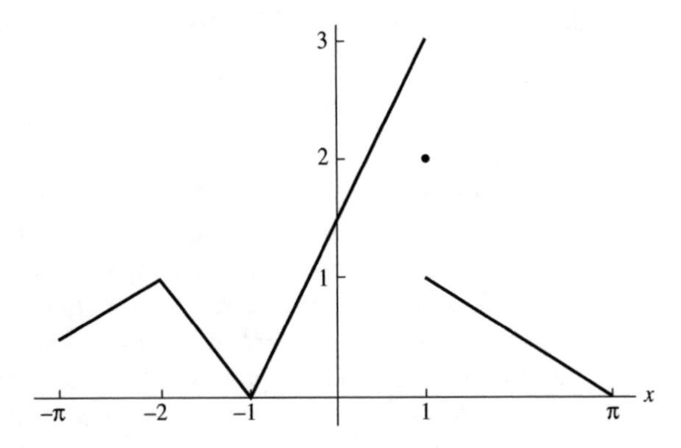

Figure 2.11 At the point of discontinuity $x_0 = 1$ the Fourier series of f converges to 2.

qualitative, for the infinitely many oscillations of $y = x \sin x^{-1}$ in proximity of the origin cannot be rendered).

Nevertheless, Fourier's intuition was correct overall due to the following *Dirichlet theorem*.[9] If a function f has at most a finite number of maxima and minima and of discontinuities, then its Fourier series at all points x where f is continuous converges to $f(x)$; at a point of discontinuity $x = x_0$ it converges to

$$f^*(x_0) = \frac{f(x_0+) + f(x_0-)}{2}$$

which is the average between the value from the right $f(x_0+)$ and from the left $f(x_0-)$ (Fig. 2.11). Note that in the last case $f^*(x_0)$ might be different from the value f takes at the point of discontinuity x_0.

The assumptions of the theorem are usually satisfied: They surely are by any graph that might be drawn on a sheet of paper (Fig. 2.12), eventually subdivided in pieces corresponding to functions $y = y(x)$ and $x = x(y)$.

---

[9]P. L. Dirichlet, "Sur la convergence des séries trigonométriques qui servent à représenter une fonction arbitraire entre des limites données," *Journal für die reine und angewandte Mathematik* 4 (1829), 157–169.

Figure 2.12 Leonardo, *Portrait of Isabella d'Este.* (The Louvre, Paris.) The profile, from the nose to the chin, can be expressed as a Fourier series.

It is worth mentioning that, under the broad assumption of absolute integrability alone, convergence can fail at any x (Kolmogorov [1926]). Nevertheless, in a celebrated proof, Carleson [1966] showed that if $|f|^2$ is integrable, then the Fourier series converges almost everywhere, that is except possibly on a set of Lebesgue measure zero.

2.5 Lord Kelvin, Michelson, and Gibbs phenomenon

Tides are an oscillatory phenomenon to which it is natural to apply Fourier analysis. Being primarily due to the combined gravitational effects of the moon and of the sun upon the oceans, the simplest model accounts only for these two forces, additionally assumed to be periodic. The periods are that of the rotation of the Earth with respect to the Moon and that of the rotation of the Earth with respect to the Sun. Take for instance the solar tide, which is the least complicated of the two. The fundamental harmonic (2.5) has a frequency of one cycle per day and the $n = 2$ harmonic — which is stronger in any given month — has a frequency of two cycles per day.

(Actually neither the solar tide nor the lunar tide are exactly periodic and so the corresponding Fourier coefficients a_1 and a_2 are not exactly independent of time, as periodicity would imply. Coefficients slightly varying over the course of the year are used to obtain a better mathematical model. In Section 4.9 another example can be found.)

Lord Kelvin (1824–1907), who began his scientific career with articles[10] on Fourier's work — and was the first to advance the idea that Fourier's mathematics could be used as well to study the flow of electricity in conductors — invented a machine for the purpose of predicting tides that computed a periodic function $h(t)$, the tide at time t, from its Fourier coefficients as well as another machine, the harmonic analyzer, that could compute the Fourier coefficients of a past height $h(t)$ on record (W. Thomson and P. G. Tait *Treatise on Natural Philosophy* (1867)). (A tide machine can be seen at the Smithsonian Museum in Washington, D.C.)

Albert Abraham Michelson (1852–1931), whose ability in building equipment to new standards of accuracy is legendary, constructed one of Kelvin's machines. It had to involve many more terms than previous models to reach a higher accuracy. To test it, Michelson fed in the first eighty Fourier coefficients of the sawtooth wave. To his surprise the machine produced a graph very close to that of the sawtooth, except for two blips on either side of the points of discontinuity. The effect of increasing the number of Fourier coefficients fed in was to move the blips closer to the points of discontinuity, but they remained there and their height remained about 18% above the correct value. After making every effort to remove mechanical defects that could account for the blips, their existence was confirmed by hand calculation (Michelson [1898]).

Josiah Willard Gibbs (1839–1903), one of America's greatest physicists, whose main field of interest was theoretical physics and chemistry (his formulation of thermodynamics transformed a large part of physical chemistry from an empirical to a deductive science) had spent almost three years in Europe and studied with Karl Weierstrass (1815–1897). In two letters to *Nature* — the second, Gibbs [1899], a correction of the first — showing an appreciation of mathematical fine points, he clarified the above phenomenon that ever since has gone under his name.

A careful examination of Dirichlet's theorem detects its pointwise nature: The convergence of a Fourier series at any point, those of discontinuity included, is described. The procedure is that of fixing a point and then summing all the infinitely many terms of the series. If only a finite number of them is summed, then an approximate value is obtained. Such an approximation will now be examined in a neighborhood of a point of discontinuity.

[10]"On Fourier's expansions of functions in trigonometric series," *Cambridge Mathematical Journal* 2 (1839–1841), 258–262; "Note on a passage in Fourier's Heat," ibid., 3 (1841–43), 25–27; "On the linear motion of heat," ibid., 3 (1841–1843), 170–174 and 206–211.

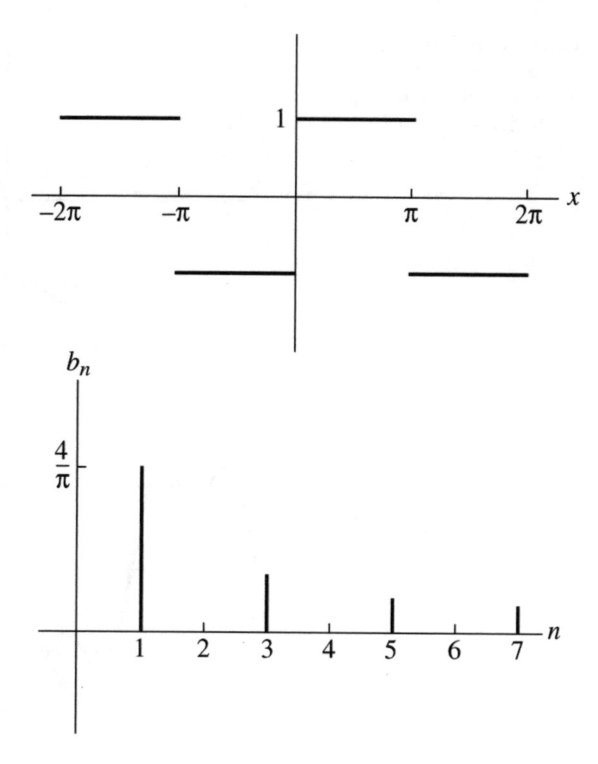

Figure 2.13 Above, the square wave. Below, its Fourier coefficients b_n.

Let us consider the simplest example, the (asymmetrical) *square wave* defined by

$$g(x) = \begin{cases} -1 & \text{if } -\pi < x < 0, \\ 0 & \text{if } x = 0, \\ 1 & \text{if } 0 < x < \pi, \end{cases}$$

clearly discontinuous at $x = 0, \pm\pi, \ldots$ (Fig. 2.13). The coefficients having been computed, the related Fourier series can be written. The following rather striking equality, which Fourier calculated himself, holds

$$g(x) = \frac{4}{\pi}(\sin x + \frac{1}{3}\sin 3x + \frac{1}{5}\sin 5x + \cdots) \tag{2.7}$$

On the right only smooth functions appear; on the left there is a function with jumps. Euler never believed it was possible. Trigonometric series do account for that, but as shall be seen, it is not an easy job. Writing $S_1(x) = 4\pi^{-1}\sin x$, $S_2(x) = 4\pi^{-1}(\sin x + \frac{1}{3}\sin 3x)$, and more generally, $S_n(x)$ for the sum of the first n terms of the series, let us fix x_0 as in Figure 2.14. There S_1 and S_2 take a small value, rather far from the value of unity for the square wave. If many more terms are added, then the approximation improves (Fig. 2.15). Nevertheless, independently of how many terms one might sum, it is impossible to obtain a good approximation over an entire interval containing

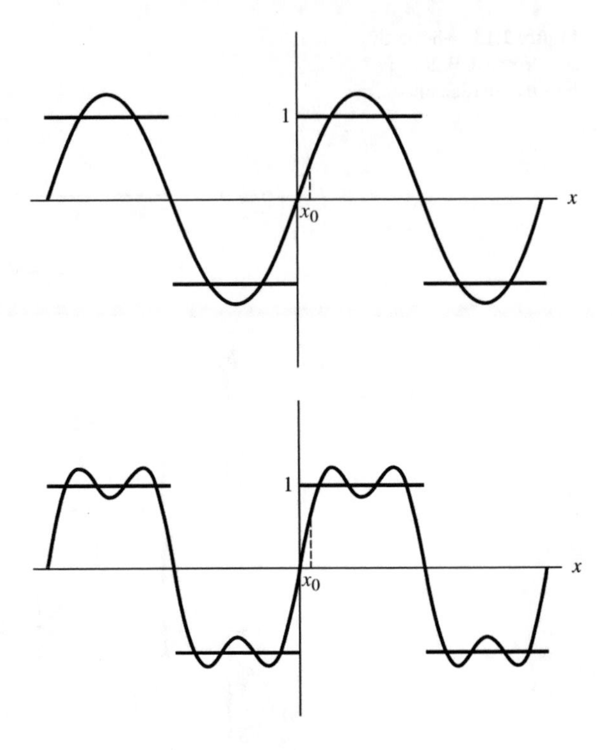

Figure 2.14 Superimposed on the square wave: above $S_1(x)$ and below $S_2(x)$, the first and second partial sums.

the origin: There are always two blips next to the origin that are easily pinpointed, being visibly different from the flat graph of the square wave. The height of the blips is about $y = 1.18$ against $y = 1$ of the underlying wave. In other words the approximation of the square wave, given by any partial sum, is not uniform over the whole interval $(0, \pi)$.

This phenomenon of *overshooting*, of practical relevance in automated computation of Fourier series, had originally been discovered by an English mathematician Wilbraham [1848]. (The property in question, discussed by both Wilbraham and Gibbs for a particular series, characterizes the behavior of the Fourier series of a function in the neighborhood of a point of jump discontinuity. This was showed by Maxime Bôcher [1906], who gave it the name the Gibbs phenomenon (see also H. S. Carslaw [1930] for the history)). Computers were not around yet and Wilbraham's point, regarded as a mere curiosity, was likely forgotten fifty years later. Unfortunately timing is important in scientific discoveries too.

2.6 The construction of a cellar

The problem of heat diffusion attracted Fourier's attention as well as that of other contemporary mathematicians. For instance Poisson published a large part of his results, which originally appeared in the *Journal de l'École Polytechnique*, in the book *Théorie mathématique de la chaleur* [1835] (*Mathematical Theory of Heat*).

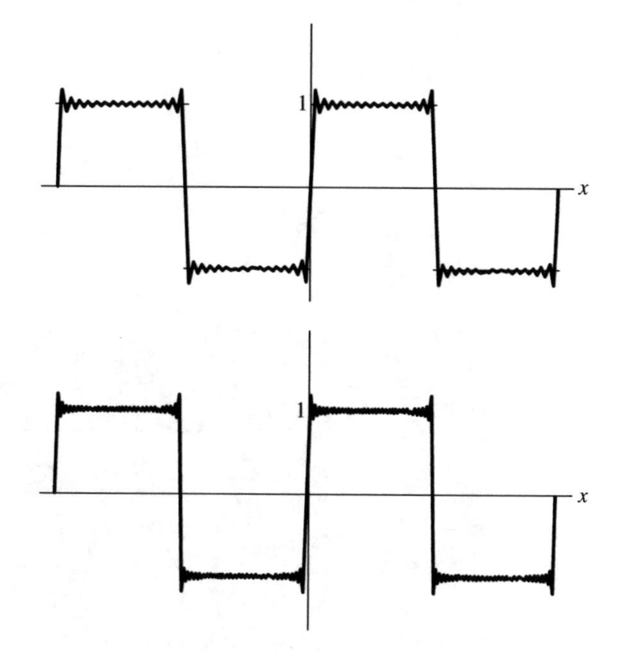

Figure 2.15 The partial sum $S_n(x)$, of the Fourier series of the square wave, for $n = 15$ above and $n = 22$ below.

The topic was interesting not only from a theoretical viewpoint, relative to the determination of the temperature in the interior of the Earth, but also from a practical viewpoint, such as metal working.

To illustrate the principle in a simple way, suppose a cellar has to be built. It is natural to ask what its depth should be in order for it to be cool and with only small variations of temperature as the seasons change (ideally it ought to keep the same temperature at all times).

As time goes by, heat propagates through the earth's crust as a wave so that the hottest day of the year at the surface is not the hottest day in the interior because the peak of temperature there will occur with a delay (*change of phase*). Think for instance of springs in the mountains: During summer their water is cool, often so cool indeed that it can be drunk only in small sips, whereas during winter it is found at an ideal temperature. Also it can be expected that both heat and cold, when they reach deep inside, are attenuated (*damping*).

If the problem is formulated and solved mathematically, as below, the preceding qualitative considerations can be quantified. It turns out that the heat wave reaches 4.5m below the Earth's crust with a change of phase of π, which is equivalent to a six-month delay: If at the surface summer is starting, at that depth winter is starting. The damping factor is one sixteenth; that is, the winter cold in the cellar is one sixteenth of the winter cold at the surface, and the same for the summer heat. Thus at 4.5m summer and winter are switched around and if at the surface the temperature change during an entire year is 40 degrees, in the cellar it is only two and half degrees (Fig. 2.16).

Figure 2.16 Fourier descending into a cellar. (Drawing by Enrico Bombieri.)

The mathematical model that proves what is stated assumes a flat Earth. This is a good approximation if one is not interested in knowing the temperature at great depths. For the same reason, heat sources in the interior of the Earth can be disregarded. Depth will be denoted by x, time by t, and the unknown temperature by $u(x,\ t)$. Only at the surface is the temperature known, that is $u(0,\ t) = f_0(t)$ is given (boundary condition). The partial differential equation of heat diffusion in this case is

$$\frac{\partial u}{\partial t} = K \frac{\partial^2 u}{\partial x^2} \tag{2.8}$$

where K is a constant depending upon the soil. The temperature is assumed periodic in time, one year being the period. To a large period T there correspond densely packed frequencies: the fundamental frequency $1/T$ accompanied by the multiples n/T. It is useful to work with *angular frequencies* $n\omega$, with $\omega = 2\pi/T$ the *fundamental angular frequency*. The method for solving the problem begins by expanding the unknown function $u(x,\ t)$ in Fourier series with respect to time t. The Fourier coefficients depend on the choice of x

$$u(x,\ t) = \sum_{n=-\infty}^{\infty} u_n(x)\, e^{i\omega nt}. \tag{2.9}$$

Note that the time variable t and the space variable x are now separated and that the dependency upon t is explicit. Differentiating term by term — assuming that is permissible — and upon substitution, (2.8) becomes

$$\sum_{n=-\infty}^{\infty} i\omega n u_n(x)\, e^{i\omega nt} = K \sum_{n=-\infty}^{\infty} \frac{d^2u_n}{dx^2}\, e^{i\omega nt}.$$

By the *uniqueness* of Fourier series (that can be proved), the coefficients on both sides have to be equal. Thus the heat equation gets transformed into infinitely many equations, one for every coefficient

$$K\frac{d^2u_n}{dx^2} = i\,\omega\, n\, u_n. \tag{2.10}$$

Time t has disappeared: (2.10) is an ordinary differential equation in the x variable, easy to solve, being linear with constant coefficients. The solution is

$$u_n(x) = C_n\, e^{\alpha_n x} + D_n\, e^{-\alpha_n x} \tag{2.11}$$

where C_n and D_n are arbitrary constants and $\alpha_n = (1 \pm i)\sqrt{|n|\omega/2K}$, depending upon $n > 0$ or $n < 0$. Because the real part of α_n is positive, $C_n = 0$. Otherwise the solution in (2.11) would go to infinity as x tends to infinity, whereas it has to be bounded for lack of interior sources (boundary condition at infinity). Therefore $u_n(x) = D_n\, e^{-\alpha_n x}$. Finally back to (2.9): When the boundary condition $u(0,\ t) = f_0(t)$ is imposed and developed in Fourier series $f_0(t) = \sum_{n=-\infty}^{\infty} c_n\, e^{i\omega nt}$, it is found that $D_n = c_n$. Therefore the problem is solved by

$$u(x,\ t) = \sum_{n=-\infty}^{\infty} c_n\, e^{i\omega nt - \alpha_n x}.$$

The result is easier to decode if it is written in real terms

$$f_0(t) = \frac{A_0}{2} + \sum_{n=1}^{\infty} A_n\, \cos(n\omega t + \varphi_n)$$

and

$$u(x,\ t) = \frac{A_0}{2} + \sum_{n=1}^{\infty} u_n(x)\, \cos(n\omega t + \phi_n(x))$$

where

$$u_n(x) = A_n \exp\left(-\sqrt{\frac{n\omega}{2K}}x\right),$$

$$\phi_n(x) = \varphi_n - \sqrt{\frac{n\omega}{2K}}x.$$

The amplitude $u_n(x)$ of the nth harmonic decreases exponentially as the depth increases. It suffices to consider the first harmonic, the least damped. Both phase $\phi_1(x)$ and damping $\exp(-\sqrt{\omega/2K}x)$ depend upon the quantity $\sqrt{\omega/2K}$, which is approximately $2\pi \times 900^{-1}\text{cm}^{-1}$, with $T = 365 \times 24 \times 3600$ s and $K = 2 \times 10^{-3}\,\text{cm}^2\text{s}^{-1}$ for ordinary soil. Therefore at the depth $x = 450$ cm the solution shows a change of phase with respect to the initial data of $\sqrt{\omega/2K}x$, which is approximately π: The heat wave gets there with a delay of $T/2$ (six months). The damping factor is about $e^{-\pi} \cong 1/16$. The results are even more striking if the period T is chosen to be one day. With $T = 24\times 3600$ s, a change of phase of π — that is midnight instead of midday — and 1/16 as the concomitant damping factor of the first harmonic are found to take place at 23 cm. Thus, the daily temperature fluctuations penetrate the ground with noticeable intensity only for a few centimeters (*skin effect*).

2.7 The Fourier transform

In the general case of a function f defined over $(-\frac{1}{2}T, \frac{1}{2}T)$ the formula for the complex Fourier coefficients,

$$c_n = \frac{1}{T}\int_{-T/2}^{T/2} f(x)\, e^{-i2\pi nx/T}\, dx \tag{2.12}$$

where c_n stands for $c_{2\pi n/T}$, can be formally extended to include functions defined on the entire real line by letting T tend to infinity. Then the frequencies n/T invade all the reals and give rise to a *continuous spectrum*. Thus the case of a nonperiodic function is dealt with as a limiting case of periodicity, the period being infinite. Accordingly the *Fourier integral* is defined by

$$F(\omega) = \int_{-\infty}^{\infty} f(t)\, e^{-i\omega t}\, dt \tag{2.13}$$

or alternatively by (2.15) or (2.16). (Note that the missing multiplicative factor $1/2\pi$ in (2.13) will appear in the inversion formula (2.14).) F is called the *Fourier transform* of f and is also commonly denoted by $\hat{f}$. Under the assumption that f is absolutely integrable, (2.13) is well defined.

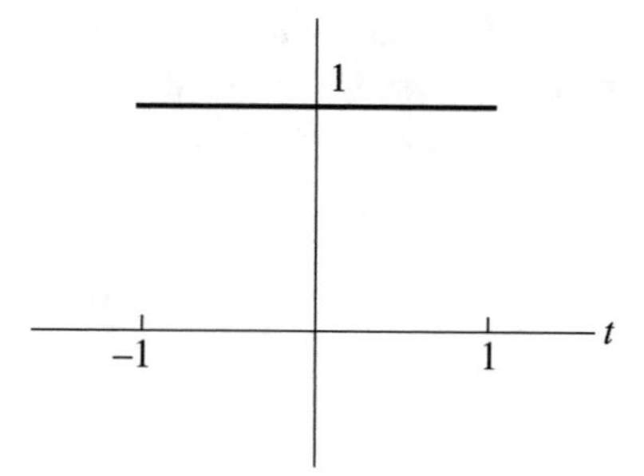

Figure 2.17 The rectangular pulse $r(t)$.

The Fourier transform of the *rectangular pulse* $r(t)$, equal to 1 on the interval $(-1, 1)$ and zero otherwise (Fig. 2.17), is found by an immediate integration to be equal to $2\omega^{-1}\sin\omega$ (Fig. 2.18). The geometric meaning of $R(\omega) = \int_{-1}^{1}\cos\omega t\,dt$ for a fixed ω is that of an area that varies continuously with ω. So $R(\omega)$ is continuous even though $r(t)$ is not. Continuity is a general property of the Fourier transform F, defined in (2.13), under the basic assumption that f is absolutely continuous.

Similarly, the Fourier series corresponding to the coefficients in (2.12) leads to the so-called *Fourier integral theorem* or *inversion formula*

$$f(t) = \frac{1}{2\pi}\int_{-\infty}^{\infty} F(\omega)\,e^{i\omega t}\,d\omega, \tag{2.14}$$

which holds under the assumption that $f,\ F$ are absolutely integrable and continuous. A more general statement that takes into account functions f showing a finite number of "jumps" discontinuities holds (Körner [1988]). Thus, functions satisfying the stated assumptions can be expressed as a "sum" of infinitely many harmonics

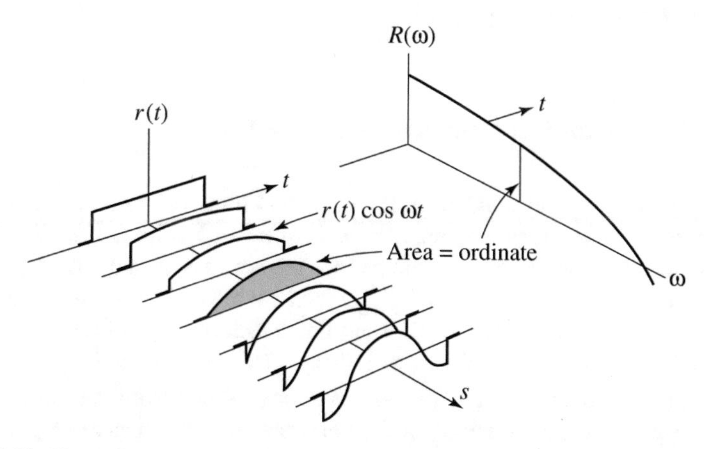

Figure 2.18 The surface $r(t)\cos\omega t$ sliced in one of two possible ways and the geometric interpretation of the Fourier transform $R(\omega)$ of the rectangular pulse $r(t)$, (R. N. Bracewell, *The Fourier Transform and Its Applications*, McGraw Hill [2000]), with permission.

$F(\omega)\, e^{i\omega t}$, where ω is any real number. The above (2.14) first appeared in a paper[11] by Fourier that was a summary of his book yet to be published. It is common to find (2.13) and (2.14) written in two other equivalent ways, one being

$$F(\omega) = \int_{-\infty}^{\infty} f(t)\, e^{-i2\pi\omega t}\, dt,$$
$$f(t) = \int_{-\infty}^{\infty} F(\omega)\, e^{i2\pi\omega t}\, d\omega \tag{2.15}$$

and the other

$$F(\omega) = \frac{1}{\sqrt{2\pi}} \int_{-\infty}^{\infty} f(t)\, e^{-i\omega t}\, dt,$$
$$f(t) = \frac{1}{\sqrt{2\pi}} \int_{-\infty}^{\infty} F(\omega)\, e^{i\omega t}\, d\omega. \tag{2.16}$$

The integrability hypothesis on the entire real line, both for f and F, is rather strict. For instance $\sin x$ does not satisfy it and so it has no Fourier transform (but it has a Fourier series, namely itself, being integrable on $(-\pi,\ \pi)$). Also $R(\omega)$ in Figure 2.18 is not absolutely integrable (it does not die off suitably fast at infinity) and the inversion formula (2.14) does not hold for every t. These restrictions can be lifted to include a larger class of functions and more, as in Section 3.4 where the notion of Fourier transform will be defined on *generalized functions*.

It is worth mentioning that, even if f is real-valued, its Fourier transform might be complex-valued. In other words it is characterized by *modulus* $|F(\omega)|$ and *phase* $\phi(\omega)$ and both have to be known to reconstruct f by the inversion formula (see Chapter 6 where lack of information on the phase constitutes a major problem in crystallography). Two basic properties of the Fourier transform follow.

2.8 Dilations and the uncertainty principle

With a any positive real number, $f_a(t) = f\,(at)$ turns out to be a dilation or a contraction of $f(t)$, depending upon $a < 1$ or $a > 1$ (Fig. 2.19). The Fourier transform evaluated at the origin accounts for the area associated with the original function, as in the formula $\hat{f}(0) = \int_{-\infty}^{\infty} f(t)\, dt$, which follows directly from the definition (2.13) by setting $\omega = 0$. Then the transform of a dilated $f\,(at)$ will present a multiplying factor of a^{-1}, as indicated by the formula $\int_{-\infty}^{\infty} f\,(at)\; dt = a^{-1} \int_{-\infty}^{\infty} f(t)\, dt$, to account for the change of area. Aside from that, to a dilation there corresponds a contraction of the Fourier transform and vice-versa (Fig. 2.19). Indeed

[11] J. B. J. Fourier, "Théorie de la chaleur," *Annales de chimie et de physique*, 3 (1816), 350–375.

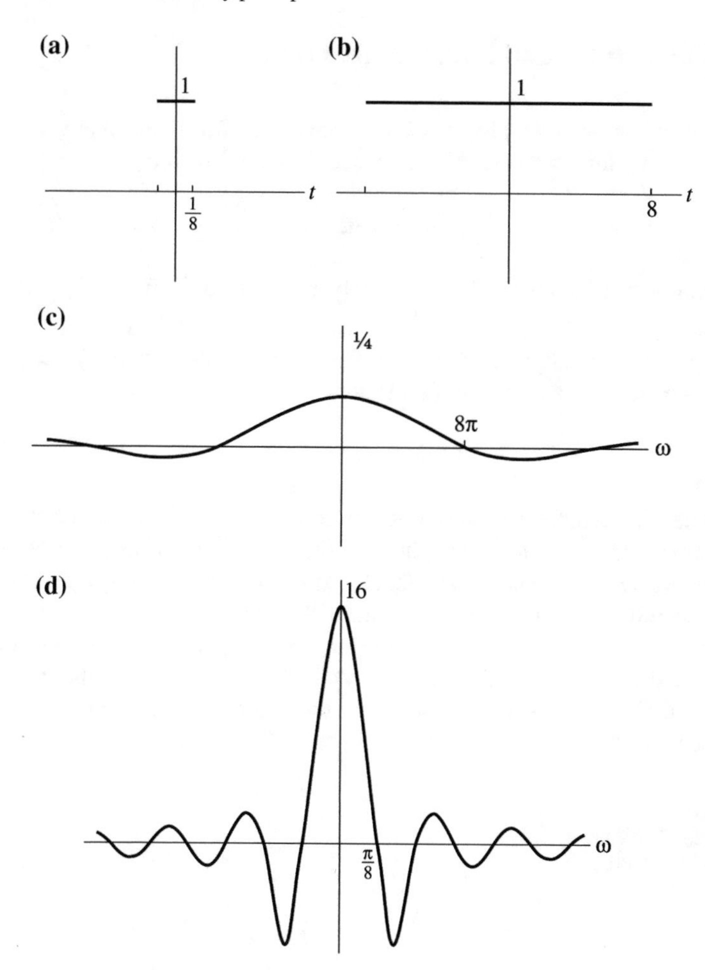

Figure 2.19 a) The rectangular pulse contracted by 1/8, and dilated by 8 in b); in c) and d), respectively, the Fourier transforms.

$$\hat{f}_a(\omega) = a^{-1}\hat{f}(\omega/a). \tag{2.17}$$

This formula is proved by the change of variables $u = at$ in the integral $\hat{f}_a(\omega) = \int_{-\infty}^{\infty} e^{-i\omega t} f(at)dt = \int_{-\infty}^{\infty} e^{-i(\omega/a)u} f(u)a^{-1}du = a^{-1}\hat{f}(\omega/a)$.

Formula (2.17) makes clear that a function and its Fourier transform cannot be both of short "duration." On the other hand, if details of f as small as a are to be reconstructed (Fig. 5.6), then $\hat{f}$ has to be known on an interval as big as $1/a$. Formula (2.17) is known as the *uncertainty principle*. Its various interpretations depend on the context and the meaning of the term "duration." For instance in quantum mechanics this principle involves position and momentum.

2.9 Translations and interference fringes

To the multiplication of $f(t)$ by the exponential $e^{i\omega_0 t}$, with ω_0 real, there corresponds a translation of $\hat{f}(\omega)$, or a delay if ω stands for time. Indeed if $g(t) = e^{i\omega_0 t} f(t)$, then

$$\hat{g}(\omega) = \hat{f}(\omega - \omega_0). \tag{2.18}$$

This formula can be checked immediately by writing $\hat{g}(\omega) = \int_{-\infty}^{\infty} e^{-i\omega t} e^{i\omega_0 t} f(t)dt = \int_{-\infty}^{\infty} e^{-i(\omega-\omega_0)t} f(t)dt = \hat{f}(\omega-\omega_0)$. The graphical illustration in Fig. 2.20 involves the rectangular pulse once more. The opposite holds too. If $h(t) = f(t-t_0)$, then it can be proved, as easily as (2.18), that

$$\hat{h}(\omega) = e^{-it_0\omega} \hat{f}(\omega). \tag{2.19}$$

Formula (2.19) matters in questions of interference. For instance a function like the one in Figure 2.21 can be thought of as the sum of two rectangular pulses, one translated to A and the other to $-A$. Hence, it is given by the formula $r(t-A)+r(t+A)$. The corresponding Fourier transform is $\hat{r}(\omega)(e^{iA\omega} + e^{-iA\omega}) = 2\hat{r}(\omega)\cos A\omega$, where the oscillating factor $\cos A\omega$ accounts for the so-called *interference fringes* (Fig. 2.22). If A is very big, then $\cos A\omega$ oscillates sharply making the fringes very dense and difficult to distinguish one from the other; if on the contrary A is very small, then the fringes are very sparse. In Figure 2.22, A is of the order of magnitude

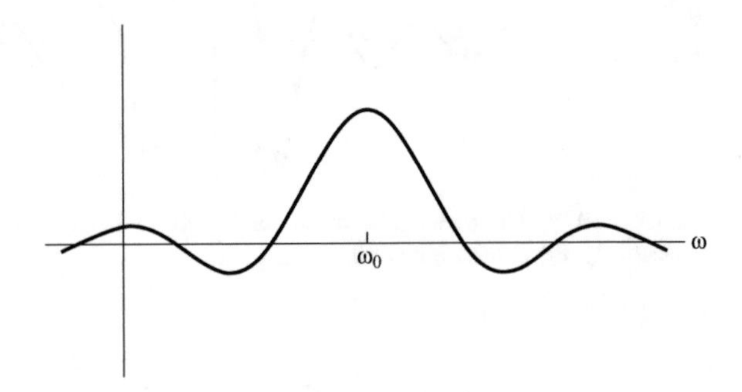

Figure 2.20 The Fourier transform of $e^{i\omega_0 t} r(t)$.

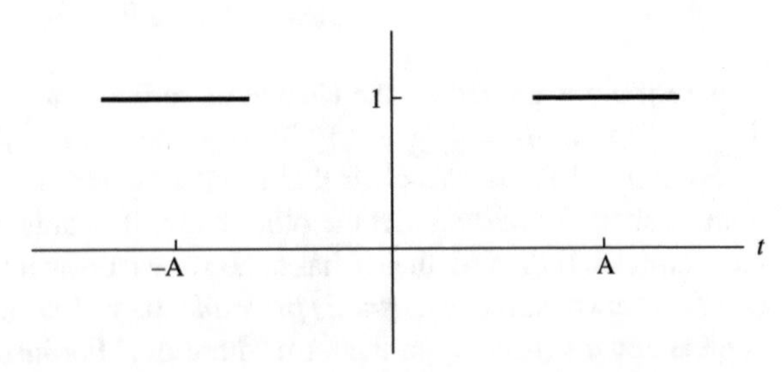

Figure 2.21 The rectangular pulse translated to A and $-A$.

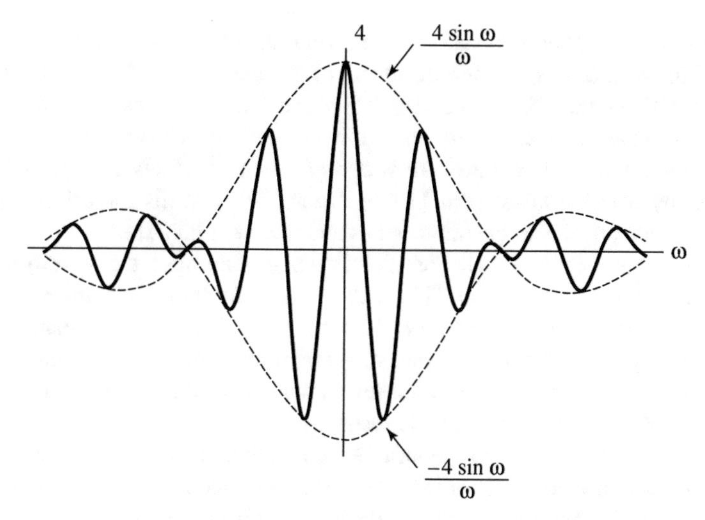

Figure 2.22 Graph of the Fourier transform of the function in Figure 2.21 with A=2.

of 1 — so A is comparable to the pulse duration — and the fringes stand out nicely (see also Figures 5.7 and 5.8 pertaining to the same phenomenon in two dimensions).

The phenomenon of interference is relevant to many different fields. One example is the *principle of interferometry* (Chapter 9), which greatly improved the resolving power of radiotelescopes, so allowing the development of radioastronomy. This in turn changed our conception of the universe.

2.10 Waves, a unifying concept in science

Since Fourier's time the theory has been greatly extended in dimension 1 (Zygmund [1968], Edwards [1967]), greater than 1 (Stein and Weiss [1971]), and more generally on groups and even on discrete structures. The decomposition in waves typical of Fourier analysis finds applications to partial differential equations, which lie at the origin of the theory. The wellknown mathematical physicist Arnold Sommerfeld (1868–1951) opens his classical text *Partial Differential Equations in Physics* [1967] with the statement "Fourier's *Théorie Analytique de la Chaleur* is the bible of the mathematical physicist. It contains not only an exposition of the trigonometric series and integrals named after Fourier, but the general boundary value problem is treated in an exemplary fashion for the typical case of heat conduction."

Audrey Terras at the very beginning of her book *Harmonic Analysis on Symmetric Spaces and Applications* [1985] writes "Since its beginnings with Fourier (and as far back as the Babylonian astronomers), harmonic analysis has been developed with the goal of unravelling the mysteries of the physical world of quasars, brain tumors, and so forth, as well as the mysteries of the nonphysical, but no less concrete, world of prime numbers, diophantine equations, and zeta functions." (As for the reference to Babylonian astronomers, it was a discovery of Neugebauer [1952] that Babylonians used a primitive kind of Fourier series for the prediction of celestial events.)

The role of harmonic analysis in quantum mechanics has been hinted at in Section 2.8. Moreover the Fourier transform is one of the most widely used techniques in quantum physics (Abrikosov et al. [1975], Dirac [1947], Walls and Milburn [1994]) from solid state physics to quantum optics and quantum field theory.

This book takes off in yet another direction that relies upon the waveform model of the electromagnetic spectrum. This leads to major fields of contemporary research that make use in a fundamental way of harmonic analysis in one dimension, such as electrical signals and computer music, in two dimensions, such as Fourier optics, computerized tomography, and radioastronomy, and in three dimensions, such as crystallography and nuclear magnetic resonance.

In this we might be following Fourier's own inclination, as reported in a letter by Karl Gustav Jacob Jacobi (1804–1851) to Legendre. Jacobi — himself thinking that "the only goal of science is the honor of the human spirit and in this respect a question in the theory of numbers is as valuable as a problem in physics" — refers in the same letter to a rather different belief, when he writes: "It is true that Fourier was of the opinion that the chief end of mathematics is the public good and the explanation of the natural phenomena" (Jacobi [1846]).

Chapter 3
Telecommunications and Space Exploration

3.1 The great electrical revolution

Technically speaking, signals such as the train's whistle, the blinking of a car's beams,... are quantities, varying in time, to which an informative content is assigned. The signals of smoke by day and fire by night, used by American Indians in early colonial times, are just another example, common to ancient civilizations from China to Greece. In the play *Agamemnon* by the Greek dramatist Aeschylus, quantitative information can be found: Nine relay points on natural prominences covered a distance of 800 km, by means of beacons of fires, to bring to Queen Clytemnestra in Argos the news of the imminent homecoming—from the Trojan war—of her victorious husband and to prompt her desire for vengeance. It was more than a thousand years before Christ. These may be regarded as early forms of *telegraphy*, which means "to write far" from the Greek. In telegraphy a message, consisting of a set of words, is transmitted by ascribing to each letter a certain coded signal.

The alphabet can be encoded into a matrix. The 25 entries of a 5 by 5 matrix suffice for the 24 letters of the ancient Roman alphabet. The ancient Greeks had a somewhat similar idea already by 300 B.C., as recorded by the historian Polybius. By ten vases, any letter could be identified: for instance omega, the 24th and last letter of the Greek alphabet, was represented by five vases placed on a low left wall and four on the right wall. Indeed (5,4) is the position in the matrix of its 24th element, counting by rows.

The Italian Gerolamo Cardano (1501–1576)—physician of fame all over Europe and major mathematician of his time—replacing space configuration by time configuration, suggested in 1551 the use of five torches set on towers to spell out letters and therefore messages. Each torch would figure in the code as "light" or "dark" allowing 32 combinations.

After the invention of the telescope in the seventeenth century, two "optical telegraphs," based on Cardano's idea, were built in 1794 and 1795. The first one in France, under and for the government of the First Republic, consisted of relay

E. Prestini, *The Evolution of Applied Harmonic Analysis*,
Applied and Numerical Harmonic Analysis,
DOI 10.1007/978-1-4899-7989-6_3

stations equipped with operators and telescopes, placed on hilltop towers spaced five to ten kilometers apart (3 to 6 miles). On the top of the tower a vertical timber supported a beam with an arm at both ends that could assume seven angular positions, accommodating in all 49 letters and symbols. A first line, from Lille to Paris, was built for the needs of the war against the Austrians. Under good visibility conditions, it provided a way of communication tens times faster than mounted couriers. Its inventor, the cleric Claude Chappe (1760–1829), gave it the name of *telegraph* and within a decade variants appeared in Europe, in Russia, and eventually in India and Egypt. The second such device was established for the Admiralty in England. Made with six shutters that could be open or closed, it allowed 64 combinations. It spread to the United States where a line, mainly used by merchants, connected New York to Philadelphia until 1846.

At the University of Bologna the Italian physician and physicist Luigi Galvani (1737–1798) held lectures on the anatomy of frogs and, in the late 1770s, on the entirely new subject of electrophysiology. He made numerous ingenious experiments and obtained muscular contraction in a frog by a metallic contact between leg muscles and nerves leading to them. He held that as a manifestation of a new form of electricity that he termed "animal" electricity, to be added to the known "natural" electricity (lightning) and static electricity obtained from friction.

Alessandro Volta (1745–1827), professor of physics at the nearby University of Pavia, on the one hand thought that Galvani's work "contains one of the most beautiful and most surprising discoveries," on the other hand held a different explanation, thinking that the contact of dissimilar metals in a moist environment was the true source of stimulation. While partisan groups were taking shape, the two scientists went further ahead, each following his own line of thought. Galvani succeeded in causing a muscle to contract by touching the exposed muscle of one frog with the nerve of another, thus establishing for the first time the existence of bioelectric forces within living tissues, while Volta provided a dependable source of continuous current with his battery, named after him *voltaic pile*. The threshold of the great electrical revolution was in sight.

In 1777 Volta proposed to string an iron-wire signaling line, supported on posts, from Como to Milan, but the way to the true future telegraph was opened by the Danish Hans Christian Ørsted who observed, in 1819, the deflection of a pivoted magnetized needle by an electric current. In Göttingen the great mathematician Carl Friedrich Gauss (1777–1855)—after whom the unit of magnetic induction is named—and his friend, the physicist Wilhelm Eduard Weber (1804–1891)—after whom the unit of magnetic flux is named—studied terrestrial magnetism. In 1833 they constructed a two-wire telegraph line of copper, carried 2.3 kilometers over housetops, a small project for limited finances. Working alone Weber connected the astronomical observatory, directed by Gauss, to the physics laboratory by a double wire that broke "uncountable" times as he strung it over houses and two towers.

In 1835 Samuel F. B. Morse (1791–1872), American painter and inventor, drew his Morse Code of dots and dashes. In connection he had a device, something like a single typewriter key, which when depressed, sent an electric signal to a distant receiver. This embossed a series of dots and dashes on a paper roll. In 1872 Jean

Maurice Emile Baudot (1845–1903) made another major innovation by inventing a device that could send many messages altogether. By the end of the century numerous telegraph lines and cables across the Atlantic Ocean were available for this purpose and by about 1960 the receiver had developed into a machine, known as a *teleprinter*, capable of typing almost a thousand characters per minute. The telegraph remained the most important system of telecommunications until the advent of the telephone.

The English scientist Michael Faraday (1791–1867) stands at the origin of both telephone and radio techniques. One of four children of a poor blacksmith, Faraday, who could recall times in his childhood when he was given one loaf of bread that had to last him for a week, received rudiments of formal education at church Sunday school. At an early age he began earning money by delivering newspapers and at fourteen became apprentice to a bookdealer and bookbinder: Faraday read the books brought in for binding. In 1812 he became assistant of the prominent chemist Sir Humphry Davy (1778–1829) and held the post until 1820. Then a series of discoveries that astonished the scientific world followed, in chemistry first and then in physics pertaining to electricity and magnetism.

Unlike his contemporaries, Faraday thought of electricity as a vibration or force rather than a material fluid flowing through wires. Having worked on the theory of sound, he was familiar with the patterns formed by light powder spread on iron plates when these are thrown into vibration, for instance, by a violin bow, and impressed by the fact that such patterns could be induced in one plate by bowing another near by, he attempted a similar experiment—his most famous one—involving electricity. It had been known for more than a decade that the flow of an electric current through a wire produces a magnetic field around the wire when, on August 1831, Faraday wound a thick iron ring, on one side with insulated wire connected to a battery and, on the opposite side, with wire connected to a galvanometer (instrument for measuring a small electrical current). When he closed the primary circuit the galvanometer needle jumped, signaling that a current had been induced in the secondary circuit by a current in the first one. That was what Faraday expected but, to his surprise, when he opened the circuit he saw the needle jump in the opposite direction, providing evidence that a changing magnetic field creates an electric field (*Faraday's law of induction*). The quantitative relation between the two was later formulated by the Scottish physicist James Clerk Maxwell (1831–1879) by means of the classical equations named after him.

In the preface to his *A Treatise on Electricity and Magnetism* (1873) Maxwell stated that his major task was to convert Faraday's physical ideas into mathematical form. His interest in Faraday's work was already apparent in his first electrical paper "On Faraday's Lines of Forces" read to the Cambridge Philosophical Society in December 1855. In his second paper "On Physical Lines of Forces," which appeared in 1861 in the *Philosophical Magazine*, he was lead to the theoretical conclusion that a changing electric field would induce a magnetic field—the converse of Faraday's experimental discovery—for it became apparent in his model that the two were interchangeable. Furthermore he was to conclude that field changes could be transmitted through dielectric media (insulators), including air and free space (which Maxwell considered to be permeated by ether). He wrote: "A conducting body may be compared to a porous membrane which opposes more or less resistance to the passage of a fluid, while a dielectric is like an elastic membrane which may be impervious to the fluid, but transmits the pressure of the fluid on one side to that on the other." Any electrical disturbance would propagate in the form of transverse waves through his model dielectric medium with finite velocity, which he calculated using electrical measurements for air made by R. Kohlrausch and W. Weber. It turned out to be close to the velocity of light. From this "he could scarcely avoid to infer that light consists in transverse undulations of the same medium which is the cause of electric and magnetic phenomena." The discovery of the electromagnetic spectrum had started (Fig. 5.9), even though in Maxwell's time and for eight years after his death—until Hertz's famous experiment—no one ever detected electromagnetic waves.

In *telephony*, the human voice, which generates sound waves propagating in the air, induces the vibration of a diaphragm located in the telephone mouthpiece. It is the diaphragm, working as a microphone, that transforms the conversation into an electrical signal: The diaphragm presses against an assembly of carbon particles and causes their electrical resistance to vary, thus altering an electric current—flowing through the particles—in accord with the pressure on the particles. At the receiver the current flows through an electromagnet whose power of attraction, on an adjacent steel diaphragm, varies accordingly. The vibrations of the diaphragm move the air and reproduce the sound. The Italian inventor Antonio Meucci (1808–1889)—who emigrated to the United States in 1850 and worked in Staten Island, New York—was the first to believe electrical transmission of voice to be possible and commercially practical as well. In December 1871 he submitted a *patent caveat* titled "Sound Telegraph" to the U.S. Patent Office. After 1874 he could not maintain it, being led to poverty by some fraudulent debtors. It was the Scottish born Alexander Graham Bell (1847–1922)—whose family had migrated from London to Canada out of concern for the health of their last remaining son after two others had died of tuberculosis—to patent the telephone in March 1876 after settling in Boston. Then the dispute and legal battles over whom was the inventor of the telephone—a device until then regarded as a joke by most—started and went so far that the Canadian Parliament even voted a motion on the issue in 2002.

A student of Hermann von Helmholtz, the German physicist Heinrich Hertz (1857–1894) was the first in 1887 to generate, transmit, and receive electromagnetic waves in the laboratory. He installed a spark gap at the focus of a parabolic metal mirror and another spark gap placed at the focus of another parabolic metal mirror placed at a distance of 1.5 meters, in line with the first one. A spark across the first gap caused a smaller spark across the second gap 1.5 m away. Thus Hertz showed that radio waves travel in straight lines and that they can be reflected by a metal sheet, as light is by a mirror, proving Maxwell's predictions at least over small distances. In his honor the unit of frequency of electromagnetic waves is named *hertz* (Hz), which corresponds to one cycle (complete oscillation) per second.

Electromagnetic waves are periodic both in space and time, just as waves in a pond. In space the period is λ, the *wavelength* or distance between two crests, as it would appear by freezing time as in a pond's picture. In time the *period* is T, whose inverse is *frequency* (Tab. 3.1). This last periodicity, in our comparison, would be

Table 3.1 Frequency terms

1 hertz (Hz) = 1 cycle per second
1 kilohertz (kHz) = 1,000 Hz
1 megahertz (MHz) = 1,000,000 Hz
1 gigahertz (GHz) = 1,000,000,000 Hz

that shown by the oscillations of a cork on the surface of the pond: The cork holds the same location in the pond and just moves up and down as time goes by. The relation

$$\lambda = vT$$

holds with v denoting the velocity, which is 300, 000 km per second in vacuum for all electromagnetic waves. Therefore the smaller the wavelength, the smaller the period and the higher the frequency.

The Italian physicist Guglielmo Marconi (1874–1937) started experimenting with Hertz's ideas as early as 1894 at his family estate near Bologna. He first increased the range of signaling to about 2.4 kilometers using a vertical aerial. Receiving little encouragement in Italy, he moved to England where he succeeded in sending signals across the Bristol Channel (about 14 kilometers). That was still insufficient to convince skeptics about the usefulness of this means of communication. Everything changed in 1899 when Marconi equipped two ships to report to newspapers in New York City the progress of the yacht race for the America's Cup. In 1900 Marconi filed his second patent, the famous No. 7777, for "Improvements in Apparatus for Wireless Telegraphy."

But his even greater success was still to come. In spite of the prevailing opinion that the Earth's curvature would limit communications by means of electric waves to a maximum of about 300 kilometers, in 1901 Marconi succeeded in transmitting across the Atlantic Ocean. That created a sensation in every part of the civilized world. With Marconi—Nobel prize in 1909—playing an important role, radio communications used in broadcasting and navigation services made phenomenal progress. In 1915 the first significant demonstration of speech transmission by the American Telephone & Telegraph Company took place between Arlington, Va., and Paris.

A simple radio echo device for use in navigation was first patented in 1904 by the German engineer Christian Hülsmeyer, thus putting to work the principle of radio reflection discovered by Hertz, which is at the base of the *radar* (from "radio detecting and ranging"). After Marconi elaborated the principle of radio reflection for detection in 1922, the height of the Earth's ionosphere—whose existence had been predicted as early as 1902—was measured in 1925 by Gregory Breit and Merle Tuve by bouncing radio pulses off the ionized layers of air and determining the time taken by the echoes to return. During the 1930s several countries started research on radar systems whose design was perfected during World War II in England, spurred by defensive needs.

The development of *satellite communications* is also linked to World War II. Rockets were at first considered as a means to send humans into space by the Russian

Konstantin E. Tsiolkovsky (1857–1935), a withdrawn schoolteacher with permanent hearing problems due to scarlet fever, in the small town of Kaluga located two hours south of Moscow. Tsiolkovsky started laying the theoretical foundation of rocketry as early as 1896, while the American university professor Robert H. Goddard (1882–1945) was first in achieving a brief lift-off of a liquid-propelled rocket engine in Auburn, Mass., on March 16, 1926. Then the war pushed the field ahead, with the V-2 rockets used on air assaults.

Early proposals envisioned manned satellites. For instance in 1945 the British scientist and author Arthur C. Clarke proposed three geostationary space stations with living quarters for a crew—built of materials flown up by rockets—to provide radio communications among all locations on Earth. He filed a patent that was denied in 1946 by the U.S. Patent Office on the ground that the orbital rocket to test his hypothesis did not then exist. Progress accelerated and ten years later the reason for denying him the patent once more had changed to "eminent domain."

Unmanned satellite communications, analyzed early in 1955 in a paper by the U.S. engineer J. R. Pierce, were the ones to achieve success. The first experiment was on December 18, 1958 when the U.S. government's project SCORE launched a satellite that was to function for thirteen days, until the batteries ran down.

3.2 More and more hertz

In *telephony* the electric signal is usually transmitted to the receiving site by cable, that is by a couple of electrical wires, but it may also be transmitted by satellite, using radio waves in part of the transmission. Figure 3.1 shows the signal associated with the uttered vowels "a" and "e" and with "sc." From these extremely simple examples the complexity of the signal generated by a common conversation can be imagined. It is well known that the telephone is not an example of high-fidelity transmission. Even though the human voice, above all in the "cantato," might reach rather high frequencies and the human ear can perceive from about 30 Hz up to 18000 Hz, the width of the range of frequencies—the *bandwidth* of the signal—used in telephony is 4000 Hz. The band itself is referred to as a *telephone channel*. Higher frequencies, which may be present in the initial conversation, are cut off by suitable filters for economical reasons. At the same time the result is adequate since, for intelligible speech, frequencies from 300 Hz to 3400 Hz are sufficient. If we were to imagine, just for fun, that telephones were used by dogs, who can hear up to 38000 Hz, would they need a larger frequency band to recognize their master's voice?

A *radio signal* consists of a radio-frequency sine or cosine wave, called the *carrier wave*, that has undergone modulation. In amplitude modulation (AM), the carrier amplitude is varied by the modulating signal (Fig. 3.2 a)). In frequency modulation (FM), it is the carrier frequency that is varied by the modulating signal (Fig. 3.2 b)). Acceptable audio quality may be achieved with a bandwidth as small as 5 kHz, as in AM radio, but when the modulating signal consists of musical sounds the frequency bandwidth extends to about 10 kHz in commercial broadcasting. For the perfection

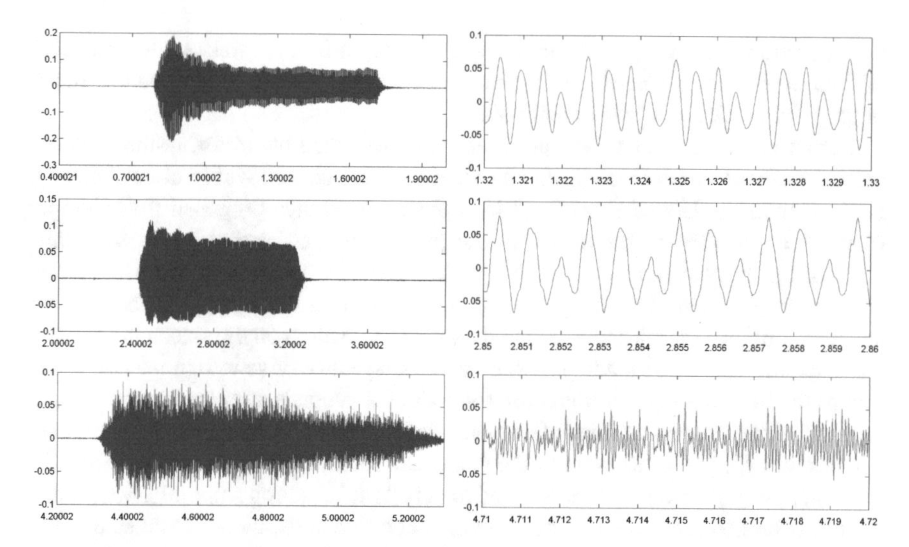

Figure 3.1 To the left the waveform associated with the vowels "a" as in father (top), "e" as in get (middle), and to "sc" as in scion (bottom); to the right an enlargement of the central part. (*Courtesy of P.E. Giua, Istituto di Acustica, CNR*).

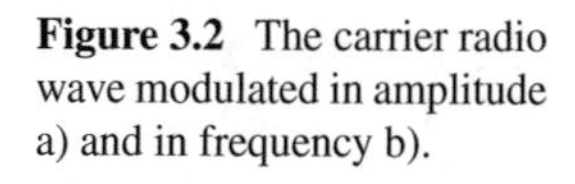
Figure 3.2 The carrier radio wave modulated in amplitude a) and in frequency b).

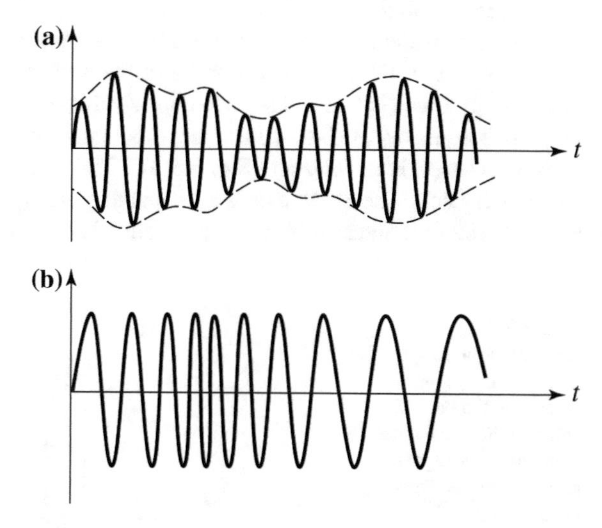

of high-fidelity music transmission, the range ought to be wider: from about 30 Hz, which is the lowest frequency of a large organ pipe, to 15 kHz for cymbal and triangle sounds. Frequency modulation is used in this case because wide band FM signals are more effective than the equivalent AM signals in combatting noise and interference.

The *television system* has to take into consideration not only the ability of the ear but also the ability of the human eye to distinguish the brightness, color, size, shape, details, and position of objects. A picture can be broken up into elementary details by superimposing a fine grid. Thus a single picture, or frame, can be realized as a set of lines each made by a string of elementary details and associated to each a shade of gray, for black and white transmission.

Earlier ideas for the realization of television assumed the transmission of every picture element simultaneously, each over a separate circuit. But in about 1880 W. E. Sawyer in the United States and M. Leblanc in France proposed the principle of rapidly scanning each element in the picture in succession line by line, thus establishing the possibility of using only a single wire or channel of transmission. The elementary pieces are transmitted sequentially, one after the other, over the channel and reassembled at the receiver end in their correct position on the screen to form the electrical image. This is made possible by the persistence of vision. Indeed the brain retains the impression of illumination for about 0.1 seconds after the source of light goes off. If the process of image synthesis, described above, occurs within less than 0.1 seconds, the eye is unaware of the piecemeal reassembling. Commonly from 25 to 30 pictures per second are transmitted to depict rapid motion smoothly. Since each picture is broken into 300, 000 or more elementary details, their transmission over the television system exceeds the rate of 4, 000, 000 per second. The high frequencies required for picture details are limited by the frequency-allocation authorities to 6 MHz; most of that signal is occupied by the picture signal, part by the color signals, and part by the audio signal.

Also large is the band of frequencies used by *radars*, extending from about 400 MHz to about 40 GHz. To detect and localize a distant object the *pulsed radar*—most widely used type of radar—usually transmits a short signal made by a carrier wave, of frequency 1 GHz for instance, modulated by a train of rectangular pulses with a low *repetition interval* (Fig. 3.3 a)). The target, hit by the signal, reflects some of it back. The radar receiving apparatus calculates the distance of the target from the time interval between the transmitted signal and the detection of the

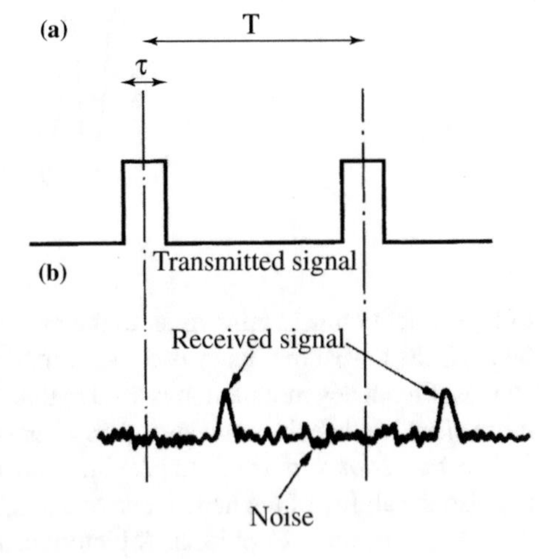

Figure 3.3 a) The modulating pulse typically transmitted by a pulsed radar: τ is the pulse duration and T is the pulse repetition interval; b) the signal reflected by the target (figure not to scale).

time-averaged reflected one (*echo*). The radar echo-ranging technique is also used at optical frequencies—sometimes called "lidar" (li for Light).

A reasonable accuracy is obtained by employing very short pulses of the order of microseconds or less. Here is an example in which the *pulse duration* is assumed to be one microsecond and the repetition interval one millisecond. Thus the 1 GHz carrier wave oscillates a thousand times during a single pulse duration and the pulse duration itself is one thousandth of the pulse repetition period.

The usual rule of thumb is that a transmitter-pulse duration of one microsecond requires a receiver bandwidth of one MHz, that is just the reciprocal. This is because the frequency content of a rectangular pulse of one μs duration is mostly confined to a one MHz interval, centered on plus or minus the modulation frequency (see Section 3.5 for more details).

In one millisecond, radio waves travel 300 km. A reflecting aircraft 150 km away will return an echo one millisecond after the pulse was emitted. This echo will return to the receiver just as the next pulse is being emitted. If aircrafts are to be detected out to say 50 km, then it is desirable that the transmitter-pulse power and antenna gain be proportioned so that a plane as far away as 150 km should ordinarily be lost in the receiver noise.

But suppose that an object 160 km away did return a detectable signal; then the received echo could be interpreted as coming from a small plane only 5 km away (Section 10.14). This would be a serious situation for airport traffic control. The situation can occur under certain atmospheric conditions or with certain orientations of a distant large reflecting object. In any case, these are the considerations that determine the choice of repetition interval.

The signal received back by the radar is always disturbed by noise (Fig. 3.3 b)) of various origins that might give rise to false alarms. This is a problem of utmost importance for radar performance. Filters allow some noise to be eliminated, but not all.

The International Telecommunication Union in Geneva allocates bands of the radio spectrum on the basis of usage, including *mobile radio* (aeronautical, marine, land), *radio navigation* (aeronautical, marine), *broadcasting* (AM, FM, TV), *amateur radio, space communications* (satellite communications, communications between space stations), and *radioastronomy*. For instance in the U.S., AM broadcasting is allocated the 535–1,605 kHz band and FM the 88 – 108 MHz band, TV has four bands, the first being the 54 – 72 MHz and the last being the 470 – 890 MHz.

The study of signals, those of electrical nature in particular, originated a field rich both in theory and applications, used not only in telecommunications but also in telemetry, astronomy, oceanography, optics, crystallography, geophysics, bio-engineering, and medicine to mention a few. As will be seen, there are several reasons Fourier series and transforms are of fundamental importance for the mathematical theory of signals.

3.3 Tables of Fourier transforms

In Section 2.3 it has been described the procedure by which a function $f(t)$, periodic of period 2π, may be realized as a sum of a Fourier series by three equivalent formulas (2.3), (2.4), (2.5). In case of a generic period T, (2.3) reads

$$f(t) = \frac{a_0}{2} + \sum_{n=1}^{\infty} a_n \cos\left(\frac{2\pi nt}{T}\right) + b_n \sin\left(\frac{2\pi nt}{T}\right). \tag{3.1}$$

The signal $f(t)$ is realized as a sum of infinitely many *harmonics* whose *frequencies* are integer multiples n of the *fundamental frequency* $1/T$. The Fourier coefficients measure the "weight" of the corresponding harmonic in the signal and are given by the formulas $a_0 = \frac{2}{T}\int_{-T/2}^{T/2} f(t)dt$, $a_n = \frac{2}{T}\int_{-T/2}^{T/2} f(t)\cos\frac{2\pi nt}{T}dt$, and $b_n = \frac{2}{T}\int_{-\frac{T}{2}}^{\frac{T}{2}} f(t)\sin\frac{2\pi nt}{T}dt$.

In case of a signal the variable t denotes time, usually measured in seconds and so frequencies are measured in s^{-1} or Hz. Thus a list of numbers, namely the *amplitudes* a_n and b_n of the harmonics, completely determine a periodic signal of known period T. For, once this information is acquired, the signal can be fully reconstructed by means of (3.1).

An element of simplicity can be added. Even though the *spectrum* (set of all frequencies) can be infinite and generally is, it can be proved that the amplitudes a_n and b_n get small as n increases if f is absolutely integrable. So a finite number of harmonics is sufficient to obtain the signal with good approximation, adequate for communications. Consequently the use of the *bandwidth* (total range of frequencies) in communication systems can be made effective and economical at the same time. Figures 2.8, 2.9, and 2.13 show some of the waves commonly used and their spectrum. If the function is not periodic, as the *rectangular pulse* of Figure 2.17, then it can still be interpreted as periodic with period $T = \infty$. The corresponding frequencies cover the entire real axis, as seen by making $T \to \infty$ in the discrete spectrum $2\pi n/T$. The formula for the Fourier coefficients and the inversion formula (3.1) melt into (2.13) and (2.14), respectively the Fourier transform of $f(t)$ and the inverse Fourier transform (or equivalently (2.15) and (2.16), or (2.17) and (2.18)). Some signals and their Fourier transforms can be found in Figure 3.4. The proof is immediate in the first case and it amounts to an integration by parts in the second case, as follows:

$$2\int_0^1 (1-t)\cos(\omega t)\,dt = (2/\omega^2)(1-\cos\omega) = (4/\omega^2)\sin^2(\omega/2).$$

In the fourth case, with $F(\omega)$ denoting the Fourier transform, an integration by parts gives

$$F(\omega) = 2\int_0^{\infty} e^{-t}\cos(\omega t)\,dt = 2 - \omega^2 F(\omega).$$

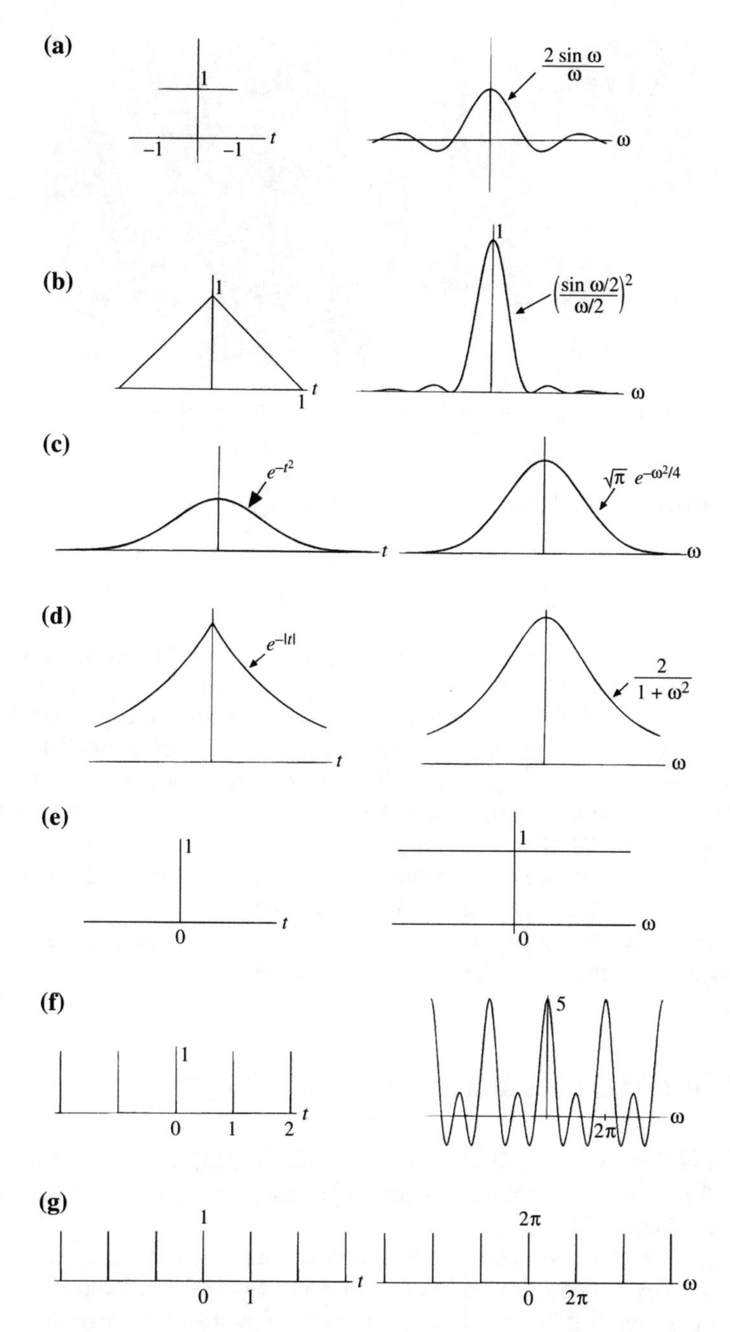

Figure 3.4 On the left from a) to g), signals; on the right, their Fourier transforms.

Figure 3.5 Gauss and the gaussian on the former ten mark note of the Bundesbank.

In the third case, from the known formula $\sqrt{\pi} = \int_{-\infty}^{\infty} e^{-x^2}\, dx$ follows

$$\sqrt{\pi} = \int_{-\infty}^{\infty} e^{(\omega/2-it)^2}\, dt = e^{\omega^2/4} \int_{-\infty}^{\infty} e^{-t^2-i\omega t}\, dt.$$

The first four examples on the left of Figure 3.4 show the *rectangular pulse*, the *triangular pulse*, the *gaussian* e^{-t^2} (Fig. 3.5), and the $e^{-|t|}$ (whose transform goes by the name of *Poisson kernel*); on the right are the corresponding Fourier transforms. All these tend to zero as ω tends to infinity—in analogy with the Fourier coefficients—according to a property (*Riemann–Lebesgue lemma*) that holds under the very general assumption that allows the Fourier transform to be defined, namely integrability of the absolute value of f. In particular the Fourier transform of the gaussian, in the fourth example, is a gaussian itself while the third example is central in the theory of nuclear magnetic resonance (Chapter 7).

Of different nature are the last three cases—to be dealt with in the next section—which also appear in Section 3.7 and in later chapters.

3.4 The Dirac delta function and related topics

Topics here, among which are the Dirac delta function and its Fourier transform, are presented in a heuristic fashion. A rigorous presentation can be found for instance in Gelfand, Shilov [1964].

The graph of $r_a(t)$ in Figure 3.6 is obtained from the rectangular pulse $r(t)$ in Figure 3.4 a) by a contraction and by changing its constant value in such a way that the resulting underlined area is 1. Thus the value of the Fourier transform $R_a(\omega)$ at the origin is 1 since, by the very definition (2.13), $F(0) = \int_{-\infty}^{\infty} f(t)\, dt$.

To define heuristically the *Dirac delta function, a* has to be made smaller and smaller, and in the limit as a tends to zero, $r_a(t)$ defines a *generalized function* or *distribution* that is commonly denoted by $\delta(t)$, following the notation introduced by

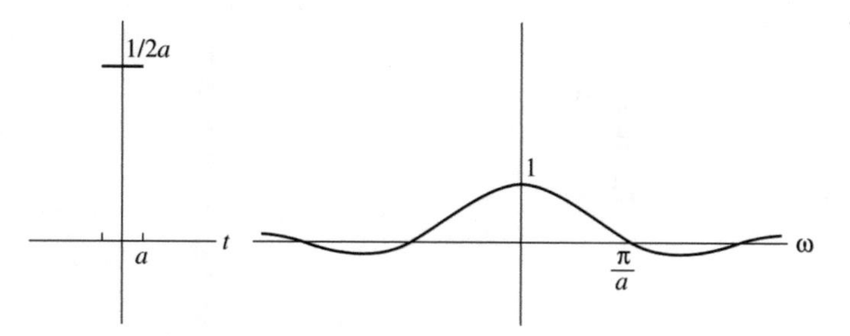

Figure 3.6 On the left, the rectangular pulse $r_a(t)$; on the right, its transform $R_a(\omega)$.

Paul Adrien Maurice Dirac (1902–1984) in *The Principles of Quantum Mechanics* (4th. ed., 1958). The properties of the Dirac delta function show that it is not a function: It is zero for every t with the exception of $t = 0$ where it is infinite and its integral over the entire real axis is 1. Indeed this property, common to all the r_a, is inherited by the delta.

The computation of $R_a(\omega)$, the Fourier transform of the rectangular pulse r_a, gives $(a\omega)^{-1}\sin(a\omega)$. The corresponding graph is shown in Figure 3.6. As a tends to zero, the abscissa π/a tends to infinity, and in the limit, R_a becomes identically equal to 1: the Fourier transform of the delta (Fig. 3.4 e)) has been reached.

The Dirac delta is used in electronics to represent the *unit impulse* and sampled signals, but it is also employed to represent a unit mass ideally concentrated at the origin. If concentrated elsewhere, for instance at the integer n, then it is denoted by δ_n, the delta translated at n. Precisely $\delta_n(t) = \delta(t - n)$. Its Fourier transform is equal to $e^{-in\omega}$, by (2.19).

Much more is true. A sequence of complex numbers $\{c_n\}_{n=-\infty}^{\infty}$ can be represented by $\sum_{-\infty}^{\infty} c_n\,\delta_n(t)$ and its Fourier transform is $\sum_{-\infty}^{\infty} c_n e^{-in\omega}$, going from "discrete" to "periodic."

On the contrary, a periodic function $f(t)$ with Fourier series $\sum_{-\infty}^{\infty} c_n e^{-int}$ does not have a Fourier transform for lack of integrability at infinity (unless f is identically equal to zero), but considered as a distribution, it does. Its Fourier transform is equal to $\sum_{-\infty}^{\infty} c_n\delta_n(\omega)$, going from "periodic" to "discrete." In the space of distributions, the notion of Fourier transform includes that of Fourier series. A unified theory is thus derived.

For instance, from Euler's formula $\sin t = [e^{it} - e^{-it}]/2i$, it is easily calculated that $\sin t$ has Fourier transform equal to $(-i/2)\delta_{-1}(\omega) + (i/2)\delta_1(\omega)$. Other examples follow.

A *finite lattice* made by unit masses concentrated at the integers n, with $-N \le n \le N$, is represented mathematically by $\sum_{n=-N}^{N} \delta_n(t)$ and graphically by a "comb" as in Figure 3.4 f), where $N = 2$. The inverse Fourier transform $\sum_{n=-N}^{N} e^{in\omega}$, manifestly periodic of period 2π, is equal to

$$\frac{\sin(N + \frac{1}{2})\omega}{\sin(\omega/2)}.$$

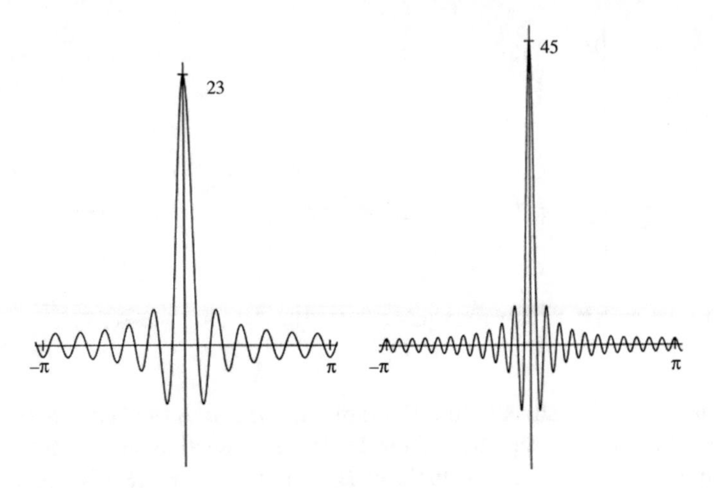

Figure 3.7 The Dirichlet kernel D_N for $N = 11$ and $N = 22$.

This can be seen by writing

$$\sum_{n=-N}^{N} e^{in\omega} = e^{-iN\omega}[1 + e^{i\omega} + e^{i2\omega} \cdots + e^{i2N\omega}] = e^{-iN\omega}\,\frac{1 - e^{i(2N+1)\omega}}{1 - e^{i\omega}}.$$

Then by collecting $e^{i\omega/2}$ both at the numerator and denominator we obtain

$$\frac{e^{-i(N+\frac{1}{2})\omega} - e^{i(N+\frac{1}{2})\omega}}{e^{-i\omega/2} - e^{i\omega/2}}$$

and using Euler's formula $\sin\omega = [e^{i\omega} - e^{-i\omega}]/2i$, the stated result. The *Dirichlet kernel* $D_N(\omega) = \sum_{n=-N}^{N} e^{in\omega}$ is drawn for different values of N in Figure 3.7. Then it is clear that the Dirichlet kernel focuses at the origin as N increases. Actually, due to the observed periodicity, at $0, \pm 2\pi, \pm 4\pi, \ldots$ it shows pronounced peaks of height $2N + 1$ and width of the order of $2\pi(2N + 1)^{-1}$, so that their underlined area is approximately 2π. It is worth remarking that the height of the peaks is equal to the number of elements of the lattice.

To deal heuristically with the case of an *infinite lattice* it suffices to let $N \to \infty$ in the above formulas. In the t variable $\sum_{n=-\infty}^{\infty} \delta_n(t)$ is obtained, graphically represented by an "infinite comb" (Fig. 3.4 g)). In the ω variable the periodicity is maintained, the peak heights tend to infinity, the width tends to zero, while the underlined area keeps approximately the same value 2π for any N and therefore in the limit. This explains the last case of Figure 3.4: to an "infinite comb" in time t (both periodic and discrete) there corresponds an "infinite comb" $2\pi \sum_{n=-\infty}^{\infty} \delta_n(\omega/2\pi)$ (periodic and discrete) in the frequencies. A formal derivation can be found in Bracewell [1998] or Papoulis [1987].

3.5 Convolution

The convolution of two absolutely integrable functions f and g, denoted by $f * g$, is itself an absolutely integrable function. Therefore $h = f * g$ has a Fourier transform. The definition of convolution, to be given further below, might appear complicated at first, but its action is simple on the Fourier transform side: It consists of pointwise multiplication, according to the formula

$$\hat{h}(\omega) = \hat{f}(\omega)\hat{g}(\omega).$$

From this, the importance of convolution can be inferred, for instance in describing and dealing mathematically with *filters* (Section 3.6) or with the *resolving power* of observing instruments, like radiotelescopes (Section 9.5).

Here comes the definition

$$h(x) = \int_{-\infty}^{\infty} f(y)g(x-y)dy.$$

To compute the value of the above convolution at a point x one has to start by folding $g(y)$ to obtain its reflection $g(-y)$, as in Figure 3.8, then proceed to translate such

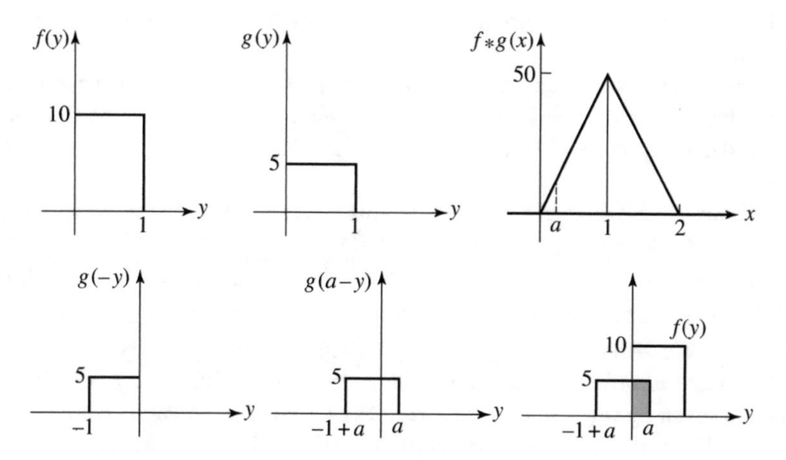

Figure 3.8 Graphical description of convolution. Ten times the shaded area is $50a$, the value of $f * g(a)$.

a reflection at x obtaining $g(x-y)$, at last to be integrated against $f(y)$. If g is even, then the above convolution is just an average of f, against g translated at x.

For instance, given a periodic function f of period 2π with Fourier coefficients c_n, $-\infty < n < \infty$, and given the function whose Fourier coefficients are 1 for $-N \leq n \leq N$ and zero otherwise, multiplication on the Fourier transform side gives c_n, $-N \leq n \leq N$, and zero otherwise or equivalently $\sum_{n=-N}^{n} c_n \delta_n(\omega)$. Back in the time domain we have $\sum_{n=-N}^{n} c_n e^{int}$ which turns out to be equal to the convolution $f * D_N(t)$. It has been observed that D_N focuses at the origin, therefore in the convolution, evaluated at t, D_N enters translated at t where it focuses. Hence it can be expected that, as N tends to infinity, $f * D_N(t)$ tends to $f(t)$. This is actually true, under some assumptions, namely those of the Dirichlet theorem (Section 2.4).

Convolution can be thought as a kind of multiplication in the class of absolutely integrable functions, being commutative, associative, and distributive under the pointwise addition. It is natural to ask whether there is a unit, that is an element that plays a similar role to the number 1 for the multiplication of real or complex numbers. The delta partially works that way: If f is continuous, in addition to being absolutely integrable, then

$$f * \delta(x) = f(x).$$

For precision it has to be noted that the search ended up outside the class of absolutely integrable functions. Indeed the delta is not a function (while 1 is a number among numbers, in the preceding comparison).

Now the Fourier transform of a radar signal is going to be computed. We assume for simplicity that the transmitted signal (Fig. 3.3) has unit power and moreover that it is a pulse, repeated periodically infinitely many times (a true radar signal is a pulse repeated only a few times, so it is a truncation of the one we are going to consider. More on truncations and their effects on the Fourier transform can be found in Section 3.8).

Let us assume the frequency of the carrier wave $v = 1$ GHz, the modulating pulse to have duration $\tau = 1$ μs and repetition period $T = 1$ ms. These figures can be considered representative of airport surveillance radars with a range of about 100 km.

The Fourier transform of a rectangular pulse $R(t)$ of width τ is given by $\hat{R}(\omega) = 2\omega^{-1}\sin(\tau/2)\omega$ according to formula (2.17). If a periodic signal $p(t)$ is obtained from $R(t)$ by repetition with a period of T, that is $p(t) = \sum_{n=-\infty}^{\infty} \delta_n(t/T) * R(t)$, then $\hat{p}(\omega)$ is discrete and given by the formula $\hat{p}(\omega) = \hat{R}(\omega)\omega_0 \sum_{n=-\infty}^{\infty} \delta(\omega - n\omega_0) = \omega_0 \sum_{n=-\infty}^{\infty} \hat{R}(\omega - n\omega_0)$, where $\omega_0 = 2\pi/T$ (Papoulis [1987], p. 44).

The radar transmitted signal $m(t)$, being $\cos 2\pi\nu t$ modulated by $p(t)$ (Section 10.14), is given by $m(t) = p(t)\cos 2\pi\nu t$. Thus, by Euler's formula $\cos t = [e^{it} + e^{-it}]/2$ and formula (2.18), we have $\hat{m}(\omega) = 0.5\hat{p}(\omega + 2\pi\nu) + 0.5\hat{p}(\omega - 2\pi\nu)$, that is $\hat{p}(\omega)$ translated by frequency shifts $2\pi\nu$ to both left and right.

3.6 Filters, noise, and false alarms

Mathematically the action of a filter is a convolution. It is easily described if the signal is given in terms of frequencies, that is by its Fourier transform. An ideal filter stops certain frequencies and leaves unaltered all the others, as in Figure 3.9, which describes the ideal *high-pass, low-pass*, and *band-pass filters*.

For instance if $x_i(t)$ denotes the input time function and $x_o(t)$ the output of an ideal low-pass filter, then the relation between the corresponding Fourier transforms X_i and X_o is stated in the formula

$$X_o(\omega) = X_i(\omega)H(\omega).$$

Here $H(\omega) = e^{-it_0\omega}r(\omega)$ where $r(\omega)$, named *amplitude characteristic*, is the characteristic function of the interval $[0, \omega_0]$, also called the rectangle function, which takes the values 1 on the interval and 0 outside as shown in Figure 3.10. The factor $e^{-it_0\omega}$ deserves a few words of explanation. Every filter has a *phase characteristic* $\varphi(\omega)$, assumed linear for simplicity in Figure 3.10, that is $\varphi(\omega) = t_0\omega$. Thus by (2.19) the filtered signal shows a delay with respect to the original signal, as one naturally expects. The true amplitude characteristic and phase characteristic are rather as in Figure 3.11, providing a good approximation of the ideal case.

Multichannel operation is of general use in signal transmission. The most common type of multichannel operation is obtained by *frequency division*, but also *time division* is employed. The *frequency division multiplex system* is made possible by band-pass filters. Messages or channels, each with a band of frequencies allocated, are mixed at the transmitting site to be sent together along the same communication path. At the receiving site the frequency band of each channel can be singled out by appropriate filters, thus ensuring a correct reception.

Noise, inevitably picked up during signal transmission and reception, is dealt with by the theory of signals and of random variables. The purpose is to get rid of most of it,

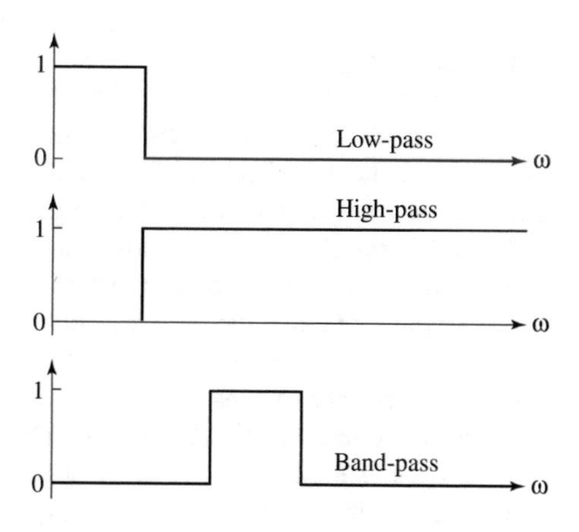

Figure 3.9 Ideal high-pass, low-pass, and band-pass filters.

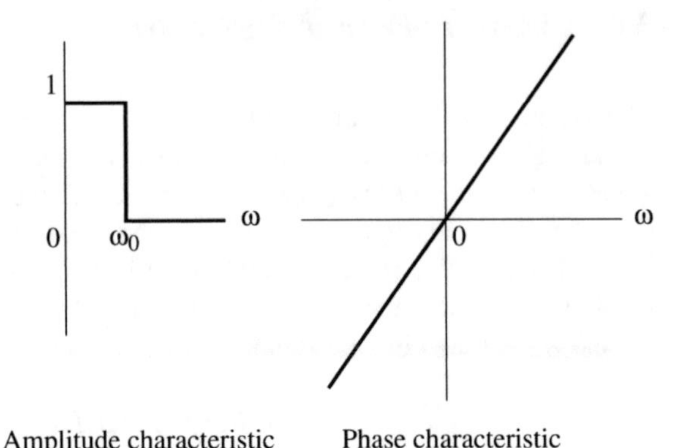

Figure 3.10 On the left the amplitude characteristic and on the right the phase characteristic of an ideal low-pass filter.

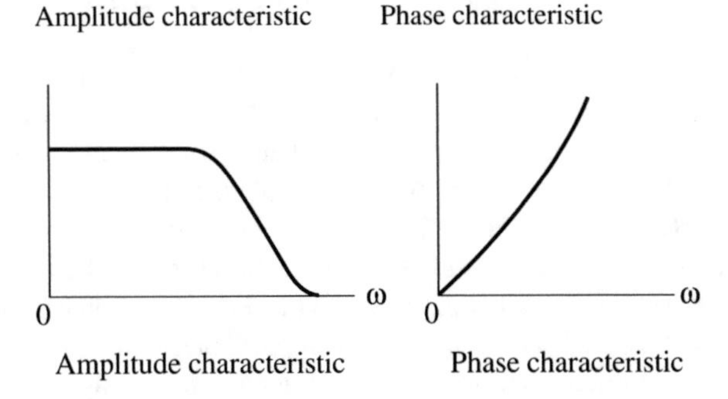

Figure 3.11 On the left the amplitude characteristic and on the right the phase characteristic of a low-pass filter.

for instance by means of suitable filters or by improved system design. Electrical noise may be defined as a signal unwanted but always present in a communication system that tends to impede the reception of the wanted signal and to limit its detection. It is important, in the evaluation of the performance of a system to carry out a general analysis of noise and in addition of *interference*, another kind of unwanted signal (such as those from neighboring stations) that may be equally important. While noise and interference have a similar effect on communication systems, their nature is different: Usually noise is made of randomly occurring voltages, sometimes very peaky in shape, while interference is usually periodic or regular in form. When viewed on an oscilloscope, noise gives a spiky impression, as in Figure 3.3. The peaks, continuously occurring, are about a microsecond in duration, on the average. Thus, by (2.17), they have high-frequency components easy to filter if the signal band is much lower. Therefore the importance of Fourier analysis, in conjunction with the study of the statistical behavior of noise.

Natural forms of noise are those due to *cosmic* and *atmospheric radiation*, usually entering the system via the receiving antenna. An example of the former type is the largely studied solar radiation—whose effects on radio reception are well known and may be reduced by modifying the directional properties of the antenna—as well as radiation from other sources. For instance in our galaxy the radio source Cassiopeia A in the Cassiopeia constellation emits a definite and regular amount of noise.

Atmospheric noise due to scattered radiation from vapor or snow has frequencies of about 22 GHz, while radiation's absorption by O_2 in the atmosphere generates noise of about 60 GHz. *Sky noise* is low only on a frequency range situated around 1 GHz (microwaves). Signals received from radiogalaxies or from space satellites are generally small, thus knowledge of the background noise received from various parts of the sky is required.

Another form of naturally occurring noise is the more troublesome *circuit noise* generally known as *thermal noise*. Due to the random motion of free electrons in a conductor and increasing with temperature, it is linked to the nature of electronic components and cannot be completely eliminated. Thermal noise makes up an important part of communication systems and is well studied: Generally it covers a wide range of frequencies in the receiver and it may be reduced by a proper choice of components and use of low temperatures. A specific example is given by a radar: Its antenna inevitably picks up cosmic and atmospheric radiation and inevitably is affected by noise generated by the electrical circuits of the radar itself. Consequently electric energy of the "right frequencies" might be occasionally present in the receiving circuits without having originated in a target, thus causing *false alarms*. Attempts to eliminate them could end up eliminating the wanted signal, therefore efforts go in the direction of keeping false alarms at an acceptable level. In case of a big radar, like the one at Fiumicino airport in Rome, whose surveillance extends over more than 300 km in radius and that might detect in one second from a hundred to more than two hundred airplanes (objects moving at 900 km per hour), an average number of false alarms of one or two per second is acceptable. The number of possible false alarms in a second being of the order of one million, it may be stated that the mentioned radar has a probability of false alarms of 10^{-6}.

3.7 Sampling and the Nyquist frequency

To make the most efficient use of time and frequency bands often, instead of transmitting a continuous signal, samples of it are transmitted without degrading reception significantly. Sampling and quantization are always present in a transmission system that makes use of computers, whose large and increasing exploitation has brought a revolution in the theory of signals and its applications.

When an airplane pilot speaks with the controlling tower, the acoustic pressure of his voice is a continuous function of time. Technically speaking it is an *analog signal*. Similarly for the electric signal generated in his transmitting radio. Computers though deal with *digital signals* which are allowed to assume only a finite number of values. Therefore one part of a transmission procedure, which makes use of computers, deals with *sampling*: Signals are taken only at certain instants, for example every 0.5 microseconds (Fig. 3.12).

Every signal's value, so obtained, is furthermore rounded off, that is restricted to a certain finite set of possible values (*discretization levels*). This procedure is called *quantization*. For instance if the signal at time $t = 0$ takes on the value 0.567, then the quantized value could be 0.6. Every set of levels is designated by a decimal number

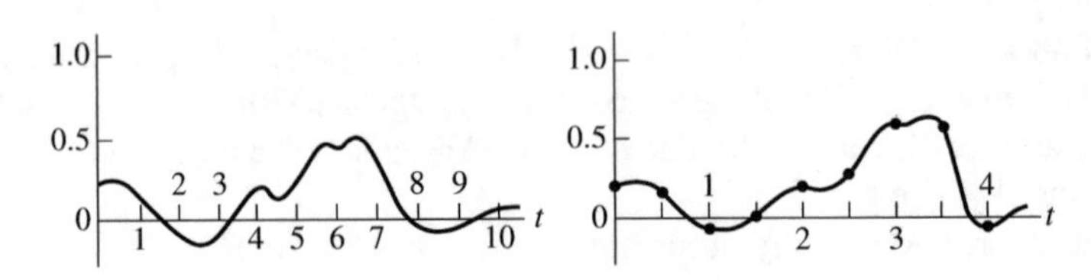

Figure 3.12 To the left an analog signal; on the right the same signal sampled every 0.5 microseconds.

0, 1, 2, ... and then expressed in binary form using the binary digits (*bits*) 0 and 1, the only digits appearing in the computer's language.

A converter from analog to digital (A/D) at the transmission site is matched by an opposite converter from digital to analog (D/A) at the receiving end where a sequence of numbers has to be transformed back into an acoustic pressure, reproducing at best the pilot's voice. The fidelity of A/D and D/A conversion depends on the sampling rate and on the number of discretization levels which have to be chosen adequately to the purpose.

Unavoidably, quantization introduces errors while the former operation of sampling poses a problem: What choice of *sampling rate* allows an accurate reconstruction of the original signal? The answer depends upon the signal: The more rapidly the signal varies, the denser the sampling has to be, in technical terms the higher the *sampling rate*. If the signal shows significant variation in a microsecond, samples have to be taken every microsecond or fractions of it. Roughly speaking two or three samples have to cover those parts of the signal that vary the most rapidly.

In Figure 3.13 a suitable sampling has been chosen for the given signal: The rate of sampling is sufficiently high to allow a faithful signal reconstruction. If a lower frequency was chosen, for instance one-fifth of the preceding one (in Figure 3.13 it would amount to taking one sample from every five) then the ensuing reconstruction could be grossly wrong (dashed curve in Figure 3.13).

In Figure 3.14 the opposite procedure is followed. The sampling is given and from that the signal has to be reconstructed. Both signals $x(t)$ and $y(t)$ are compatible with the given sampling. Which is the right one? Actually a moment of thought suggests

Figure 3.13 Signal, adequate, and inadequate sampling (see text).

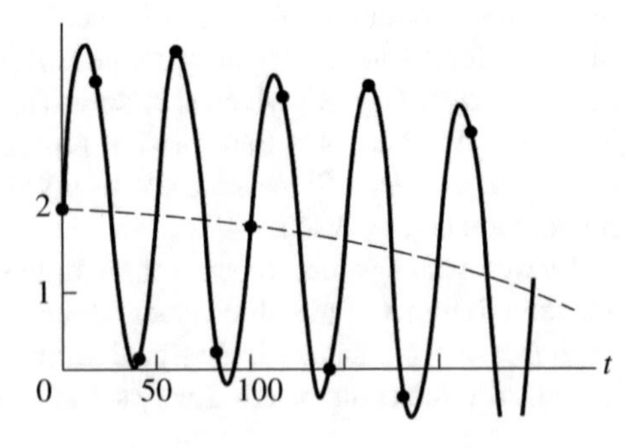

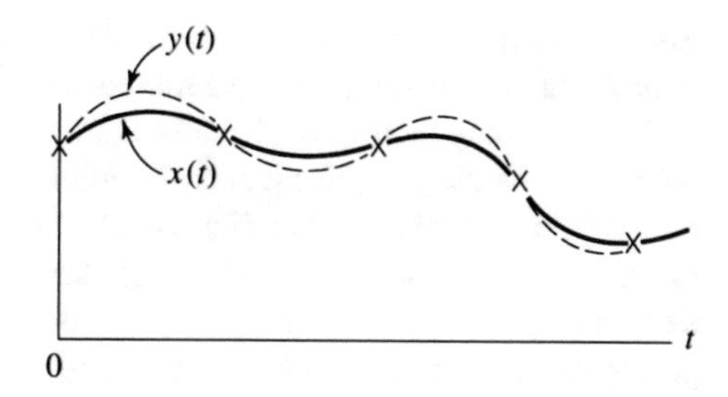

Figure 3.14 Sampling compatible with the two signals $x(t)$ and $y(t)$.

that infinitely many signals are compatible with the given sampling. This confusion goes under the name of *aliasing*. Fortunately it is possible to recover the wanted signal if it is known, beforehand, that all signals undergoing sampling are "slowly" changing in the time interval between two successive samples. This restriction is quite reasonable since a transmission system is always designed for a class of signals that share common features. For instance the telephone is designed to transmit the human voice, not the bat's scream or the sound of the fog horn.

The preceding qualitative considerations are manifestly insufficient from an operational point of view. It is again Fourier analysis, with a fundamental result known as the *sampling theorem*, that states quantitatively the assumptions that allow the reconstruction of the signal and the formula that does it. The basic assumption is for $x(t)$ not to have components beyond a certain frequency B, which implies that $X(\omega) = 0$ for $|\omega| \geq 2\pi B$. Then the signal can be recovered uniquely by choosing as sampling time $T = 1/2B$ and by applying the formula

$$x(t) = \sum_{k=-\infty}^{+\infty} x(kT) \frac{\sin \pi(tT^{-1} - k)}{\pi(tT^{-1} - k)}.$$

In Figure 3.14 we intentionally chose $y(t) = x(t) + \sin 2\pi Bt$ to have the Fourier transform $Y(2\pi B) \neq 0$, thus violating the stated assumption.

The sampling theorem is also called the Shannon theorem after Claude E. Shannon of Bell Telephone Laboratories, who proved it in 1948. The sampling frequency $T^{-1} = 2B$ is known as the Nyquist frequency (see Whittaker [1915], Nyquist [1928], Shannon [1948]).

The above theorem might appear incredible: The full values of $x(t)$ are recovered from the values of $x(t)$ on a sequence of equidistant points. The values of a function might change from point to point arbitrarily in size and sign, so it is manifestly impossible to predict the values of $x(t)$ on any interval, however small, from its values on a given sequence of points. The above formula works, contrary to expectations, because of the theorem's rather strong assumption. For any signal $x(t)$ starting at an instant t_0 and ending at an instant t_1, like any conversation over the phone, it is known that there is no B such that $X(\omega) = 0$ for $|\omega| \geq 2\pi B$.

The technique of sampling therefore might require one to set $X(\omega)$ equal to zero outside some interval that has to be chosen large enough not to destroy, in so doing, essential information. This is obtained by suitably filtering the signal. It is the filter's output that is to be sampled, transmitted, and finally reproduced at the receiving end.

Digital sound recording provides another relevant example of application of this theorem. The range of frequencies audible to the human ear has an upper bound at 20,000 Hz. Thus any audible sound can be exactly reproduced, no matter how intricate, by taking 40,000 evenly spaced samples for every second. Indeed a sampling frequency of 41.5 kHz is used in compact discs where sample values are stored as points of varying reflectance. The resulting sound fidelity is much better than in analog recordings. Besides digital processing and reproduction, Shannon's theorem similarly underlies digital generation of sound as well.

Figures 3.15, 3.16, and 3.17 describe graphically the theorem on the Fourier transform side. Figure 3.15 has the signal $x(t)$ and its Fourier transform $X(\omega)$. The next two paragraphs deal with the discrete Fourier transform and the fast Fourier transform, routinely used in applications. Being of rather technical nature they can be omitted in a first reading. It is enough to know that there exists a way to compute the Fourier transform that involves a certain degree of approximation but, on the other hand, it is very convenient since it can be implemented on a computer. If provided with suitable software, the computer will show the graph of the Fourier transform, with a certain approximation, just by punching a key. In case of adequate sampling (Fig. 3.16) the Fourier transform of the sampled signal, on the interval $[-2\pi B,\ 2\pi B]$, is exactly the Fourier transform of the original signal; otherwise aliasing occurs (Fig. 3.17 and Fig. 10.33).

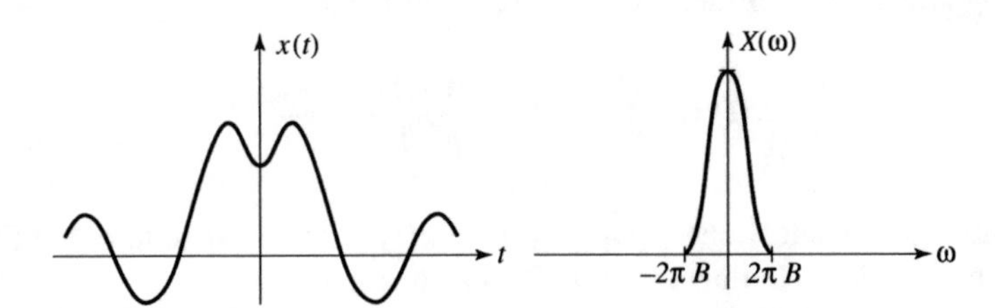

Figure 3.15 Signal and Fourier transform. (E. O. Brigham, *The Fast Fourier Transform*, Prentice Hall, [1988])

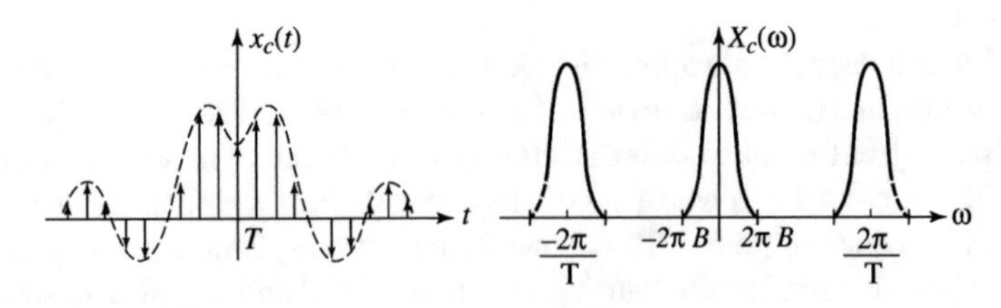

Figure 3.16 Signal adequately sampled and its Fourier transform. (E. O. Brigham, *The Fast Fourier Transform*, Prentice Hall, [1988])

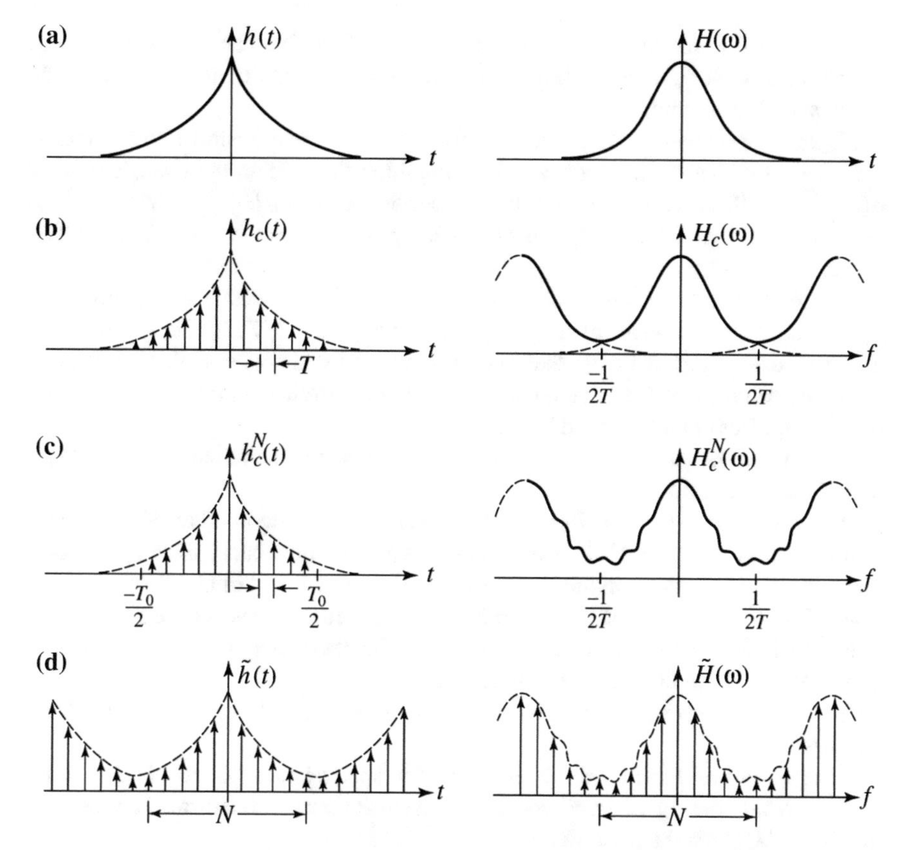

Figure 3.17 On the left the signal; on the right the DFT. (Adapted from E. O. Brigham, *The Fast Fourier Transform*, Prentice Hall, [1988])

3.8 The discrete Fourier transform

The Fourier transform deals with functions defined on the real line, so it cannot be implemented on computers that handle a possibly enormous, but nevertheless finite, number of data. Even experimental measurements, relative to a given function $h(t)$, are taken for a finite number of chosen values of the variable t. Moreover it is $h(t) = 0$ outside a finite interval—like the electric signal corresponding to a conversation—or it is set $h(t) = 0$ outside a finite interval suitably chosen so as to make negligible the error of the resulting approximation.

To meet the needs of both automated and experimental computations, the *Discrete Fourier Transform* or DFT has been introduced. It involves only a finite number of additions and multiplications. Subsequently, to compute the DFT, a suitable procedure has been found that reduces the number of operations and in turn saves computing time. It is named *Fast Fourier Transform*, abbreviated as FFT.

Going back to the DFT, which logically comes first, there are two goals: It has to approximate at best the continuous Fourier transform and moreover it has to be implementable on computers.

Figure 3.17—showing signals to the left and their Fourier transforms to the right—describes graphically the procedure for computing the DFT in the specific case of $h(t) = e^{-|t|}$. (It is worth noting that the Fourier transform $H(\omega) = 2/(1+\omega^2)$ is never zero so that the assumption of the sampling theorem is not fulfilled.) The procedure starts by

1) *sampling $h(t)$ with sampling rate T*. This produces discrete data denoted by $h_c(t)$ and a Fourier transform $H_c(\omega)$ periodic with period $1/T$. In this specific case aliasing occurs (Fig. 3.17 b)) that can be reduced by choosing a finer sampling. Sampling can be described mathematically as the multiplication by a "comb," made of unit impulses equally spaced by T.

The number of samples at the moment is infinite and no computer can operate on that. Thus there follows:

2) *a truncation of length T_0 in the time variable*. A finite number of data arises in this case, precisely N if $T_0 = NT$. This has a slight drawback, since it implies a degraded discrimination on the Fourier transform, shown by "ripples," enhanced in Figure 3.17 c). Indeed on the Fourier transform side we are witnessing the convolution of H with the Fourier transform of a finite comb (Fig. 3.4 f)). Ripples can be attenuated by increasing T_0 and in turn the samples number.

The Fourier transform is still continuous; therefore the last step to define the DFT consists of

3) *sampling on the frequency variable*. The spacing rate has necessarily to be $1/T_0$ to avoid aliasing when computing the inverse Fourier transform. The samples number is N, given the periodicity of H_c.

In Figure 3.17 the upper left shows the graph of the signal, while the lower right shows the DFT. To the left of the DFT there is its inverse Fourier transform, recognizable as the truncation in Figure 3.17 c) made periodic. The discrete Fourier transform $\tilde{H}$ of the function h is therefore given by the following formula, to which we will always refer in the sequel:

$$\tilde{H}(n/T_0) = \sum_{k=0}^{N-1} h(kT)e^{-2\pi i n k/N} \qquad n = 0, 1, \ldots, N-1 \tag{3.2}$$

for $h(t)$ defined on $[0, T_0]$ and sampling rate T, such that $NT = T_0$. Manifestly (3.2) is the discrete analog of (2.15), which reads

$$H(\omega) = \int_{-\infty}^{\infty} h(t)e^{-2\pi i \omega t}\,dt.$$

Similarly to the Fourier transform, the DFT has an inverse given by

$$h(kT) = \frac{1}{N}\sum_{n=0}^{N-1} \tilde{H}(n/T_0)e^{2\pi ink/N} \qquad k = 0, 1, \ldots, N-1$$

denoted by $\tilde{h}(t)$ in Figure 3.17.

3.9 The fast Fourier transform

The number of operations required by the discrete Fourier transform is of the order of N^2 if N is the number of samples. The fast Fourier transform (FFT) is an algorithm that reduces the number of operations to the order of $N \log_2 N$. If calculations are hand made, then N is necessarily small and the advantage is not that significant but, in case N is large, the number of operations is drastically reduced.

The best-known algorithm that goes under the name of *Fast Fourier transform* was introduced in 1965 by J. W. Cooley and J. W. Tukey (Cooley–Tukey [1965]) of the IBM Research Center, Yorktown Heights, N.Y. The discrete Fourier transform is written more conveniently as

$$X(n) = \sum_{k=0}^{N-1} x_0(k)e^{-2\pi ink/N} \qquad n = 0, 1, \ldots, N-1, \tag{3.3}$$

with k in place of kT and n in place of n/T_0 for simplicity, x_0 and X in place of h and $\tilde{H}$. Then (3.3) is a system of N equations, each one presenting a power of the complex number $W = e^{-2\pi i/N}$ with exponent nk.

In case $N = 4$, (3.3) reads as follows:

$$\begin{cases} X(0) = x_0(0)W^0 + x_0(1)W^0 + x_0(2)W^0 + x_0(3)W^0, \\ X(1) = x_0(0)W^0 + x_0(1)W^1 + x_0(2)W^2 + x_0(3)W^3, \\ X(2) = x_0(0)W^0 + x_0(1)W^2 + x_0(2)W^4 + x_0(3)W^6, \\ X(3) = x_0(0)W^0 + x_0(1)W^3 + x_0(2)W^6 + x_0(3)W^9, \end{cases} \tag{3.4}$$

or, more synthetically, as

$$\begin{pmatrix} X(0) \\ X(1) \\ X(2) \\ X(3) \end{pmatrix} = \begin{pmatrix} W^0 & W^0 & W^0 & W^0 \\ W^0 & W^1 & W^2 & W^3 \\ W^0 & W^2 & W^4 & W^6 \\ W^0 & W^3 & W^6 & W^9 \end{pmatrix} \begin{pmatrix} x_0(0) \\ x_0(1) \\ x_0(2) \\ x_0(3) \end{pmatrix}. \tag{3.5}$$

A simple examination of (3.4) shows that, in order to compute $X(0)$, four multiplications and three additions of complex numbers are required, so that to, compute all

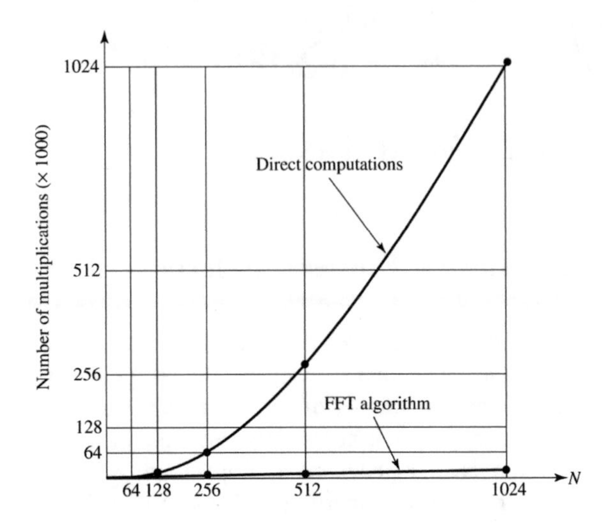

Figure 3.18 Number of multiplications required by direct computations of the discrete Fourier transform and by the FFT algorithm, as the number N of samples increases up to 1024.

four components of the vector $\mathbf{X}$, 16 multiplications and 12 additions are needed. The FFT reduces all that to four multiplications and eight additions.

In case N is generic, the FFT involves $(N/2)\log_2 N$ multiplications in place of N^2 and $N \log_2 N$ additions in place of $N(N-1)$. Since computing time mainly depends upon multiplications, in Figure 3.18 the number of multiplications required by direct computations and by the FFT algorithm is shown up to $N = 1024 = 2^{10}$. For instance in case $N = 1024$, it is $N^2 = (1024)2 = 1,048,576$ and $N \log_2 N = 10,240$ so that the ratio $(2N)^{-1}\log_2 N \approx \frac{1}{200}$. Therefore the saving in the number of multiplications is about 99%. The efficiency of the FFT is immediately perceived since its advantage over the direct method increases dramatically with N.

The algorithm of the FFT is easily described by going back to the case $N = 4$. The first simplification is based on the properties of the roots of unity in the field of complex numbers. For $W = e^{-2\pi i/4}$ is a fourth root of unit that is $W^4 = 1$. Then $W^6 = W^4W^2 = W^2$ and $W^9 = W^4W^4W = W$. Besides $W^3 = -W$ and $W^2 = -W^0$. Obviously $W^0 = 1$, yet W^0 will be used for reasons of uniformity. Then (3.5) becomes

$$\begin{pmatrix} X(0) \\ X(1) \\ X(2) \\ X(3) \end{pmatrix} = \begin{pmatrix} 1 & 1 & 1 & 1 \\ 1 & W & W^2 & W^3 \\ 1 & W^2 & W^0 & W^2 \\ 1 & W^3 & W^2 & W \end{pmatrix} \begin{pmatrix} x_0(0) \\ x_0(1) \\ x_0(2) \\ x_0(3) \end{pmatrix}. \tag{3.6}$$

The other idea ingrained in the FFT is the possibility of writing the matrix in (3.6) as a product of two matrices with many null elements. Precisely

$$\bar{\mathbf{X}} = \begin{pmatrix} X(0) \\ X(2) \\ X(1) \\ X(3) \end{pmatrix} = \begin{pmatrix} 1 & W^0 & 0 & 0 \\ 1 & W^2 & 0 & 0 \\ 0 & 0 & 1 & W^1 \\ 0 & 0 & 1 & W^3 \end{pmatrix} \begin{pmatrix} 1 & 0 & W^0 & 0 \\ 0 & 1 & 0 & W^0 \\ 1 & 0 & W^2 & 0 \\ 0 & 1 & 0 & W^2 \end{pmatrix} \begin{pmatrix} x_0(0) \\ x_0(1) \\ x_0(2) \\ x_0(3) \end{pmatrix}, \tag{3.7}$$

as can be easily checked by multiplying explicitly, row by column, in (3.7). A small price has been paid: The second and third component of the vector $\mathbf{X}$ are switched around. That is why the result has been denoted by $\bar{\mathbf{X}}$, to be distinguished from the wanted $\mathbf{X}$.

The total number of operations required to compute $\mathbf{X}$ can be inferred from (3.7). The product of a matrix times a vector occurs twice. The first time, by two multiplications and four additions the vector $\mathbf{x}_1$ is obtained, having components

$$\begin{pmatrix} x_1(0) \\ x_1(1) \\ x_1(2) \\ x_1(3) \end{pmatrix} = \begin{pmatrix} 1 & 0 & W^0 & 0 \\ 0 & 1 & 0 & W^0 \\ 1 & 0 & W^2 & 0 \\ 0 & 1 & 0 & W^2 \end{pmatrix} \begin{pmatrix} x_0(0) \\ x_0(1) \\ x_0(2) \\ x_0(3) \end{pmatrix}.$$

Indeed only the multiplication $W^0 x_0(2)$ is needed to compute the first and third component

$$\begin{cases} x_1(0) = x_0(0) + W^0 x_0(2), \\ x_1(2) = x_0(0) + W^0 x_0(2), \end{cases}$$

and only the multiplication $W^0 x_0(3)$ is needed to compute the second and fourth component

$$\begin{cases} x_1(1) = x_0(1) + W^0 x_0(3), \\ x_1(3) = x_0(1) - W^0 x_0(3), \end{cases}$$

Similar steps are required to compute $\mathbf{x}_2 = \bar{\mathbf{X}}$, since

$$\begin{cases} x_2(0) = x_1(0) + W^0 x_1(1), \\ x_2(1) = x_1(0) - W^0 x_1(1), \end{cases}$$

and

$$\begin{cases} x_2(2) = x_1(2) + W^1 x_1(3), \\ x_2(3) = x_1(2) - W^1 x_1(3). \end{cases}$$

In all, four multiplications and eight additions had been performed as stated. There remains to exchange the second and third component of $\mathbf{x}_2$ to obtain $\mathbf{X}$.

The procedure can be extended. For instance, in case N is a power of 2, say $N = 2^v$, the number of required steps becomes $v = \log_2 N$. For, the original matrix can be written as a product of $v = \log_2 N$ factors so that $\log_2 N$ is the number of the intermediate vectors $\mathbf{x}_0, \mathbf{x}_1, \ldots, \mathbf{x}_r, \ldots, \mathbf{x}_\nu = \bar{\mathbf{X}}$, with $\mathbf{x}_0$ denoting the initial data and $\bar{\mathbf{X}}$ the wanted result (up to an exchange of components). The computation of any intermediate vector $\mathbf{x}_r$, having N components $x_r(0), \ldots, x_r(N-1)$, is based on the preceding $\mathbf{x}_{r-1}$: For a fixed $\mathbf{x}_r$ and fixed component $x_r(h)$ there exists a suitable integer p such that

$$x_r(h) = x_{r-1}(h) + W^p x_{r-1}(h + 2^{-r} N).$$

The multiplication $W^p x_{r-1}(h + 2^{-r} N)$ allows to compute another component, namely

$$x_r(h + 2^{-r} N) = x_{r-1}(h) - W^p x_{r-1}(h + 2^{-r} N).$$

Therefore, to compute any intermediate vector $\mathbf{x}_r$, $N/2$ multiplications are required and a total of $(N/2)\log_2 N$ to compute $\bar{\mathbf{X}}$. Once the rule for p is given the whole procedure can be automated.

3.10 Gauss and the asteroids: history of the FFT

The publication of the FFT algorithm by Cooley and Tukey in 1965 was a turning point in digital signal processing and in certain areas of numerical analysis. The algorithm was regarded as new by many and only little by little was its intriguing history found out.

In response to the Cooley–Tukey paper, P. Rudnick [1966] of Scripps Institution of Oceanography, La Jolla, Cal., proposed a similar algorithm. It was based on the work of G. C. Danielson and C. Lanczos [1942] who had described their method in the context of X-ray scattering problems. In this field, the calculation of the Fourier transform (Chapter 6) was a serious bottleneck and kept being that way to researchers unaware of the method for years after 1942.

The Rudnick paper prompted an investigation into the history of the FFT algorithm by Cooley himself, Lewis, and Welch [1967]. They pointed out that Danielson and Lanczos, for the source of their method, referred to two papers by Carl David Tomé Runge (1856–1927) dealing with trigonometric series in real form (2.3) and deriving computational economy from the symmetries of the sine and cosine functions. The papers, Runge [1903] and [1905], were apparently overlooked even though they had been published by a well-read mathematician. Thus the roots of the FFT algorithm were traced back to the early twentieth century.

Then Goldstine [1977] discovered that C. F. Gauss had an algorithm similar to the FFT for the computation of the coefficients of finite Fourier series. Originally written in the *Theoria Interpolationis Methodo Nova Tractata*, undated but likely written in late 1805, the treatise was published only posthumously in 1866 in Volume 3 of his collected works (Gauss [1866]). This result by Gauss had remained hidden or went unnoticed for almost two hundred years. Gauss's motivation and the mathematician himself are worth a digression.

Carl Friedrick Gauss (1777–1855) was born in Brunswick, Germany, into a poor and unlettered family. He showed extraordinary precocity. Nothing like it is on record in the history of mathematics. Without the help or knowledge of others the little Carl Friedrick learned how to calculate—before he could talk, as he joked years later—and taught himself how to read. At the age of three, he corrected an error in his father's wage calculations. At the age of eight he astonished his teacher by instantly computing the sum of the first hundred integers and at fifteen, by an ingenious method of his own, he was able to rapidly calculate square roots to fifty decimal places. Later in the 1820s, relative to the project of the triangulation of Hannover, he handled more than a million numbers without assistance.

His interest and ability in computations, which he retained through his life, extended first to the theory of numbers and then to algebra, analysis, geometry, probability, and the theory of errors. The list of branches of science that have been the object of his intensive empirical and theoretical investigations is even longer, including observational astronomy, celestial mechanics, surveying, geodesy, capillarity, geomagnetism and electromagnetism (Section 3.1), mechanics, optics, and design of scientific equipment. He also had an interest in actuarial science: Having known poverty, he always maintained habits of frugality but succeeded well enough to own an estate worth nearly two hundred times his annual salary. His method was one: massive empirical investigations in close interaction with rigorous theory construction.

At the University of Göttingen, which he entered in 1795, mathematical classics and journals were finally available and he often found that his own discoveries were not new. In 1796 Gauss obtained a criterion by which it can be decided whether a regular polygon with any given number of sides can be constructed using only rule and compass. These include the regular polygon with 17 sides, the first advance in two millennia since the time of Euclid. This convinced him to abandon philological studies, which also interested him, and concentrate on mathematics alone. During the following years, from 1796 to 1801, mathematical ideas came to him so fast that he could hardly write them down.

The year 1801 saw two extraordinary achievements, the *Disquisitiones Arithmeticae*, pertaining to number theory, which quickly won Gauss recognition as a first order mathematician, and the calculation of the orbit of the newly discovered planet Ceres.

On January 1, 1801, Giuseppe Piazzi of the Palermo Observatory in Sicily discovered a body, subsequently named Ceres and identified as a *minor planet*, that seemed to be approaching the sun. This was the first success in the search for a new planet, predicted by Bode's law, in the region between Mars and Jupiter. It was to be followed by a thousand others. Ceres was briefly observed during some forty days and then lost. Where did it go? During the rest of the year astronomers tried, in vain, to relocate it. In September with the *Disquisitiones* just printed, Gauss decided to take up the challenge. He did not use telescopes—everybody else had done that—rather he set up some computations using a more accurate orbit, based on the ellipse instead of the usual circular approximation, and methods of his own. By December he had the result. And there was Ceres, found by the astronomer Franz von Zach in the predicted position.

The extraordinary achievement of locating a heavenly body, distant and tiny—Ceres, which revolves around the sun in 4.6 years and has a diameter estimated at 700 km (440 miles)—appeared to be almost superhuman, especially since Gauss did not reveal his method. It was a triumph. He had a similar success with the asteroid Pallas, detected in 1802, and second largest after Ceres. In 1809 Gauss published his second masterpiece *Theoria Motus Corporum Coelestium* in which he set the laws, at the basis of practical astronomical computations for many years to come, including a difficult analysis of perturbations. It was in conjunction with astronomical computations that Gauss hit upon the algorithm presently called *Fast Fourier Transform*.

Gauss found some of the way already paved. Euler, who pushed on the use of trigonometric series in analysis [1748], [1750], [1753], [1793], [1798], derived in [1750] the formula for the coefficients of a periodic function, expressed as a finite series of sines, given samples of the function. His work was followed by Alexis Claude Clairaut (1713–1756) [1754] who used equally spaced samples and obtained the first known explicit formula for the DFT for cosine series, and by Joseph Louis Lagrange for sine series [1759], [1762]. Both Clairaut and Lagrange were concerned with the problem of determining an orbit from a finite set of observations. Gauss presumably knew the papers of Lagrange having borrowed from the library in Göttingen, while he was a student, the volumes of *Miscellanea Taurinensia* containing them (Dunnington [1955]).

In the mentioned treatise *Theoria Interpolationis* Gauss extended the work of Lagrange. For general periodic functions normalized so to have period 1

$$f(t) = \sum_{n=0}^{N-1} C_n e^{2\pi int},$$

once the values $f_n = f(t_n)$ at N equally spaced points $t_n = 2\pi k/N$, $n = 0, \ldots, N-1$, are known, he showed that the coefficients C_n are given by

$$C_n = \frac{1}{N} \sum_{k=0}^{N-1} f_k e^{-2\pi ikn/N} \quad n = 0, \ldots, N-1 \tag{3.8}$$

(see Goldstine [1977], pp. 249–253, for an English translation). The formula might be easily verified by using the orthogonality relations

$$\sum_{k=0}^{N-1} e^{2\pi ink/N} e^{-2\pi imk/N} = \begin{cases} N & \text{if } m = n, \\ 0 & \text{if } m \neq n. \end{cases}$$

The above (3.8) is the earliest known explicit formula for the general DFT (Heideman et al. [1985]).

Then Gauss proceeds by a method that "greatly reduces the tediousness of mechanical calculations, success will teach the one who tries it." If the number N is composite, namely $N = N_1 N_2$, Gauss suitably writes

$$\begin{aligned} n &= N_1 n_1 + n_2, \\ k &= N_2 k_2 + k_1, \end{aligned}$$

where $n_2,\ k_2 = 0, \ldots, N_1 - 1$ and $n_1,\ k_1 = 0, \ldots, N_2 - 1$. Then changing notation, to adapt to the above decomposition of the n and k and writing $f_k = f(k_1,\ k_2)$ and $C(n_1,\ n_2) = NC_n$, (3.8) becomes

$$C(n_1,\ n_2) = \sum_{k_1=0}^{N_2-1} \left[\sum_{k_2=0}^{N_1-1} f(k_2, k_1) W^{n_2 k_2 N_2} \right] W^{nk_1} \quad n = 0, \ldots, N-1. \tag{3.9}$$

Here $W = e^{-2\pi i/N}$ is the Nth root of unity so that the fundamental identity $W^N = 1$ holds. For the computation of a single C_n, the formula (3.9) requires $N_1 + N_2$ multiplications, N_1 for the inner sum and N_2 for the outer one, versus $N_1 N_2$ with (3.8). Totally $N(N_1 + N_2)$ multiplications are needed to compute all of the C_n's by (3.9). Then Gauss applies the algorithm to the orbit of Pallas, with $N = 12$ and $N_1 = 4,\ N_2 = 3$ and also $N_1 = 3$ and $N_2 = 4$. He gives another example with $N = 36$ and $N_1 = N_2 = 6$. Finally he states, "Now the work will be no greater than the explanation of how that division can be extended still further and can be applied to the case where the majority of all proposed values are composed of three or more factors, for example, if the number N_1 would again be composed." He just omits to quantify the computational complexity to the now familiar expressions $N \sum N_i$ and $N \log_2 N$ in case N is a power of two.

Gauss's algorithm, which uses real trigonometric series rather than the complex exponential above, is equivalent to Cooley–Tukey's one, being as general and powerful. His treatise though is difficult to read because of the Latin language and notation. Gauss's work predates Fourier's and should have led to the Discrete Fourier transform and the fast Fourier transform being also named after him.

3.11 Unsteady combustion and space propulsion

Following the astronomical content of the preceding section now is space propulsion, to end with space exploration, a field still in its infancy, which powerfully captures the human mind. In space propulsion much scientific and technical knowledge is involved. Examples of the contribution of signal analysis will be given.

As stated by the inversion formula (2.14) a signal $f(t)$, depending upon time t, may be equivalently expressed by its Fourier transform $\hat{f}(\omega)$. In many cases, the Fourier transform allows an easier interpretation and provides a better characterization of the signal itself. For this reason, the FFT is commonly used to interpret experimental data expressed under the form of a signal, be it analog or digital.

Remarkable examples will be presented in Chapter 8 relative to the phenomenon of resonance, exploited both for diagnostic purposes (nuclear magnetic resonance imaging) and to reconstruct the spatial atomic structure of matter (nuclear magnetic resonance spectroscopy), after a preliminary identification of the resonant frequency.

The instance that follows also has to do with resonant conditions but investigated with the opposite aim, that is, to avoid them. The context is a specific topic in combustion—a wide area of research encompassing engineering, physics, and applied chemistry—and a field strategically important in its own right. It suffices to consider that between 70 and 80 percent of the energy produced in the industrialized countries is generated by burning fossil fuel: gasoline or diesel fuel in automotive engines (cars and trucks), natural gas or heating oil for heating purposes, kerosene in airplanes and heavy oil fuel, among other fuels, in power stations. The remaining 20 to 30 percent is divided between nuclear and hydroelectric power.

In most industrial applications fuel and oxidizer (air) are initially separated. They are mixed and react in a certain container, called *combustor* or *burner*. Conventional gasoline engines are an exception: Gasoline and air enter inside the engine cylinders already mixed.

In rocket motors for space launchers (motors for air to air or air to ground missiles may be somewhat different) so-called *solid propellants* are employed to generate thrust. Fuel and oxidizer are both solid and have been thoroughly mixed. As both fuel and binder, a polymer is used and as oxidizer, a salt is used, for example ammonium perchlorate.

This kind of propellant is employed to accelerate, during the initial two to five minutes of their trajectory, the huge space launchers used to orbit crew and payload—as in the case of the space shuttle—or satellites. Satellites launchers are for instance the European Ariane 5 and the U.S. Delta. In those initial minutes hundreds of tons of solid propellant are burned to provide thrust. It is of vital importance for the purpose of injecting the payload into the desired orbit that the thrust profile be exactly as planned.

The solid propellant is located in the so-called *boosters*. For instance in the case of the Shuttle they are the two big containers of cylindrical shape placed under the Shuttle at each side of the tank containing oxygen and liquid hydrogen that feed the three main engines (Fig. 3.19). The boosters are filled with *grains* (blocks) of solid propellant, cast in cylindrical shape and stacked one on top of the other. Each grain is hollow, forming a cavity (*port*). The propellant burns releasing hot gases from the port toward the exhaust nozzle. The combustion gases reach temperatures of 3000°C. For a comparison it suffices to think that, at approximately 1500°C, steel melts.

The boosters are perhaps in the memory of us all, due to the dramatic launch of the Shuttle "Challenger" that took place on January 28, 1986. The intense cold (−8°C against a minimum temperature of 11°C of the preceding launches) made the rubber

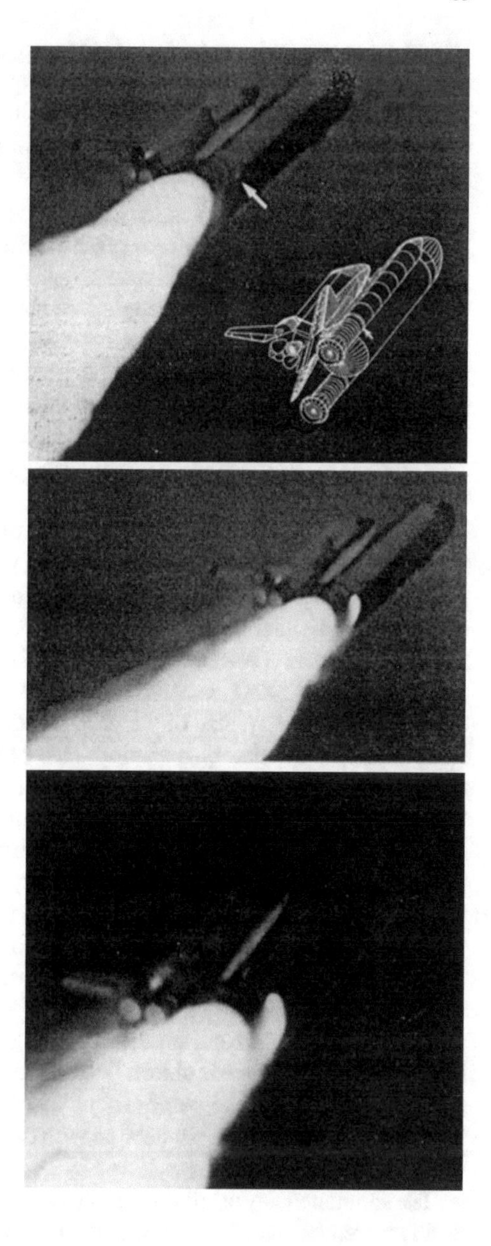

Figure 3.19 Progression of flame in the incident of the Shuttle Challenger on January 28, 1986. (*Courtesy of NASA/JPL/Caltech*)

rings that sealed the different sections of the booster case hard and brittle and unable to adapt in a few seconds to the pressure increase during the booster ignition. The seal gone, the hot combustion gases reached the outer steel casing, not designed to stand temperatures that high, and pushed against the rocket casing walls. A gap opened through which hot gases were let out and burned, eventually causing the explosion of the liquid propellant tank a few minutes after launching.

More information can be found in Feynman [1989], where the physicist Richard Feynman—Nobel prize in 1965 and member of the committee nominated by President Ronald Reagan to investigate the incident—recounts with usual frankness his experience as a member of the investigating committee and the detailed reasoning and conclusions he reached.

The cause of the Challenger failure was accidental but explosion or, for that matter, extinction may also take place for fundamental physical reasons. Much of the research on both steady as well as transient combustion (like motor ignition for instance) is motivated by the problem of combustor instability in rocket motors.

The burning characteristic of early propellants was less than ideal and explosion of motors, due to poor igniter design or unpredictable propellant burning, was common in the late 1930s and early 1940s when intensive work on solid rocket motors began. Even after propellant quality was improved, it was found that burning was often erratic sometimes going out, as if ignited incompletely, or developing violent pressure excesses.

Up to 1960, it was feasible to evaluate the performance of the relatively small motors of the time by static firing full-size motors. So after motor design and propellant had been chosen, if troubles were found in firing static tests, changes were made on a trial-and-error basis. This method became too costly for developing larger motors. Therefore a better understanding of combustion instability had to be gained and with it went along an awareness of its complexity. It became clear that small changes (in geometry, size of the propellant's particles,. . .) could cause a transition from stable to oscillatory behavior. This helped to show that many motors were operating near stability thresholds, beyond which severe oscillations were likely. "The prospect of combustion instability troubles in development programs on big solid rocket motors was a matter of real concern, because it was recognized in 1958 that it was not known what to expect in large motors, and the approach of remedial measures based on trial-and-error methods would be prohibitively expensive, both from the standpoint of cost of testing and of delays in development programs. As a result of the urging by the Polaris program managers and a U.S. Department of Defense (DOD) committee on large motors, a DOD ad hoc committee was set up to review and recommend on the combustion instability problem" (Price [1992]). Seriousness of the issue is shown by what is probably the most extreme example of direct expenditures—estimated around 25 million dollars—for combustion instability problems that arose in connection with the liquid propellant engines of the Saturn vehicle.

The spontaneously oscillatory modes of operation of a combustor that take place at different frequencies (from a few hertz up to 15,000 Hz) are studied in a given design and at a given ambient temperature, purposely to avoid them. Indeed it is of vital importance in space propulsion to anticipate exactly the performance of a burning solid propellant whose combustion cannot be modified, let alone extinguished, after ignition.

Initially the boosters are pressurized at a pressure of the order of 10 atm, the nozzle closed by a plug to be ejected as the pressure of the hot gases, produced at ignition, rises. When combustion reaches steady conditions the pressure is about

70 atmospheres. One of the causes of loss of performance, with eventual explosion of the motor, is associated with the possibility that pressure waves (longitudinal, tangential, and radial) couple with the combustion characteristic frequencies of the propellant that depend on its gasification times. Experimental and theoretical studies are made to determine if these characteristic frequencies, at operative conditions, happen to be near one of the characteristic frequencies of the motor (Zanotti and Giuliani [1993]).

Conditions of subatmospheric pressure are also investigated. This is the case of the launch of a satellite by a flying airplane—as the "Pegasus" launcher of the Orbital Sciences Corporation—and also of the upper stages (usually two) of the satellite launchers. Among these the Japanese M-V, designed to orbit scientific satellites, uses all solid propellant rocket motors stages. During ignition of the second and third stage and extinction, the pressure inside the motor is close to zero.

The phenomenon of self-sustained oscillations at subatmospheric pressure is now described quantitatively as an example of the systematic use of the FFT in this field's experimental studies (Zanotti [1992]). The combustion of a solid propellant in steady state conditions is not always time invariant. Indeed for pressure values close to the *deflagration limit* (that value of the pressure, above which fuel burns after ignition and below which extinction takes place) combustion is characterized by oscillations that are not imposed from the outside and therefore are called self-sustained. This phenomenon has been discovered rather recently and, even if limited to a brief time interval, might cause problems to the rocket engines, possibly followed by extinction or engine explosion in the case of resonance.

In the mentioned article, self-sustained oscillations are studied—in particular the dependency of their frequency upon the parameters involved and for different propellants—by means of the radiation emitted by the burning surface or through the variations of the temperature produced by the combustion. The laboratory experiment uses ammonium perchlorate as propellant mixed to a binder and takes place at ambient temperature (about 25°C).

Figure 3.20 a) shows how the temperature changes in time for a deflagration wave (the signal is amplified 500 times) as measured by a thermocouple located inside the propellant near its burning surface, the pressure being 0.08 atm. It can be seen that at a certain instant a change takes place: The phenomenon of combustion starts and keeps going on in an irregular fashion with large oscillations in the face of constant pressure and ambient temperature. A comparison could be made with a candle flame that, instead of maintaining shape and dimensions about constant as it usually happens, strongly oscillates going up and down.

The portion of the signal between 30 and 40 seconds is shown in Figure 3.20 b) and its FFT in Figure 3.20 c). Figures 3.21 and 3.22 above show the amplitude of the oscillation of the light emission as measured by a photodiode in the proximity of the combustion's surface. Pressure is 0.045 atm and 0.11 atm respectively. It can be seen with one's naked eye that the two signals are very different, but how to explain the nature of the difference? To that purpose the FFT is systematically used: The first signal has a narrow spectrum (Fig. 3.21), while the second one shows a wide spectrum (Fig. 3.22). The phenomenon of self-sustained oscillations is by now well understood.

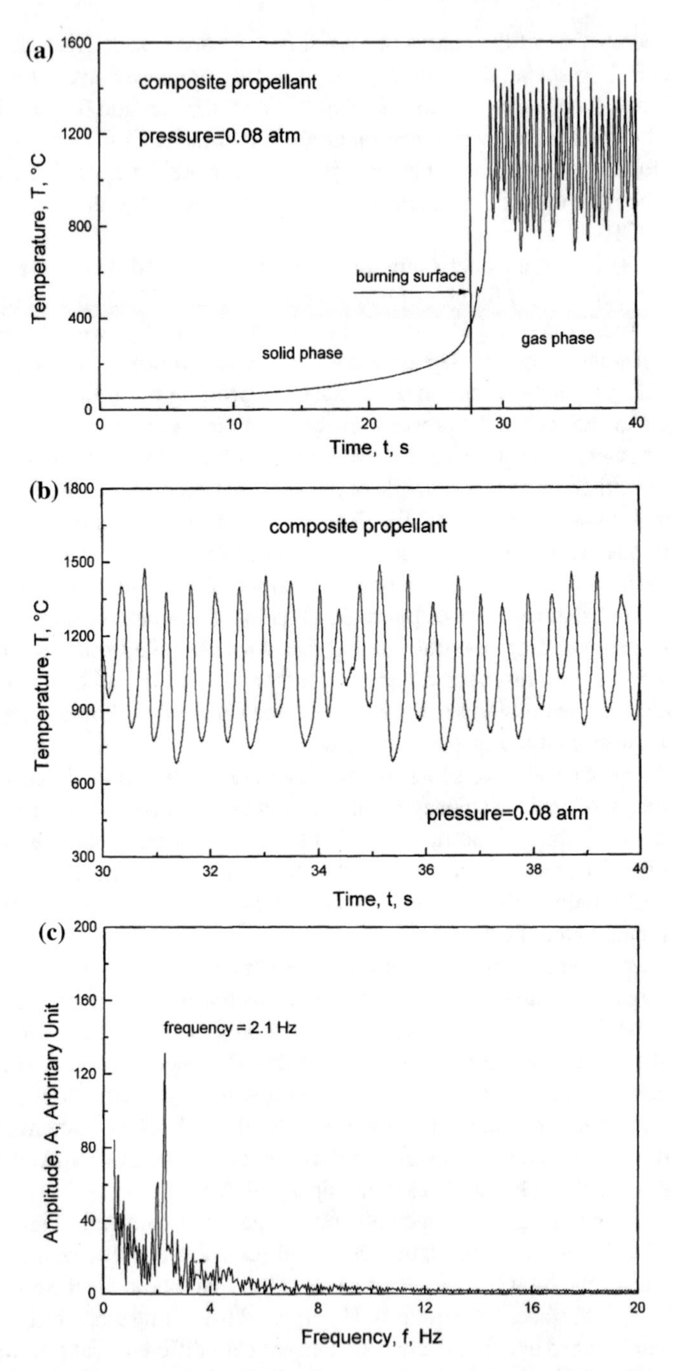

Figure 3.20 a) Temperature profile of the deflagration wave of a solid propellant at ambient temperature of 25°C and 0.08 atm; b) profile between 30 and 40 seconds; c) FFT of the signal in b). (*Courtesy of C. Zanotti, Centro Nazionale Propulsione e Materiali, CNR*)

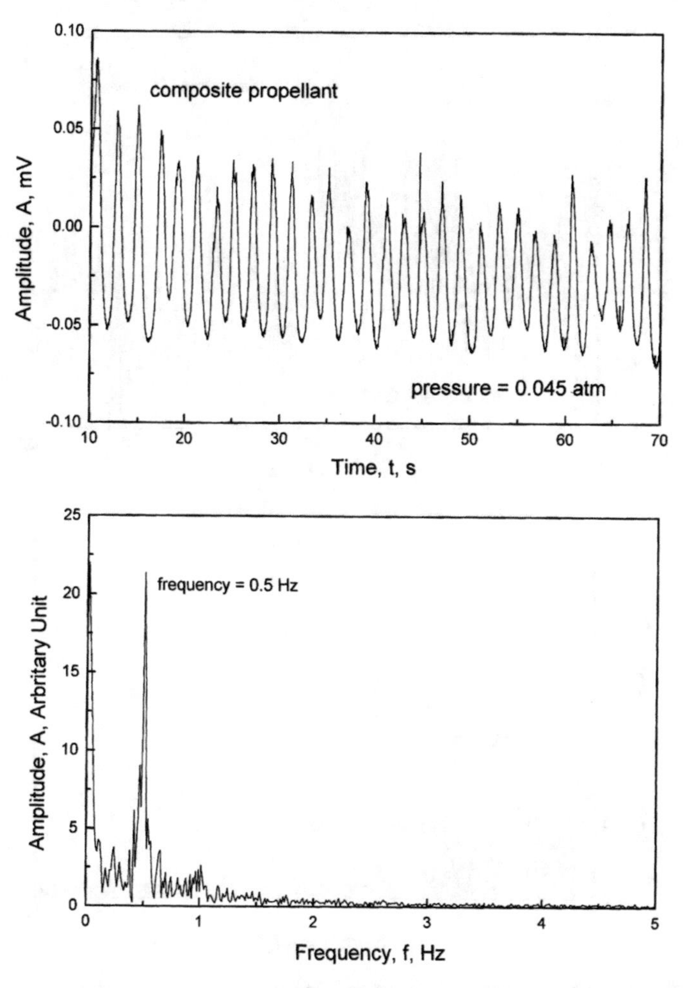

Figure 3.21 Above, the amplitude of oscillation of the light emission of a solid propellant burning at 25°C ambient temperature and 0.045 atm; below, its FFT. (*Courtesy of C.Zanotti, Centro Nazionale Propulsione e Materiali, CNR*)

It takes place only for pressure p near the pressure deflagration limit which is 0.02 atm in the case under consideration. Precisely if the pressure is between 0.02 atm and 0.3 atm then self-sustained oscillations take place: Their amplitude increases and their frequency decreases as p approaches the deflagration pressure; the reverse happens if p moves away. At the stated values of the pressure, the source of the oscillations is the heat feedback between the gas phase (the gas produced by the combustion) and the condensed phase of the propellant (solid propellant).

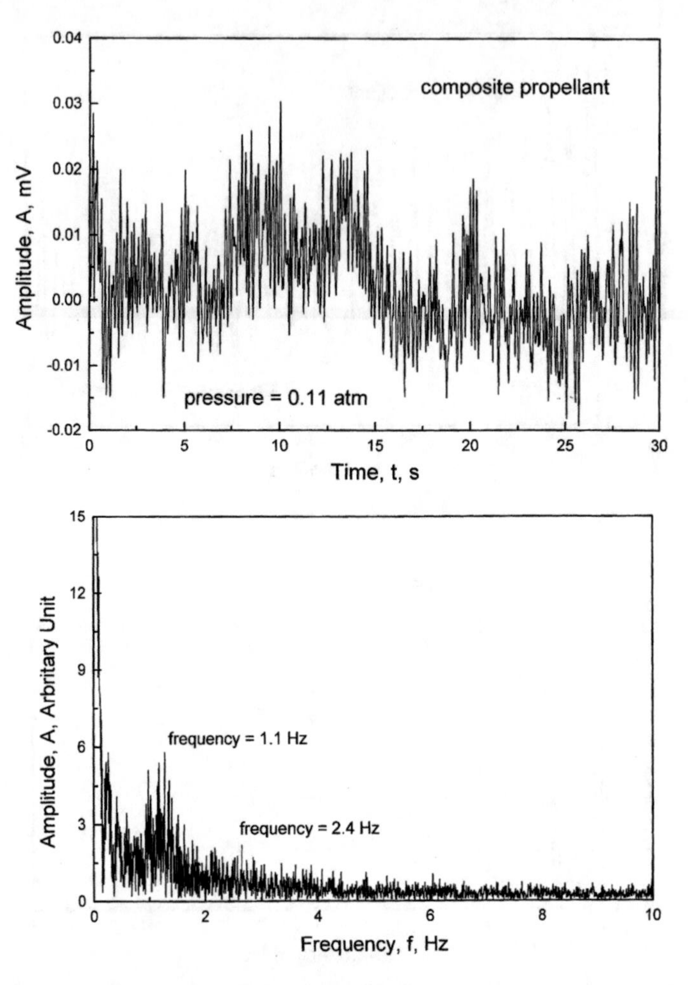

Figure 3.22 Above, the amplitude of oscillation of the light emission of a solid propellant burning at 25°C ambient temperature and 0.11 atm; below, its FFT. (*Courtesy of C.Zanotti, Centro Nazionale Propulsione e Materiali, CNR*)

3.12 The Mars project

The thread of space exploration is now picked up where it was left, with the dreams, thoughts, and theoretical work of Tsiolkovsky and the initial success by Goddard (Section 3.1). The field was greatly advanced by the German engineer Wernher von Braun (1912–1977) who played a prominent role in all aspects of rocketry and space exploration. A gifted man, a charismatic figure capable of motivating his collaborators to work even under bombings, von Braun was born into an aristocratic family. His mother, an amateur astronomer, spoke six languages and his father Baron Magnus von Braun held high positions in the government. He was minister of agriculture in the government of the Weimar Republic headed by Franz von Papen and when the Nazis seized power he retired to his estate in Silesia.

From Baroness Emmy von Braun's recounts, Wernher's childhood was happy, extremely active, showing early signs of scientific interests. "His inclination toward engineering and physics found its expression in an endless production line of vehicles, some small and powered by old clockworks, others large and driven by rockets, often to the horror of our neighbors and to the dismay of his father who had to pay for the damaged flower beds and windows" (Stuhlinger–Ordway [1994]).

Wernher's doctoral thesis of 1934, at the University of Berlin, was titled "About Combustion Test" and contained the theoretical investigation and developmental of experiments on 300- and 600-pound thrust rocket engines. By the end of the year von Braun and his group successfully launched two rockets that rose to more than 2.4 km. Soon after Hitler's advent rocket tests were forbidden and handed over to the military. von Braun became technical director of the giant military facility at the village of Peenemünde on the island of Usedom in the Baltic sea. There he developed the dreaded long range ballistic missiles A-4 (Aggregat 4) renamed by the Propaganda Ministry as V-2—meaning Vengeance Weapon 2—after September 8, 1944, when they were first used in air assaults on Paris and London. von Braun, who always retained his civilian status, had uneasy relations with the controlling SS: He lamented with them the poor living conditions of the inmates working at his project, resulting in faulty parts and poor assembly, and he himself was arrested and held in a Gestapo jail for two weeks under the accusation of sabotaging the V-2 production program by thinking too much of the future of rockets for space travel.

If, as in von Braun's words, "until 1936 Goddard was ahead of us all," by 1944 Germany with the group of Peenemünde—in the end working under allied bombs—was way ahead of any other country. "Within a period of less than five years a powerful and finely controllable rocket motor had come into existence; precision guidance and flight control systems had been invented and built as well as radio guide beams and communication links; electronic simulators—the forerunners of modern computers—to handle lengthy mathematical procedures and to help solve complicated equation systems had begun to take shape" (Stuhlinger-Ordway [1994]).

At the end of the war the highly desirable German know-how and hardware, like the remaining V-2, was split between the U.S. and Russia. von Braun and a hundred and twenty-six of his collaborators moved to the test site at White Sands, New Mexico, where initially they lived almost secluded, being after all former enemies. The scientists and technicians of his group who remained behind were soon seized by the Russians and, by the fall of 1946, gently moved to work places outside of Moscow, having being given no choice. It had all started long before with a KGB informer witnessing the rocket tests of von Braun as early as 1937, followed in 1944 by the finding of remnants of A-4 at the front line in Poland (Harford [1997]). From the nozzle dimensions the Russians calculated that the engine thrust was at least 20 tons (in fact it was 25 tons). They were shocked: At that time the limit of their dreams was one and a half. The Germans worked under Russian supervision until 1953 when it was judged that they could be let go.

From that moment the Russians were on their own and they found their leader in Sergei Pavlovich Korolev (1906–1966), a man totally dedicated to his ideas and work, who had studied in Moscow under the celebrated airplane designer Andrey N. Tupolev. One year of concentration camp in Kolyma, in far northeastern Siberia, was followed by six years in a prison for engineers and scientists during Stalin's rule; under Khrushchev and Brezhnev when in charge of systems engineering for Soviet launch vehicles and spacecraft, directing the design, testing, construction, and launching of manned and unmanned spacecraft, he was forced to anonymity: Chief Designer was his official title. At his death, signs of a feeling of gratitude from the former Soviet Union could be seen as when his coffin was shouldered by the very First Secretary Leonid I. Brezhnev, among others, to be hosted inside the Kremlin wall.

The spectacular achievements by the Soviet Union—a country considered backward in many ways—started with Sputnik 1, the first satellite to reach its orbit. "October 4, 1957, the day when Sputnik appeared in the sky" von Braun said, "will be remembered on this planet as the day on which the Age of Space Flight was ushered in." It was to be followed by Laika, first dog in space and crowned on April 12, 1961, by the Vostok 1 spacecraft carrying Yuri A. Gagarin (1934–1968), who became the first men to travel into space. Then the gigantic mobilization of manpower and money that the United States put into the Apollo program paid off: On July 16, 1969, the Saturn V rocket designed by the NASA team, lead by von Braun, with a crew of four astronauts blasted off and four days later Neil Armstrong (1930–2012) was the first man to set foot on the Moon's dusty surface.

Next is Mars, von Braun's dream. von Braun outlined a Mars expedition in *The Mars Project*, a booklet written during 1948 and 1949. It was first published in Germany in 1952 and a year later in the United States. In the 90 pages of the original German version all major phases of a manned round trip to Mars are dealt with to make the point that the project could be accomplished with the existing technology. In addition von Braun never lost one occasion to go public for, as he sometimes reminded his coworkers, "if you want to accomplish something as big as travel into space, you must win the people for your idea" (Stuhlinger-Ordway [1994]).

Mars was in the mind of Korolev too. Three months before dying of cancer surgery he told a collaborator, "I have a short time before me, maybe ten years and I want to send humans to the nearest planet" (Harford [1997]). He then promised him to double his salary if he could cut by half the time of his assignment. Korolev's plan was to launch a cosmonaut around Mars, putting the space traveler in a closed-loop life support system and an artificial low gravity environment for the lengthy round trip.

The decade 2020-2030 is presently the approximate deadline assigned to the project of sending a manned spacecraft to Mars by those who believe it can be done in the near future. It was actually President George Bush who took the initiative, back in 1989, of setting the year 2019 as a deadline for sending astronauts to Mars. Of the many problems involved, let alone the inescapable problem of the financial budget, a major one is the duration of the trip—two to three years with present technology—involving serious risks for the astronauts due to the radiation dose absorbed.

In the fall of 1998 the Italian nuclear physicist Carlo Rubbia, Nobel prize in 1984, advanced the idea for a new nuclear engine that could reduce to one year the time of a mission to Mars and back. His engine is composed by modules, each consisting of a cylinder where hydrogen is heated by the fission fragments emitted by the fission of americium 242m. The amount of energy liberated by the fission reaction has no known comparison, being about a million times greater than that obtained by chemical combustion. Rubbia worked several years on the project with his team at the Italian Space Agency (ASI). Other projects are under way (Chang-Díaz [2000]). The alternative to conventional nuclear propulsion (including that of Rubbia type) is electric propulsion. In electric propulsion a charged gas, for instance H_2 or Xenon, is accelerated to high speed by electric or electromagnetic fields by means of solar panels or by nuclear reactors. The electric thruster of Chang-Díaz has been recently refinanced by NASA as announced on August 24, 2015 on Space.com (http://www.space.com/30221-plasma-rocket-technology-nasa-funding.html). Time will tell which propulsion strategy will be chosen for a Mars mission.

Chapter 4
Sound, Music, and Computers

4.1 Understanding sound: some history

The understanding of the nature and propagation of sound has had important consequences that affect everyday life. Telephone, radio, compact disks, and sound tracks of films, often carrying spectacular sound effects, are just a few examples. Sound, the stimulus to the hearing organs produced by vibrations propagating through an elastic medium, like air, cannot be seen by observing the air. To "see" sound an oscilloscope helps, but it takes harmonic analysis to understand its structure and a repeated application of the FFT to follow in real time how it evolves (Fig. 4.6, 4.7, and 4.8).

Pythagoras and his school (c. 550 B.C.) were already interested in sound and the intriguing qualities that turn it into music. Music was part of mathematics then and musical sounds produced by plucking strings of the same thickness, tension, and material but different lengths were studied. It was discovered that strings of length 1, 3/2, 2, (C, G, C), when plucked at the same time or in succession, produce a pleasant and harmonious sound. The connection between length and pitch was the beginning of a quantitative knowledge of a complex subject that involves different fields of science such as physics, physiology, and even psychology if some understanding of why music arouses emotions is to be gained.

That motion of air is involved in sound, as a wave and not as a stream from the sound source to the listener, was understood by the Roman architect Vitruvius, first century B.C. In medieval times though, the prevailing theory explained sound as an emission of very small invisible particles moving through the air. The speed of sound being unknown and scientific method as yet unavailable, a deeper analysis could not be carried out for a long time. Also lack of adequate clocks to measure small time intervals cannot be underestimated.

Galileo (1564–1642) as well as the Abbé Marin Mersenne (1588–1648)—one of his most distinguished French disciples, member of the mendicant Order of Minims

E. Prestini, *The Evolution of Applied Harmonic Analysis*,
Applied and Numerical Harmonic Analysis,
DOI 10.1007/978-1-4899-7989-6_4

in Paris—studied the effect of a string change of mass and tension on its frequency of vibration. For a given length, an increase in mass or a decrease in tension was found to produce lower notes. This was especially important for stringed instruments such as the piano whose wide range of pitch cannot be practically obtained by just varying the length of the strings. Mersenne is also remembered for publishing the first qualitative analysis of a complex tone in terms of harmonics.

The French philosopher and mathematician Pierre Gassendi (1592–1655) who belonged to the informal group in touch with Mersenne—together with René Descartes (1596–1650), Pierre de Fermat (1601–1665), and Blaise Pascal (1623–1662)—though wrong in sharing the medieval corpuscular theory on the transmission of sound, was one of the first on record attempting to determine the velocity of sound in air. In 1635 he measured a value of 478 m/s, far too high as was recognized by other observers. Only about 1750 under the direction of the Royal Academy of Sciences in Paris a speed of 332 m/s at 0°C was measured in open air using a cannon as a sound source. Subsequent careful measurements did not change this figure significantly. The speed of sound in water of 1,435 m/s at 8°C was measured in the year 1826 in the lake of Geneva. Through steel it is much higher, about 5,000 m/s, and through the even more rigid cool glass it is about 5,500 m/s.

Galileo and Hooke (1635–1703) demonstrated that every sound is characterized by a precise number of vibrations per second, but the full understanding of even the simplest types of sound vibrators requires the power of calculus. Among the best mathematicians of the seventeenth century, Leonard Euler, Daniel Bernoulli, Jean le Rond d'Alembert, and Joseph Louis Lagrange all studied the vibrating string (Chapter 2). In 1747 d'Alembert found the mathematical expression for wave motion (the wave equation) and in 1753 Bernoulli stated the first version of the decomposition of every motion of a string as a sum of elementary sinusoidal motions. Then at the beginning of the following century Fourier developed "harmonic analysis."

The idea of applying the analytical method of Fourier to the phenomenon of sound was advanced by the German physicist Georg Simon Ohm (1789–1854) in 1843 and further elaborated by one of the greatest scientists of that century the well-known German physician, physicist, and mathematician Hermann von Helmholtz (1821–1894). In his monumental work of 1863 *On the Sensation of Tone as a Physiological Basis for the Theory of Music* (Helmholtz [1954])—in which he attempted to unveil the complete mechanism of the sensation of sound, from the ear through the sensory nerves to the brain—he also provided experimental evidence of the validity of decomposition of every sound as a combination of simple sounds whose frequencies are all multiples of one lowest frequency. He accomplished that by using special pipes, called *resonators*, of different lengths to detect different frequencies. Furthermore he could duplicate a given sound by a proper selection of electrically driven tuning forks, thus fully demonstrating experimentally the harmonic decomposition of sound.

Finally with the publication in 1877 and 1878 of the first and second volume of *The Theory of Sound* by John William Strutt (Lord Rayleigh) (1842–1919), modern acoustic began. Questions of vibrations and resonance of elastic solids and gases were studied as well as acoustical propagation in material media. In brief all aspects of the generation, propagation and reception of sound were presented in a rigorous fashion as well as experimental material. Pertaining more specifically to music, one ought to mention *The Physics of Music* by Alexander Woods, a classic in the field first published in 1913.

The physical response of the human ear to sound received attention before Helmholtz. It is worth mentioning that the *frequency limits of audibility*—which approximately cover the range from 20 Hz to 20,000 Hz, depending on the tone loudness and varying from individual to individual and with age—was established around 1830 by Félix Savart (1791–1841), who had received a medical degree at the University of Strasbourg in 1816. His early interest in the physics of the violin led him to study many phenomena involving vibrations, in acoustics as well as electromagnetism. In his earliest work he explained for the first time the function of certain parts of the violin. He also built a new violin in the form of a trapezoid with rectangular sound holes, thinking it would produce a better tone. When played before a committee that included the composer Luigi Cherubini (1760–1842), the tone was judged as extremely clear and even, but subdued. Antonio Stradivari (c. 1644–1737) still knew better.

Then in 1870 the *threshold of audibility* was measured for the first time. The resulting value was much higher than the true one. Indeed the human ear has an amazing sensitivity, being capable in the range between 500 to 4000 Hz—to which it is especially sensitive—to respond to a change of pressure of the order of 10^{-10} of atmospheric pressure. At the other end, when the change reaches about 10^{-4} of atmospheric pressure, the sensation of hearing becomes painful. In between these extremes there is a normal conversation, which induces in the vicinity of the mouth of the speaker a change of about one-millionth of the atmospheric pressure.

Inaudible sounds are of particular interest to modern physics: Infrasonic waves—below 20 Hz—naturally occur in earthquakes and tidal motions, while ultrasonic waves with frequencies above 20,000 Hz have well-known applications in medicine, for example, besides being definitely present in the animal world.

4.2 Bats, whales, and sea lions: ultrasonic reflections in air and water

Sound waves, like light waves, are spherical and propagate with a velocity v tied to the period T, wavelength λ, and frequency ν by the standard formula

$$\lambda = vT = v/\nu.$$

Sound waves are reflected by obstacles or targets in the same way that light waves are and, more generally, electromagnetic waves (Sections 3.2, 3.6). The resulting echoes are used by bats, toothed whales, and sea lions for blind hunting in the dark. It is interesting to know more about the cries emitted toward this goal.

To detect an obstacle or target, wavelengths smaller than the dimensions of the obstacle or target involved are needed. Bats are known to emit audible squeaks. These turn out to be their low tones. Indeed their cries usually fall in the range between 30,000 Hz and 80,000 Hz and are totally inaudible to people. Most species emit frequency-modulated pulses starting at a high frequency and dropping to half of that in a fraction of a second. The greatest energy goes in the middle of the range. Such extremely short and high frequency sound waves provide for the high resolution and high directionality that allows small bats, in full flight, to locate and detect the direction of motion of insects so as to catch them in dark caves or outdoors in moonless nights. Similarly bats are able to fly in the dark through threads or wires without getting entangled. This is certainly remarkable, the more so if attention is paid to the small fraction of energy that goes with the reflected signal and to the presence of disturbing noise such as the cries of many other bats and their relative echoes.

The auditory system of bats is well targeted toward that. The external ear is rather large, acting as an efficient collector and resonator of high frequency sounds. The middle ear has a structure similar to that of humans but the auditory portion of the nervous system is extraordinarily developed, a clear indication of the great predominance in these animals of hearing over the other senses. The region of greatest auditory sensitivity extends from 10,000 Hz to 70,000 Hz, the same range as for their *echolocation* cries.

Other mammals equipped for echolocation are the toothed whales which make use, for that purpose, of a rapid series of clicks containing many frequencies. Most of the energy goes in the range from 50,000 Hz up to 200,000 Hz. As mentioned, the speed of sound in water is about four times higher than in air and so the wavelength corresponding to each frequency is four times as long and significantly higher frequencies are needed to keep wavelengths small for effective echolocation. Dolphins and sea lions too use echolocation and are able to change the patterns and frequencies of their emitted clicks according to the size and distance of the prey they are pursuing, avoiding sound interference both from other species and from themselves. Sonar is ancient stuff and submarine life is not so silent after all.

4.3 Bels, decibels, and harmonic analysis of a single tone

Sound as detected by an instrument (a microphone, an oscilloscope) or perceived by a human ear is a plane wave: Locally, in a small region, that is how a spherical wave looks.

A simple sound is that produced by a tuning fork when struck around the middle of the tines or even slightly closer to the tips (Fig. 4.1). The molecules of the air in the immediate vicinity are subjected to a compression followed by an expansion again and again, so as to oscillate around a rest position. The formula that describes the movement of each molecule over time t, at a fixed point, is

Figure 4.1 Oscillations of the prongs of a tuning fork.

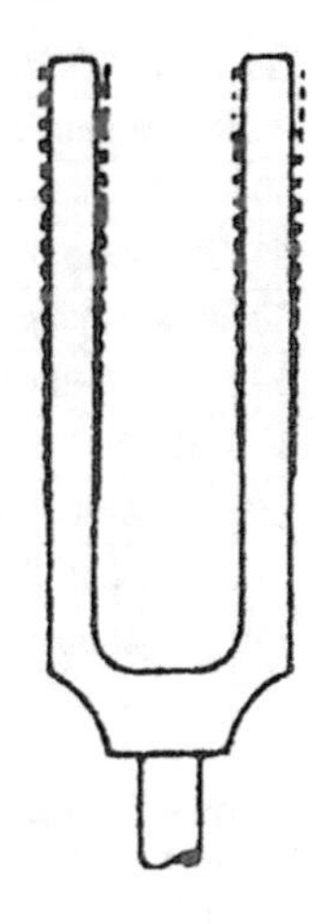

$$y = D \sin 2\pi \nu t, \tag{4.1}$$

where ν is the *frequency*, or number of oscillations (*cycles*) per second, and D is the *Amplitude* of the oscillations, that is the maximum displacement of the molecule. The oscillations take place in the direction of propagation; hence sound is a longitudinal wave.

The intensity of the sound is proportional to the square of the amplitude. However the intensity is usually not expressed in terms of displacements, which are extremely small and difficult to measure, but rather in terms of pressure. Intensity turns out to be proportional to the square of the pressure amplitude (pressure itself is a wave).

It is also common practice to measure the intensity in *decibels* (dB), the tenth part of the *bel*. The bel was devised in the 1920s by electrical engineers of the Bell Telephone Company to conveniently measure the attenuation of the telephone signals transmitted over long lines. The name was chosen in honor of the inventor of the telephone Alexander Graham Bell (1847–1922), who had just died the year before. Zero decibels is chosen as the minimum audible sound intensity I_0, as fixed by international agreements. It corresponds to a pressure of 20 micropascal (one atmosphere is about 10^5 pascal). A sound intensity I is compared to I_0 through the ratio I/I_0, which is a pure number. The logarithm to base 10 of such a ratio — another pure number—is the corresponding measure of the sound intensity I in bels. Therefore a 10-fold increase in intensity means a gain of 1 bel or 10 decibels and a 100-, 1000-, . . .fold increase means a gain of 2, 3, . . .bels or 20, 30, . . . decibels (Tab. 4.1).

Formula (4.1) can be verified experimentally. If the sound of a tuning fork in C4 is converted to an electric current and fed to an oscilloscope, the graph in Figure 4.2 is obtained. It is a simple sinusoid whose frequency ν is 261.7 Hz, meaning that there are 261.7 cycles in a second. If the same note is played by a violin, the graph

Table 4.1 Sound intensity in decibel (dB).

0 dB: minimum audibility for intact hearing
50 dB: very quiet condition
65 dB: average conversation
80 dB: annoying sound level
95 dB: very annoying sound level
117 dB: discotheque at full blast
130 dB: limit of amplified speech
135 dB: painfully loud condition
140 dB: jet operations on an aircraft-carrier deck

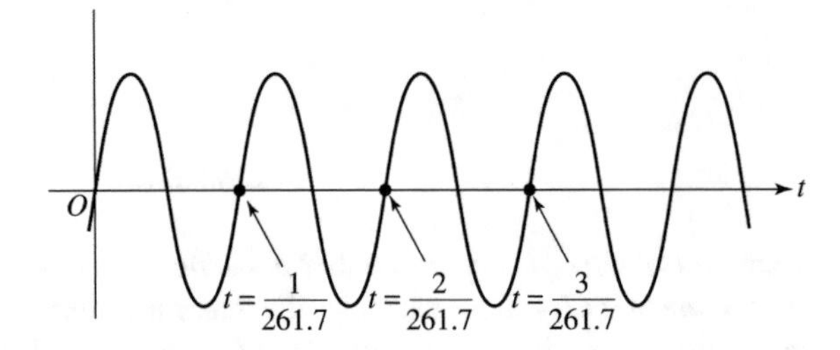

Figure 4.2 Waveform of a tone produced by a tuning fork in C4.

in Figure 4.3 (top) is obtained representing six cycles of the waveform. The lowest frequency, incorporated into the *first harmonic* and called *fundamental* because it is the frequency of the whole sound, is 261.7 Hz. The other frequencies are integer multiples of 261.7 Hz and correspond to the so-called *higher harmonics* or *overtones* (Fig. 4.3 bottom). The waveform is then decomposed as the sum of as many as 36 harmonics each with appropriate amplitude. The presence of higher harmonics is particularly noticeable in the spectrum of the violin. The procedure to obtain the sound spectrum is known as *harmonic analysis of sound* and can be performed on a computer with suitable software.

4.4 Sound processing by the ear: pitch, loudness, and timbre

A topic of great complexity, which still awaits clarification in full detail, is encountered as one moves from the physical description of sound to the human perception of it. It suffices to mention that the vibrations of the eardrum are picked up in the inner ear by some 16,000 receptor units called *hair cells* and that at the final stage, in the brain, hundreds of millions of cortical neurons, not even spatially contiguous, are active at the same time as shown by recent NMR imaging techniques.

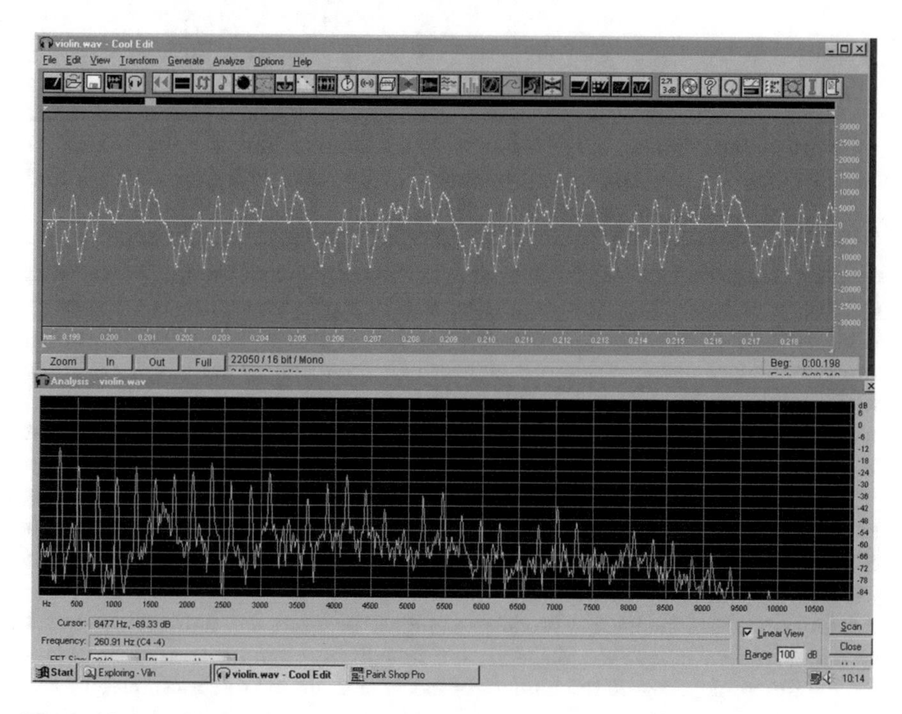

Figure 4.3 Waveform (above) and sound spectrum (below) of C4 played on a violin. (*Courtesy of G. Di Giugno*)

The considerations that follow are limited to the three essential qualities of pitch, loudness, and timbre for simple sounds.

The psychophysical perception of *pitch* in the case of a pure tone, such as that generated by an electronic oscillator, depends mainly on the frequency of the vibrations of the air (it may deviate from that at very high frequencies or at high amplitudes). The higher the frequency, the higher or sharper the perceived sound. A given frequency activates the hair cells in a limited region of the basilar membrane—in the inner ear—the so-called "resonance region." If the frequency is multiplied by a given factor the resonance region is shifted by a given amount. For instance the factor two associated with the octave corresponds to a shift of 3.5–4 mm. A complex tone—made by the superposition of different frequencies—activates hair cells in different regions but is still perceived as one entity of definite pitch by a subsequent "sharpening" process (Roederer [1979]).

Loudness, according to which sounds are described as loud or soft, is primarily a function of the physical characteristic called amplitude. The larger the amplitude of the oscillations of the sound source, the more acoustic energy is transmitted into the air and the louder is the perceived sound. Clearly loudness depends on the frequency too, since the human ear is most sensitive to a range of frequencies around 3500 Hz and is completely insensitive to frequencies less than 20 Hz or higher than 20,000 Hz, regardless of amplitude.

If two pure tones of the same frequency are played at the same time, the resulting amplitude is not necessarily the sum of the two. It is so only if the two waveforms are in phase (constructive interference), in which case a louder tone is perceived, not twice as loud though, but approximately 1.3 times as loud as a single tone. At the other extreme, when the phase difference is 180° the resulting amplitude is zero (destructive interference) and no sound is heard.

With two pure tones of closely spaced different frequencies, one tone of intermediate pitch is heard but the amplitude is more or less deeply modulated, a phenomenon referred to as *beats*. The two frequencies cannot be discriminated. If the spacing is big enough — certainly if greater than 15 Hz — then two tones are heard, the corresponding resonance regions in the ear being sufficiently separated. As far as total loudness there are complications. For instance if the frequency difference is very large, people tend to focus on one of the two tones, maybe the loudest or that of highest pitch. All this is important in music and well known to composers.

Finally we arrive at the *timbre*, the aesthetic quality par excellence belonging to a largely unknown domain. It determines whether a sound is pleasant or unpleasant. It could be said that the timbre is the recipe of the vibration: Just as any person speaks with the aid of a complex vibrating system that generates specific amounts of elementary vibrations and thereby produces a unique voice, so every musical instrument vibrates in its own unique way that makes it immediately identifiable. Broadly speaking the timbre depends on the number of harmonics, on the relationships between their amplitudes, and on the time evolution of the sound spectrum. The sounds of the human voice, the violin, and the piano are rich in harmonics, whereas the sounds produced by the recorder, the flute, and the tuning fork have few harmonics.

Special effects take place in the human ear. In the presence of a single very loud tone of frequency a, the additional frequencies $2a,\ 3a,\ 4a, \ldots$ (*aural harmonics*) are perceived. If two high intensity tones of frequencies a and b are sounded together, then the additional frequencies $a+b,\ a-b,\ a+2b,\ a-2b,\ 2a+b,\ 2a-b, \ldots$ (*combination tones*) are perceived, due to a phenomenon of nonlinear distortion in the ear. Of all these, only the first ones are loud enough to be actually heard. To delve further into these fascinating topics which focus on the understanding of the mechanical operation of the ear and on the functioning of brain in the presence of sound, the interested reader is referred to Roederer [1979].

The variety of sounds is enormous, but Fourier analysis is a powerful instrument. Every sound can be decomposed by analysis (Section 4.6, 4.7) and conversely constructed by Fourier synthesis (Section 4.9). The possibilities stagger the imagination. This is what brings Davis and Hersh to write emphatically in Davis, Hersh [1981] "To give a performance of Verdi's opera *Aida*, one could do without brass and woodwinds, strings and percussion, baritones and sopranos; all that is needed is a complete collection of tuning forks, and an accurate method for controlling their loudness." If from a theoretical point of view this statement cannot be questioned, the path toward a practical implementation is strewn with severe difficulties mostly due to the very demanding ear of listeners of classical music and opera. Nevertheless instruments exist that are able to produce music "artificially," that is without acoustic instruments and players. These devices are called synthesizers. They provide an experimental proof of the validity of Fourier analysis and synthesis in acoustics.

4.5 The tempered scale

In the following the scale of equal temperament will be used for reference. It is the scale to which all pianos are tuned. Synthesizers, however, could be tuned to other scales just as well.

A short digression, in part historical, is worthwhile. As late as 1939, in the course of an international conference, the A above the middle C was assigned the current frequency of 440 Hz. In the seventeenth century the A had been 402.9 Hz, Handel in the eighteenth century adopted an A of 422.5 Hz, and in France in the nineteenth century it took the value of 425 Hz. Naturally these variations caused problems both to instrument players and singers. The present day uniformity is mostly theoretical. Major recording companies use an A of higher frequency—445 Hz for instance—and so do major orchestras when playing in a big concert hall. Even tuners of home pianos might leave the A at a slightly higher or lower frequency if the keyboard is mostly "in line" with that value. Indeed the ratio between frequencies is what matters most.

The *diatonic scale* derives directly from the so called "Pythagoras' law," which states that the ratios between the fundamental frequencies must be equal to simple ratios of small integers for the corresponding sounds to be musical. An octave corresponds to the ratio of 2/1. A fifth (C-G), a fourth (C-A), and a third (C-E) have frequency ratios of 3/2, 4/3, 5/4 respectively (Tab. 4.2). Then the frequencies of the *diatonic scale* (Tab. 4.3) are uniquely defined.

The diatonic scale, however, suffers from a serious limitation. In music it is a frequent practice to raise or lower all notes of a piece by, for instance, one tone. It is

Table 4.2 Ratios between the frequencies of the notes in the diatonic scale.

C	D	E	F	G	A	B	C
1,000	1,125	1,250	1,333	1,500	1,667	1,875	2,000
1/1	9/8	5/4	4/3	3/2	5/3	15/8	2/1

Table 4.3 Frequencies of the diatonic and equal temperament scales.

Note	Diatonic scale (Hz)	Equal temperament scale (Hz)
C	264	261.7
D	297	293.7
E	330	329.7
F	352	349.2
G	396	392
A	440	440
B	495	493.9
C	528	523.3

easily seen that the frequency ratios of the third, fourth,... intervals are modified in the process. For instance to maintain the 3/2 ratio of the C-G fifth while raising both notes of one tone to the D-A fifth, given a D of 297 Hz, the A should have a frequency of 445.5 Hz. If the transposed piece were to be played as such, it would sound out of tune since the human ear is very sensitive to the ratios between frequencies. Correcting this problem by adding notes would entail a piano keyboard with more than 500 keys. Instead instruments like the violin and the human voice could theoretically be able to cope.

Keeping the 2/1 ratio of the octave, as imposed by the ear, the problem has been solved by making identical in all cases the ratio between consecutive keys—black or white—on the piano keyboard. This ratio, called *semitone*, is obviously equal to $\sqrt[12]{2} = 1.05946\ldots$ since there are 12 semitones in one octave. The result is the *equal temperament scale* (Tab. 4.3). The price paid for this simplification is for instance in the thirds that are slightly sharp and in the fifths which are slightly flat. This dissonance is not too offensive to the human ear, possibly because it is intrinsically "right," or more likely because one gets used to it.

4.6 Analysis of complex tones: cello, clarinet, trumpet, and bass drum

The tones of many orchestral instruments are customarily subdivided for convenience into *attack, steady state*, and *decay*. Figure 4.3 shows six cycles of the steady state of the C of a violin. Figures 4.4 and 4.5 show on top the full waveform of the E flat above middle C (about 311 Hz) of a cello, a clarinet, and a trumpet (the white areas in the middle are artifacts); at the center an enlargement of a 10 ms segment of the waveform taken from the steady state portion and showing about three periods; at the bottom the fast Fourier transform (FFT) of a sample taken again from the steady state portion of the sound and lasting 80 ms.

It is seen that the clarinet places greater energy on the odd numbered harmonics, has a more rapid attack, and a more complex steady state and that the trumpet tone has a pulse-like waveform. All three of these are *harmonic* sounds.

The last plot in Figure 4.5 is an isolated tone of a bass drum. It is classified as *inharmonic*. It has the most irregular waveform of all showing no steady state region. To have a better time resolution, the FFT of a 160 ms segment taken in the middle of the note has been computed.

Many of the characteristics of the sound of an instrument are found in the transient regions of attack and decay, mostly the attack. The explanation is simple: Initially the ear receives a complex sound made by all typical frequencies. The higher frequencies, however, have a much faster decay than the lower ones and therefore after a short time only the latter are heard. Church bells present a clear example of this phenomenon. The initial sound is loud and can be even unpleasant— resulting from the simultaneous excitation of all vibration modes of the bell—but it is quickly replaced

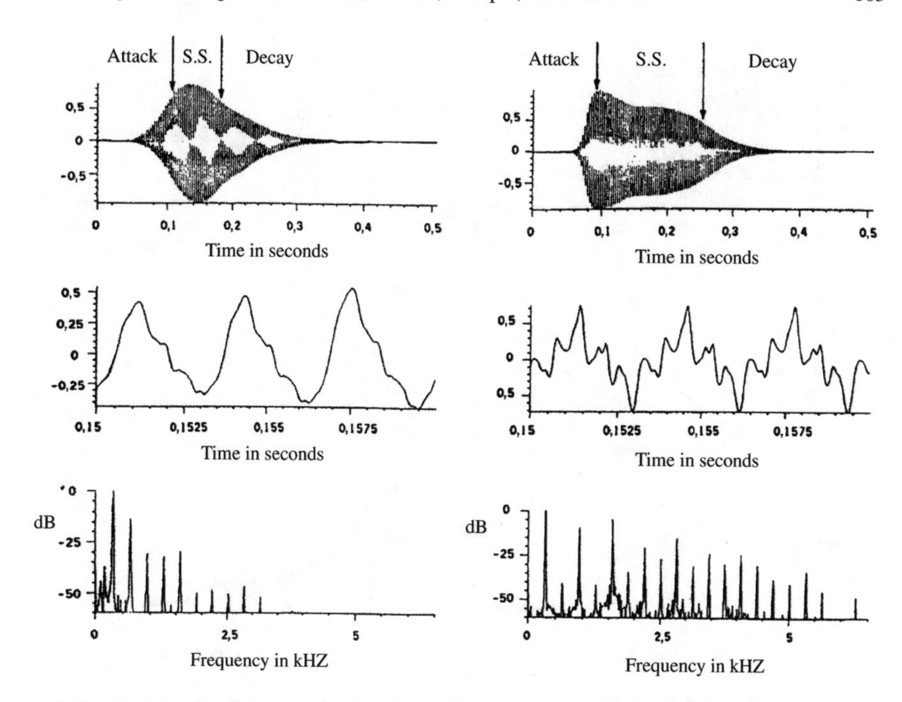

Figure 4.4 On the left the E flat of a cello, on the right of a clarinet. On top the waveform, at center an enlargement of the steady state, at the bottom the FFT (J. A. Moore [1977], "Signal Processing Aspects of Computer Music — A Survey," *Computer Music J.*, (July), pp. 4–37, ©19xx IEEE.)

by a deep, harmonious and long lasting buzz generated by the fundamental mode of vibration. Attacks are classified into families as clarinet-like, brass-like, string-like attacks and so on since they share some characteristics that lead to this type of grouping.

4.7 Three tenors in Verdi's *Rigoletto*: dynamic spectra

Dynamic spectra allow quantitative analysis of the style of a singer or a music director. To characterize each performance, various aspects need to be analyzed such as timing microstructures, dynamics, intonation, timbre, registers, and vibrato. Figures 4.6, 4.7, and 4.8 show a few seconds of the interpretations by three tenors of *Ella mi fu rapita* (She was stolen from me), an aria from Verdi's *Rigoletto*.

To decode the pictures one has to keep in mind that the FFTs of the sound waveforms are calculated for every 16 ms segment and so many graphs are obtained. As the time goes by, imagine the graphs of the FFTs set vertically, one below the other. The resulting picture, rotated by 90° counterclockwise, is shown in the figures

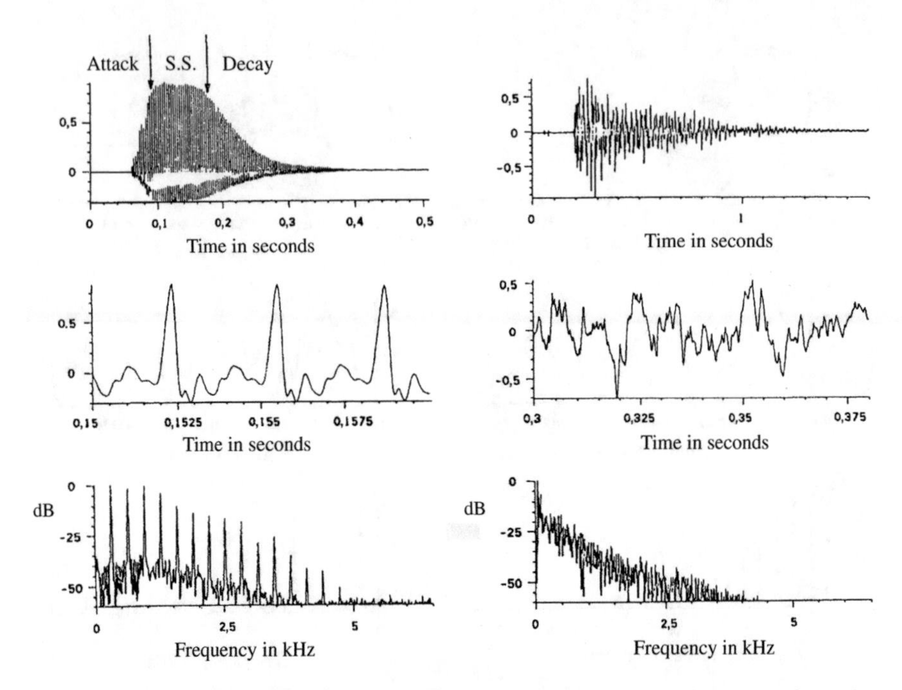

Figure 4.5 On the left the E flat of a trumpet. The waveform is on top, an enlargement of a portion of the steady state is at the center, the Fourier transform is at the bottom. On the right the tone of a bass drum (J. A. Moore [1977], "Signal Processing Aspects of Computer Music — A Survey," *Computer Music J*, (July), pp. 4–37, ©19xx IEEE.)

(*dynamic spectra*). Hence time goes left to right. The lower frequencies are at the bottom and increase toward the top, just as with musical staff notation.

In Figure 4.6 is Luciano Pavarotti, in Figure 4.7 the Spanish tenor Alfredo Krauss, and in Figure 4.8 Ferruccio Tagliavini. In all cases the tenor voices are seen to extend over the high frequency range from about 2000 Hz to 4000 Hz, while the orchestra takes up the bottom frequencies. The choice of separate ranges is, incidentally, the reason the singing is not obscured by the sound of a full orchestra.

It may be observed that the *vibrato*, relative to the word "rapita," is more agitated and shorter with Krauss than with Pavarotti, while the sound of the orchestra lasts longer. As with Tagliavini, a tenor of the 1950s, the whole sentence lasts longer—eight seconds instead of seven—as does the sound of the orchestra.

Tagliavini's interpretation expresses the ire of a master, Krauss shows a moderate irritation, and Pavarotti reaches little by little the strongest irritation of the three. The acoustical definition of these effects is found in the dynamic spectra (Födermayr, Deustsch [1993]; Bengtsson, Gabrielsson [1983]).

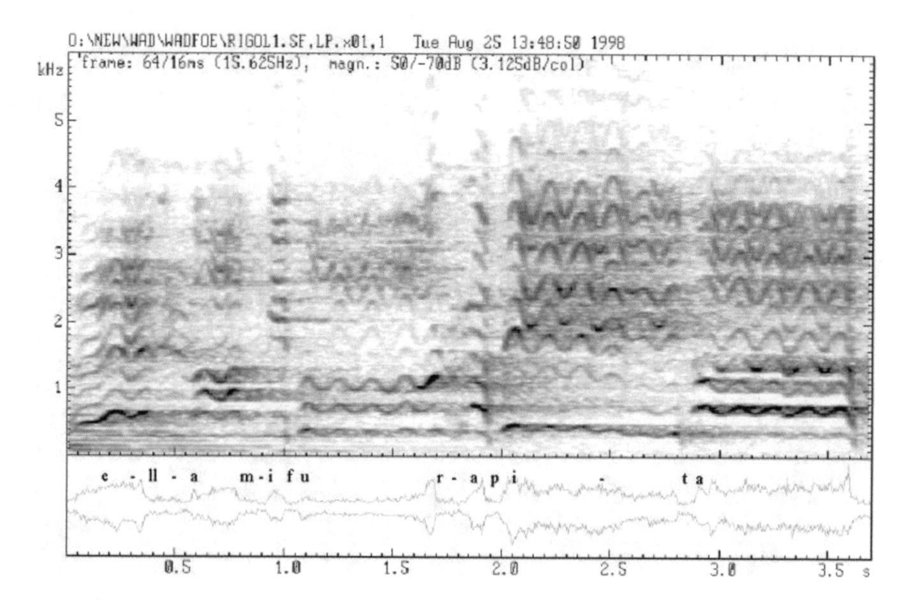

Figure 4.6 Spectrogram of a few seconds in the performance of Luciano Pavarotti [1989], Decca 425 864-2. (*Courtesy of F. Födermayr, Institut für Musikwissenschaft, Universität Wien*)

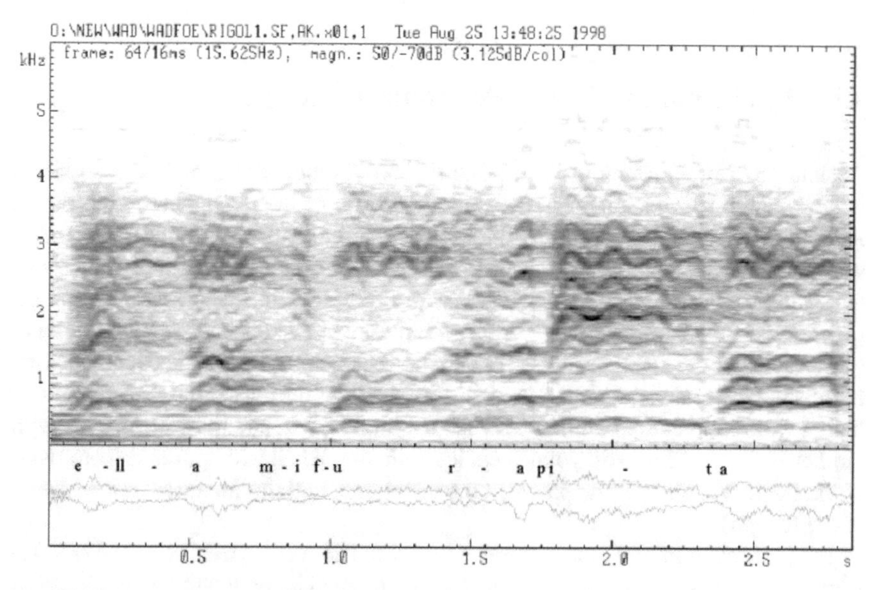

Figure 4.7 Spectrogram of a few seconds in the performance of Alfredo Krauss [1963], RCA GD 86506. (*Courtesy of F.Födermayr, Institut für Musikwissenschaft, Universität Wien*)

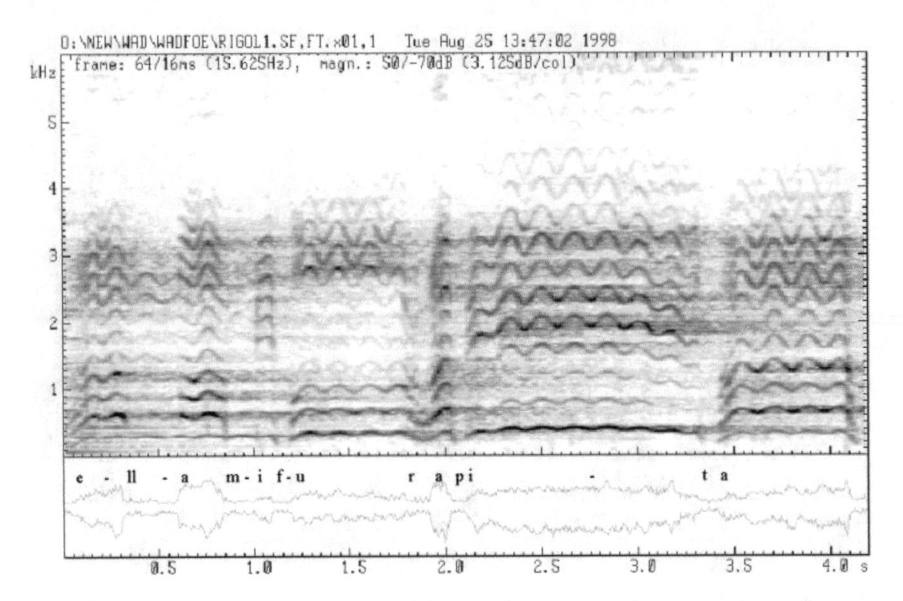

Figure 4.8 Spectrogram of a few seconds in the performance of Ferruccio Tagliavini [1954], Dino Classics LC 8410.1991. (*Courtesy of F. Födermayr, Institut für Musikwissenschaft, Universität Wien*)

4.8 The beginnings of the electronic music synthesizers

Around 1875 two inventions took place that would radically expand the possibilities of manipulating sound. First was the invention of the *telephone* by the Italian Antonio Meucci and the Scottish-born Alexander Graham Bell (Section 3.1). Bell was interested in sound because his profession—and that of his family since two generations—was to teach speech to the deaf. Emigrating to Canada with his parents (Section 3.1), he later moved to Boston where in 1872 he opened his own school to train teachers of the deaf. Soon afterwards he became professor of vocal physiology at Boston University. Attending to his institutional duties during the day, at night he would work long hours building a device for transmitting sound by electricity, with the help of an enthusiastic young repair mechanic. His health suffered from overwork, but in March 1876 he patented the telephone, initially intended as a variation of the telegraph.

In the summer of 1877 the American inventor Thomas Alva Edison (1847–1931) started working on a machine that was meant to encode the telephone messages at reception for later delivery, as done with telegraph messages. Edison built the first automatic sound recording using a carbon transmitter (Section 3.1) that ended with a stylus and left marks on a rolling strip of paper coated with paraffin. To his great surprise the faint indentations, barely visible to the naked eye, reproduced a sound that vaguely resembled the original when the strip of paper was pulled back beneath the stylus. Ten years later this original invention, which quickly made Edison popular worldwide, had its first commercial realization in the *phonograph*.

After recording and reproduction came the generation of sound by electrical means. This requires amplifiers and loudspeakers because at the origin there is no acoustic sound but rather oscillations of electric currents. Possibly the most advanced of the first electric instruments was the *dynamophone*, built by Thaddeus Cahill at the turn of the twentieth century. A huge machine resembling a power plant used dynamos to produce electric vibrations, as well as telephone receivers to convert them into sounds (Risset [1991]). Complicated and impractical, it is the ancestor of the modern synthesizers.

Electronics made the difference. Around 1920 the first electronic musical instruments were realized: The electric vibrations could be shaped by modifying the elements of an oscillating circuit. The first musical work to exist solely as a recording might be the 1939 composition *Imaginary Landscape Number 1* by the American John Cage.

After World War II, musical exploration through electroacoustic techniques became more systematic. In Paris in 1948 the group *Musique concrète*, initiated by Pierre Shaeffer, experimented with acoustically generated sounds recorded on disk or tape, to be used in various ways by composers. In 1951 in Cologne the group *Electronic music*, pioneered by Karlheinz Stockhausen among others, intended to use exclusively electronically generated sounds to realize a given score.

The last revolutionary development took place with digital sound synthesis and processing made possible by the advent of computers and by the invention, in the 1950s, of devices to convert analog signals into digital ones (Bode [1984]). Max V. Mathews and John R. Pierce at the Bell Telephone Laboratories were among the first to use computers systematically for signal processing and in particular to perform digital synthesis of musical sounds (David et al. [1958], Mathews [1969]): "We were originally drawn to the computer as a sound-analyzing and sound-producing device while investigating the factors that contribute to the efficient transmission of speech through telephone lines. It soon became clear to us that the quality of sound is of great importance not only in speech but also in music, and we enthusiastically began to study the production of musical sounds."

The results were initially disappointing. Sounds were dull, without life and identity; the imitation of existing instruments (strings, brass,. . .) was a failure, their timbrical qualities not even remotely recognizable. In short, sounds of pleasing timbre proved to be difficult to generate, also because the quantitative knowledge of musical sounds was poor. Nevertheless the pioneers of computer music did not loose faith, recognizing the infinite wealth of possibilities embodied in the new medium. And eventually the first synthesizers were built: complex instruments, designed specifically for the generation of musical sounds and depending on computers, electronic oscillators (analog or digital) and amplifiers.

In modern music synthesizers are a new element—application of modern technology to one of the oldest human activities—that overcome the limitations of traditional musical instruments and even dependency on performers. The endless mine of all possible sound combinations lies there and the golden nuggets just have to be found. Rather than the reproduction of sounds of existing instruments, it is the creation of new sounds and the modification and extension of natural sounds in ways not easily achievable otherwise that makes synthesizers attractive for composers. Besides there is awareness that the material provided by electronic media can lead to new musical structures.

In Italy the section of musicology of the CNR (Consiglio Nazionale delle Ricerche) in Pisa has to be counted among the pioneers. It was started in 1969 by maestro Pietro Grossi, professor of cello and electronic music at the "Conservatorio" in Florence. Grossi composed a musical piece, described by a program based on random processes, that can be made to last indefinitely without ever repeating itself. In 1972 the Center for Computational Sonology at the University of Padua followed. It was part of the School of Engineering and its leading figures G. B. Debiasi and Giovanni De Poli took the opposite track and strove towards maximum quality, precision, and repeatability of sound to allow the composer to achieve exactly the desired effect. Finally, around 1974 the physicist Giuseppe Di Giugno, mainly interested in real time (no appreciable delay must lapse between the commands and the sound), started to build a system for the production of music at the Institute of Experimental Physics of the University of Naples. In 1976 he moved to the IRCAM (Institut de Recherche et Coordination Acoustique/Musique) in Paris where he built the 4X, a "music station" made by several hundreds of oscillators, filters, modulators, and other units, all dynamically controllable by a computer program. In addition, at the time of execution a certain number of parameters could be modified manually by the performer.

Currently the fast paced evolution of the microprocessor industry is already making it unnecessary to build the dedicated hardware embodied in the old synthesizers, now replaced by suitable software for the analysis or the synthesis of sounds or by inexpensive synthesizers.

In traditional music the instrument player reads the music from the score and acts on the instrument to produce the sound. In this process two conceptual steps can be identified, the first being some written, symbolic representation of the music—the notes on the score—and the second one being the instrument that generates sounds in a predictable way. In computer music the instrument is the technique used for the synthesis, while the score is the collection of parameters used to identify the sound.

The synthesis technique is based on a mathematical model of the sound waveform. It can be predetermined (additive, subtractive, ...) and in this case the composer is only allowed to choose the parameters, or it can be defined by the composer itself to adapt it better to the requirements of the composition. Musical scores are computer programs, written using Music V—the software developed by Mathews in the early 1960s—or in the C language or other machine languages, with some of the parameters possibly modifiable in real time to allow an interpretation by the performing musician. It may be added that these languages, while suitable for sound synthesis, are not really fitted for composition of high-level musical structures. In other words there remains a question of notation and interface from the point of view of the composer. No such universally accepted thing exists and personal solutions flourish (Sica [1997]).

Synthesizers or softwares dedicated to synthesis are not purposefully employed to reproduce the sound of traditional musical instruments, yet their reproduction of simple natural sounds—like a single note—can be tested in order to assess their "quality." This test, deceptively simple at first sight, requires a fine analysis of the characteristics of musical notes (Section 4.6) which must be followed by an adequate synthesis, if the richness and color of the natural sound have to be retained.

4.9 Additive synthesis

In 1965 the composer and physicist Jean-Claude Risset, working at Bell Laboratories, was able to produce, by additive synthesis, trumpet tones that an average listener could not distinguish from recorded tones of a true trumpet. Risset's method led to a deeper understanding of the timbre of sounds produced by traditional instruments. In particular it made clear that the steady state is insufficient to fully characterize the tone's timbre, attack and decay being critically important as well.

If the prime concern is efficiency, then many musical notes can be synthesized by assembling an attack, a steady state of suitable duration and a decay. The problem becomes much more difficult if high fidelity is sought. To this purpose, musical sounds are classified as *harmonic* or *quasi-periodic* (Fig. 4.3 and 4.4) and *inharmonic* or *aperiodic* (as the bass drum in Figure 4.5).

Harmonic sounds, if represented by Fourier series, would require large periods and so very many significant Fourier coefficients. They are more conveniently represented as small variations of periodic sounds (Fig. 4.9) such as

$$X(t) = \sum_{k=1}^{M} A_k(t)\ \sin\{2\pi t[k\nu + F_k(t)]\}. \tag{4.2}$$

The amplitudes $A_k(t)$ of the harmonics are not constant but change slowly with time while the frequencies display small *frequency deviations* $F_k(t)$ with respect to the integer multiples of the fundamental ν, slowly changing with time. In particular different overtones are allowed to follow different time courses and this turns out to be important for the listener to recognize a timbre.

Aperiodic sounds are produced by a more limited range of instruments such as bells and drums and have a more complicated structure. They could be represented by (4.2), but the number of required spectral components would be very high. For this reason Fourier analysis is not used as a basis for the synthesis and other techniques are used that yield fairly convincing percussion instrument sounds.

After the mathematical model (4.2) has been selected, it is necessary to assign proper values to all parameters involved. The waveform is therefore analyzed by a bank of devices that detect the frequency and amplitude of each harmonic as functions of time to yield the following discrete version of (4.2):

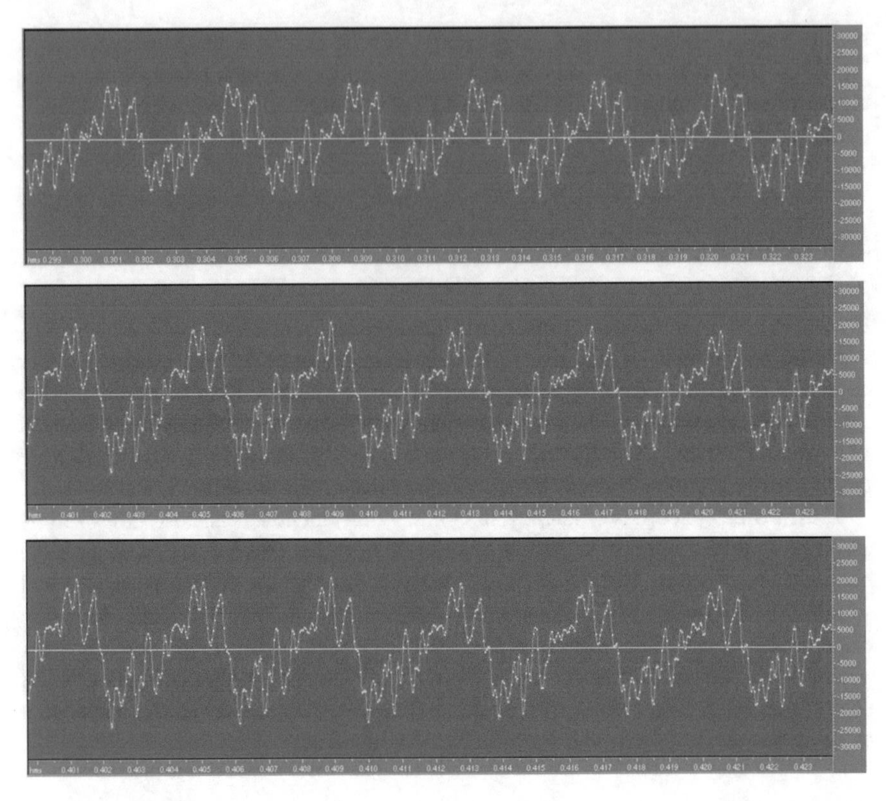

Figure 4.9 A C4 violin tone. Each line corresponds to 25 ms. A careful examination shows that the waveform gradually changes. (*Courtesy of G. Di Giugno*)

$$X(n) = \sum_{k=1}^{M} A_k(n) \sin\{2\pi n T[k\nu + F_k(n)]\}, \quad n = 0, \ldots, N-1 \qquad (4.3)$$

where T is the interval between subsequent samples and N is the number of samples. In practice the isolated tone is first digitized; then the pitch υ, which is assumed to be about constant over time, is determined very accurately; subsequently the amplitudes $A_k(n)$ and the frequency deviations $F_k(n)$ of the harmonics are computed. The number M of the harmonics is implicitly determined by the $A_k(n)$: When the amplitudes become so small as to be negligible, the summation is terminated.

At this point formula (4.3) can be used to reproduce the sound by driving a number of oscillators. It is immediately apparent the tremendous amount of data that goes with the sampling rate of 25600 Hz for each of the about fifteen harmonics. Figure 4.10 a), b) and c) show the amplitudes of the first harmonics — of the sounds of cello, clarinet, and trumpet shown in Figures 4.4 and 4.5 and lasting about 0.4 s— "fitted" by piecewise-linear functions. This approximation compress the tens of thousands of samples initially collected for each waveform to less than 200: Namely fourteen amplitude functions (and as many frequency deviations, not shown),

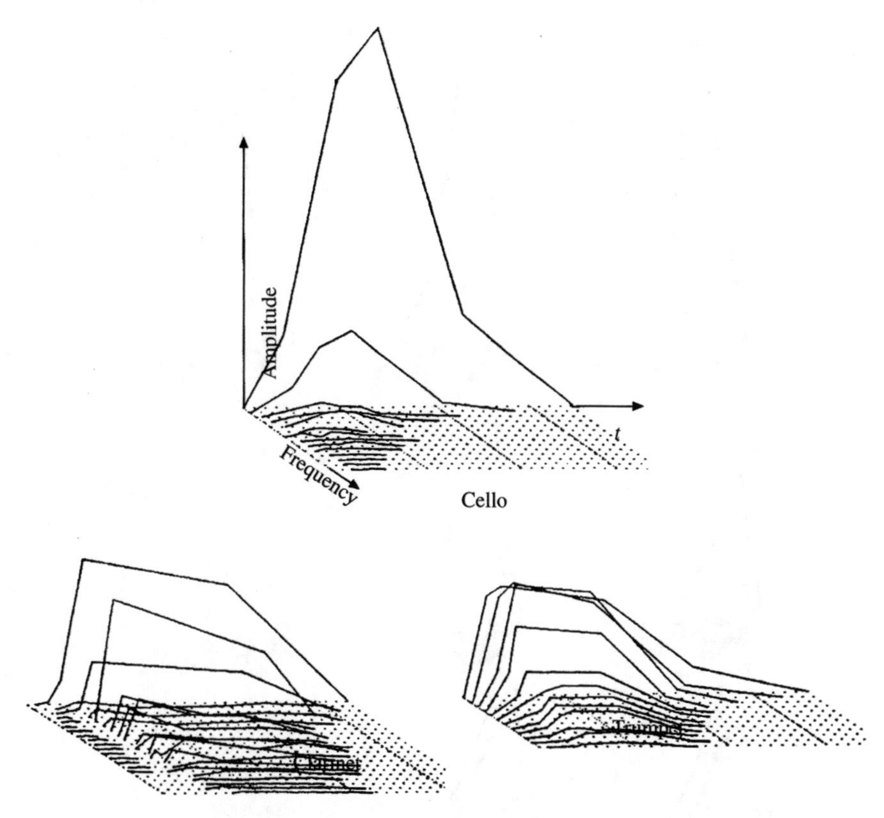

Figure 4.10 The compressed amplitudes of the sounds of cello, clarinet, and trumpet in Figures 4.4 and 4.5. The reference frame shows time (left to right), amplitude (bottom to top), and frequency (back to front). The amplitude of the fundamental frequency is in the background, while in the foreground is the amplitude of the highest harmonic. (J. A. Moore [1977], "Signal Processing Aspects of Computer Music — A Survey," *Computer Music J.*, (July), pp. 4–37, ©19xx IEEE.)

each one represented by six points suitably chosen. This amount of information is sufficient to synthesize the timbre of the three instruments, meaning that the sounds so obtained are very similar to the originals and in no way can be described as synthetic or electronic (Moore [1977]). The tones have been equalized in loudness, pitch and duration by a perceptual matching experiment so as to leave timbre as the only perceptual difference. The strong disparity in the amplitudes goes a long way in showing how much our perceptions can differ from the physical data. Additive synthesis, which is a powerful technique for computer music, and psychoacoustics as well, leads naturally to modified tones. For instance Figure 4.11 shows the amplitude functions of a cello slowly changing into those of a French horn. When the start and end instruments are similar, this "interpolation" procedure gives rise to a sequence of sounds that gradually change from one timbre to the other; when the instruments are highly dissimilar the intermediate sounds are often new and very unusual. The

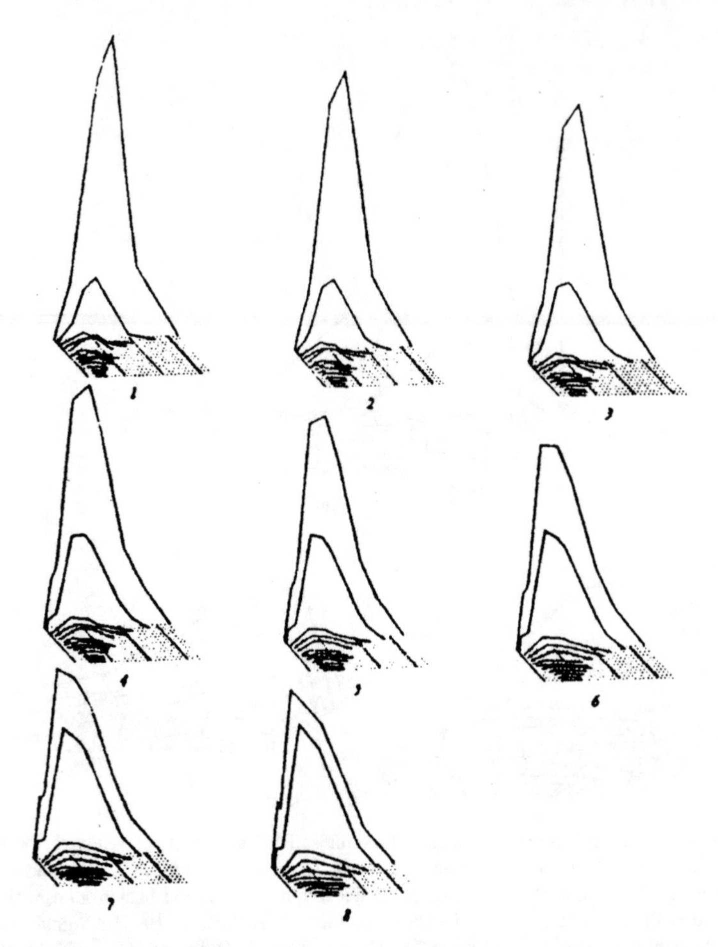

Figure 4.11 The amplitude functions of a cello gradually changing into those of a french horn. (J. A. Moore [1977], "Signal Processing Aspects of Computer Music — A Survey," *Computer Music J*, (July), pp. 4–37, ©19xx IEEE.)

method originally devised by Risset is more general, more powerful, and easier to understand for the synthesis of timbres. Yet it is also heavy on computations; thus musicians quickly started looking for short cuts that performed similarly but with less toil.

4.10 Subtractive synthesis and frequency modulation

Subtractive synthesis is a technique in which a periodic signal is generated so as to have many harmonics and then is filtered to eliminate unwanted components and to emphasize others. Many musical instruments operate on this principle. In wind instruments the player's lips or a vibrating reed generate a sound rich in harmonics. Some of them are attenuated and others amplified by the resonating cavities or more generally by the shape of the particular instrument. This happens for the human voice too, made the richest and most interesting of musical instruments by the great adaptability of the nose, throat, and mouth cavities.

While subtractive synthesis is frequently used for the human voice, frequency modulation (FM) is among the most commonly used technique for music synthesis because of its high computational efficiency and versatility. Starting in the 1930s frequency modulation was used to transmit radio signals and as such it has been taught to engineering students, but few people thought of listening to the modulated waveform itself. At Stanford University John M. Chowning, among the first musicians to make use of computers, had the idea of exploiting frequency modulation to generate sounds (Chowning [1963]). For this innovation, a landmark in the evolution of electronic musical instruments, Chowning obtained a patent that was soon bought and used commercially by Yamaha. Later Chowning himself used frequency modulation in conjunction with additive synthesis to reproduce the singing human voice, with notable results (Chowning, Bristow [1980]).

Frequency modulation (Fig. 4.12) consists in changing a simple sound $\sin \omega_c t$, called the *carrier*—of angular frequency ω_c—by a *modulating signal* as in

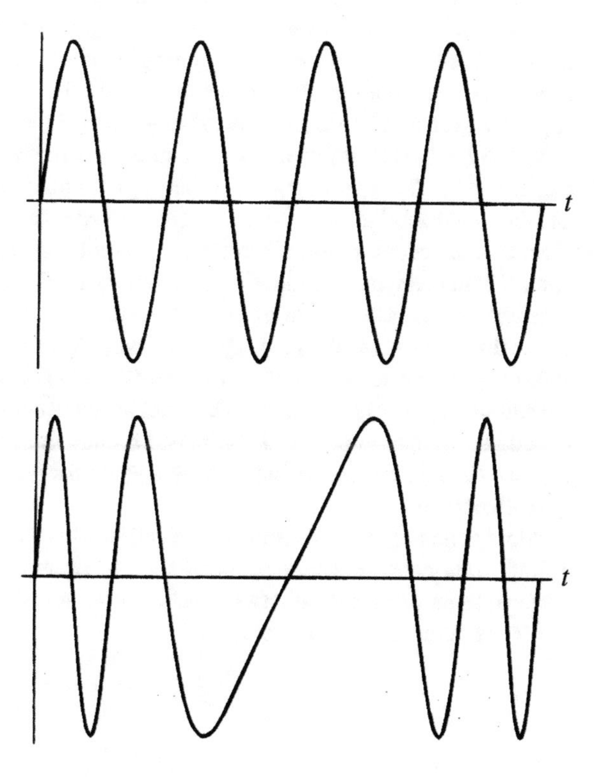

Figure 4.12 Carrier (above) and frequency modulated waveform (below).

$$s(t) = \sin(\omega_c t + I \sin \omega_m t).$$

The resulting sound has a frequency sum of the carrier frequency and the continuously varying contribution of the modulating signal. I is called the *modulation index* and ω_m the *modulation frequency*. By making use of some trigonometric identities, it can be shown that

$$s(t) = \sum_{k=-\infty}^{+\infty} J_k(I) \sin[(\omega_c + k\omega_m)t]$$

where $J_k(t)$ is the *Bessel function* of order k (Section 5.3). So the *modulated sound* is the sum of harmonics having frequencies $\omega_c + k\omega_m$, $k = 0, \pm1, \pm2\ldots$ centered at ω_c and spaced ω_m.

Chowning chose the carrier and modulator to have frequencies that were identical or of the same order of magnitude. This was avoided in FM radio transmission for it spreads the signal over a very large frequency bandwidth with no purpose.

It takes quite a bit of acquaintance and intuition to have a feeling of what change in the sound is induced by what change in the parameters. Examples help to clarify somewhat this technique. In the trivial case $I = 0$ since all $J_k(0)$ are zero, except for $k = 0$ when $J_0(0) = 1$, the modulated signal reduces to the carrier wave itself. Now let I be different from zero. For instance if $\omega_m = \omega_c$, then a harmonic spectrum is obtained. Indeed from $\sin(-\theta) = -\sin\theta$ it follows that negative frequencies can be turned into positive ones just by changing the sign of the amplitude. This done it is easily seen that the frequencies $(1 + k)\omega_c$, for k greater than or equal to zero, appear with an amplitude given by $J_k(I) - J_{-k-2}(I) = J_k(I) + (-)^{k+1} J_{k+2}(I)$ since $J_{-k}(t) = (-)^k J_k(t)$. It remains to take into account the value $k = -1$ that gives the constant term. As I gets larger more energy goes into the higher harmonics (Fig. 4.13) and the sound gets "tinnier." Moreover, for a fixed I, the values $J_k(I)$ decrease rapidly as soon as k becomes larger than the modulation index. The richness of the method comes from changing the modulation index dynamically. This allows to imitate the time variant aspect of natural spectra. For instance if the modulation index has a flatter attack than the carrier, then the high frequency overtones raise slowly to their steady state amplitude values producing a brass-instrument timbre. For a comparable sound the additive synthesis would take ten times as many computations. If $\omega_m = 2\omega_c$ only the odd harmonics appear and clarinet-like sounds are obtained. Instead the choice $\omega_m = \sqrt{2}\omega_c$ gives rise to inharmonic sounds useful to imitate bells, drums, and gong-like tones.

More generally when the ratio of the carrier and modulation frequencies ω_c/ω_m is a simple rational number, the resulting sound is harmonious, the simpler the ratio the more pleasant to the ear. Otherwise the resulting sound is inharmonious and suitable to imitate percussion instruments.

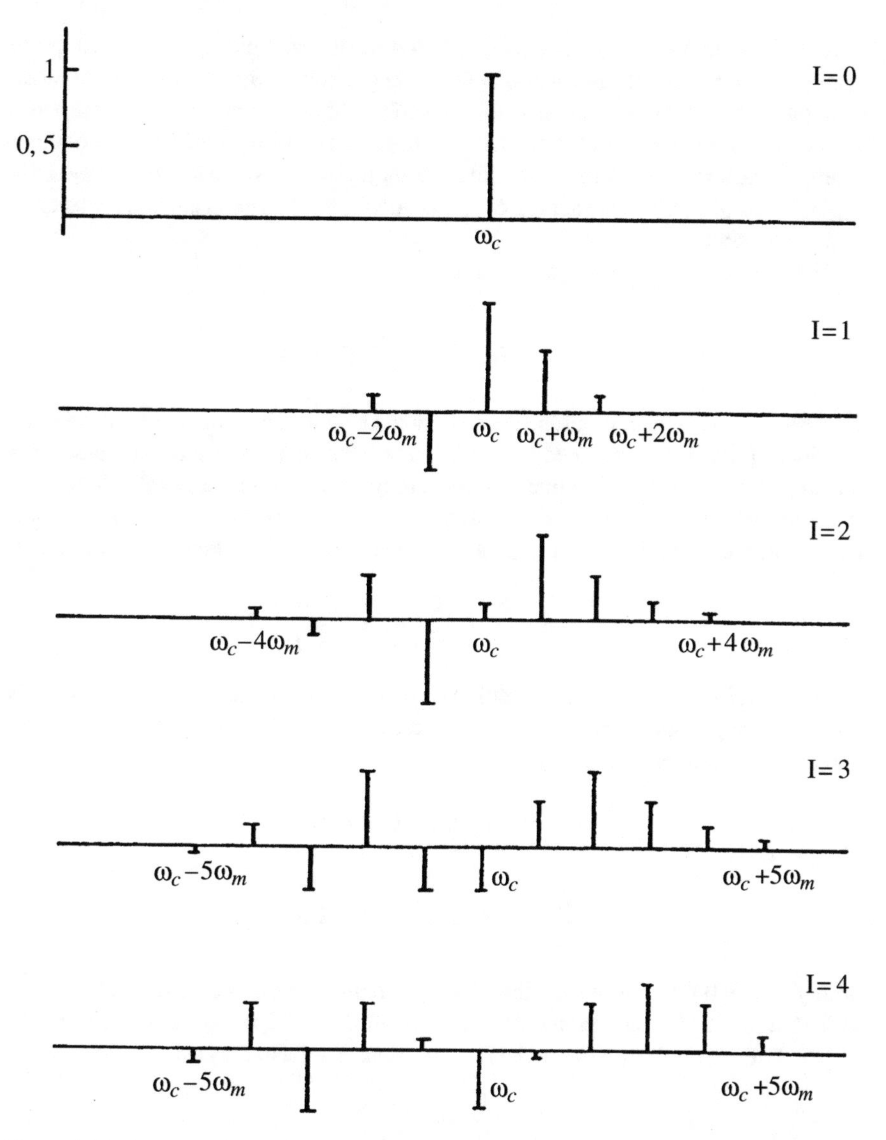

Figure 4.13 Spectrum of a frequency modulated sound, with $\omega_m = \omega_c$, as the modulation index I increases. (G. De Poli et al. "Musica e calcolatore," *Quaderni dell'Istituto di Matematica*, Università di Lecce, no. 12, [1981])

4.11 Bessel functions and drums: the vibrating membrane

A surface of negligible stiffness, with respect to the restoring force due to tension, is called a *membrane*. Taking stiffness into account clearly makes the theory of vibrations more involved. If stiffness is an important factor the surface is called a *plate*.

The vibrating membrane is a natural two dimensional analogue of the vibrating string (Section 2.1). Bessel functions play a major role in this theory, providing an example of series expansions based on a different set of functions from the sines and cosines of Fourier series. Wavelets (Section 4.12 and Appendix) are still another example and the point is made that the technique of series expansions, essentially begun by Fourier, is a far reaching one, provided the expansions are chosen so as to "suit" the phenomenon.

The wave equation for the membrane is

$$c^{-2}\frac{\partial^2 u}{\partial t^2} = \frac{\partial^2 u}{\partial x^2} + \frac{\partial^2 u}{\partial y^2},$$

where $u(x, y, t)$ is the membrane displacement and c has the dimensions of velocity. In Morse [1948], an extended analytical treatment can be found. The case of a rectangular membrane is not that different from the string: The general solution can be expressed as a double series involving sines and cosines. Instead in case of a circular membrane it is natural to write the wave equation in polar coordinates (r, θ)

$$\frac{1}{r}\frac{\partial}{\partial r}(r\frac{\partial u}{\partial r}) + \frac{1}{r^2}\frac{\partial^2 u}{\partial \theta^2} = \frac{1}{c^2}\frac{\partial^2 u}{\partial t^2}$$

and then apply the method of separation of variables. If the membrane (Fig. 4.14) is clamped along a boundary circle of unit radius r_0 — so that $u(r, t) = 0$ for $r = r_0$ and $t \geq 0$ — and the initial conditions are

$$u(r, 0) = f(r), \quad 0 \leq r \leq r_0,$$

$$\frac{\partial u}{\partial t}(r, 0) = 0, \quad 0 \leq r \leq r_0,$$

with $f(r_0) = 0$ for consistency, then it is only natural to expect u to be independent of θ, due to the circular symmetry of the equation and of the above constraints. In Boyce, DiPrima [1969] it is proved that solutions can be expressed as

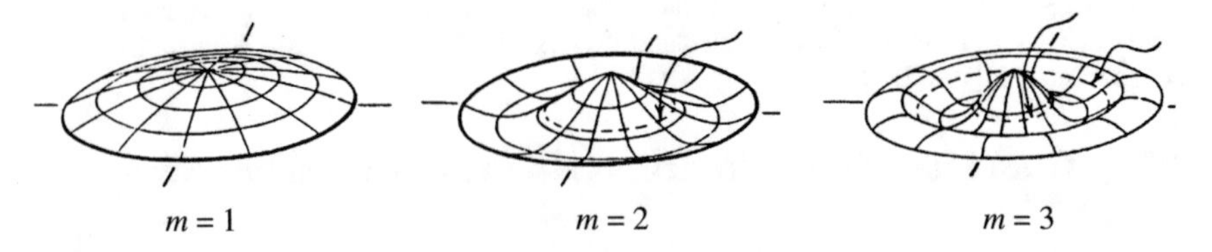

Figure 4.14 Radial vibrations of a circular membrane: shapes of the first three normal modes of vibration. Arrows point at the nodal lines. (Reprinted with permission from P. M. Morse, *Vibration and Sound*, The Acoustical Society of America, [1948])

$$u(r,\ t) = \sum_{n=1}^{\infty} c_n J_0(\lambda_n r) \cos \lambda_n ct$$

where λ_n ranges over the positive zeros of $J_0(\lambda r_0)$. These are tabulated (Jahnke, Emde [1945]) and make up an infinite discrete set. The frequency λ_1 is the fundamental. It may be observed that none of the overtones is harmonic. For instance the frequency ratio of the second and third mode with the fundamental is 1.593 and 2.135.

The above model has to be suitably changed in case the circular membrane is stretched over one end of an airtight vessel, as for the kettledrum. This musical instrument consists of a metal shell, usually copper, in the form of a hemisphere over which a membrane of calfskin about 0.2 mm thick is mounted. The pitch is tuned by adjusting the overall tension of the membrane. The reaction to the compressions and expansions, which the motion of the membrane forces on the air in the vessel, has to be taken into account. As a result the natural frequencies and general behavior of motion change (Morse [1948], Benade [1990]).

4.12 Convolution and some special effects

Convolution, already presented in the mathematical description of filters, has a profound acoustical significance: Sound waves convolve all around us, meaning that the sounds we hear are the result of convolutions.

Convolution "marries" two signals. In the case of sounds, convolution combines in a certain way the properties of two of them into a single one. Among the most important effects, easily obtained by convolution, are echoes, time smearing, and reverberation.

If a signal is convolved with the unit impulse, the result is the signal itself (Section 3.5). If a signal is convolved with two unit impulses, set one second apart, the result is the original signal and an *echo*. For a multiple echo effect, several impulses spaced at the desired delay times are needed. Lowering the amplitude of each successive impulse results in a decaying multiple echo.

If several impulses are spaced close together and if the original signal is rather long, then the multiple echoes, obtained by convolution, overlap with the result of blurring every temporal landmark. This effect is called *time smearing*.

Equally interesting are *room simulation* and *reverberation*. When a signal is convolved to the response signal of a room to the unit impulse (in the terminology of filtering this is called the "input response"), the result is as if the original sound has been played in that room. In the case of large spaces with high ceilings and many reflecting surfaces, like churches and concert halls, the impulse response is rather long resulting in all the myriad of closely spaced signals bouncing off the ceilings, walls, and floors that fuse together in a halo accompanying the original unit impulse. This effect is called *reverberation* and can be obtained by convolving with the long impulse response of the reverberant space.

More generally any two signals can be convolved, like a clarinet and the speaking voice, but the result may not be musically interesting. More experience has to be gained to map the full scope of this versatile technique (Roads [1997]).

Musical signal processing provides other techniques for sound transformation and synthesis. Indeed there is a large variety of them. For instance *granular synthesis* is another recent one. Dating from the 1970s and originating in the pioneering work of Dennis Gabor [1946], [1947] connected with the wavelet transform (Appendix), it has been employed experimentally by musicians both for analysis and for synthesis of novel sounds, especially since 1990 (Cavaliere–Picciali [1997]).

4.13 Computer music and art music

Very broadly speaking classical music may be described as experimenting with pitch and contemporary *art music* as experimenting with timbre. Presently "a new generation of composers, versed in signal processing techniques, are creating a highly refined form of musical art" (Roads [1997]). Their work might be little known to the public. This is not especially surprising. A historical comparison with the piano can be drawn. Invented in 1709 by the harpsichord builder Bartolomeo Cristofori the piano was subject to significant improvements until in 1855 when Henry Steinway designed the grand piano with a cast iron frame. Since then no fundamental change, neither in the design nor in the construction technique, has taken place, even though small improvements have been continually introduced. Considering the Chopin sonatas as the most famous pieces written for piano one cannot fail to notice that they were composed between 1827 and 1847, a hundred years after the appearance of this instrument.

The limited impact of art music so far has hardly hampered the popularity of *computer music*, which is now predominant in the production of sound tracks for television and film works. Indeed it has many advantages to offer. To mention one it is readily synchronized with the action depicted in a motion picture. Popular music orchestras are by now a thing of the past, mainly for purely economical reasons.

Chapter 5
Fourier Optics and the Synchrotron Light

5.1 The long search on the nature of light

The nature of light, which has given rise to speculations since antiquity, proved to be a difficult subject. The first on record is the Greek philosopher, Sicilian born, Empedocles (c.490–c.435 B.C.) whose writings have survived only in fragments so that his thinking is mainly known from citations in the work of Plato and Aristotle. Apparently the originator of the long-standing notion that all things are made of four elements — air, fire, water, and earth — Empedocles thought of light as a stream of small particles, emitted from a visible body, that were to enter the eyes and return to the originating body, traveling with finite speed. At other times he thought of light as originating in the eyes and returning to the eyes after reaching the object.

The Greek astronomer Ptolemy (c. 150 A.D.) measured angles of refraction, for angles of incidence from about 0° to 90°, for both the air-glass and the air-water interfaces and compiled tables accordingly.

With Galileo (1564–1642) physical matters were put on solid ground by the experimental method and mathematical language he introduced in mechanics. In 1621, the law of *refraction* was discovered from experimental observations by the Dutch Willebrord Snell (1580–1626). The phenomenon of *diffraction* — the minute deviation from rectilinear propagation, caused by small apertures and revealed by the presence of light in geometrical shadows — was first observed by the Italian priest Francesco Maria Grimaldi (1618–1663), son of a silk merchant of a wealthy family. His observations are to be found in *De Lumine*, a comprehensive treatise on light — to which he worked from 1655 until the very end — published posthumously in Bologna in 1665.

The English physicist Robert Hooke (1635–1703), known for the law of elasticity that carries his name, marked a significant point by advancing a wave model in order to explain diffraction, observed by him in 1672. In his model light consists of short vibrations "exceedingly quick," propagating every way by straight lines. Moreover — in the same manner as the waves on the surface of the water generated by a sinking

E. Prestini, *The Evolution of Applied Harmonic Analysis*,
Applied and Numerical Harmonic Analysis,
DOI 10.1007/978-1-4899-7989-6_5

stone — every vibration generates a sphere that continually increases and cuts the rays at right angles. To explain "mechanically" what was vibrating, the existence of some kind of elastic homogeneous medium pervading all space was postulated: the ether.

Whether light propagated instantaneously or not was unknown until 1675. The young Danish astronomer Olaf Römer (1644–1710), who was to become one of the greatest practical astronomers of all times, while working in Paris for the Royal Academy of Sciences discovered that light has a finite speed. The eclipses of Io, Jupiter's first moon, observed earlier by Galileo, to Römer's surprise showed irregularity in their occurrences that he was able to explain by the changed distance between the Earth and Jupiter, if the velocity of light was finite. He then proceeded to calculate such a velocity and obtained an estimate 25% too low, mainly due to an incorrect value for the change in distance between the Earth and Jupiter. His achievement was nevertheless impressive, for the great velocity had defeated many, including Galileo who had made experiments over too short a distance.

Hooke's work was continued by the Dutchman Christian Huygens (1629–1695) who advanced a principle, named after him, according to which every point of a wave front of light may be regarded as a new source of spherical waves (Fig. 5.1), the envelope of which determines the wavefront at any subsequent time (Huygens [1690]). In this framework, he was able to derive the law of reflection and, assuming

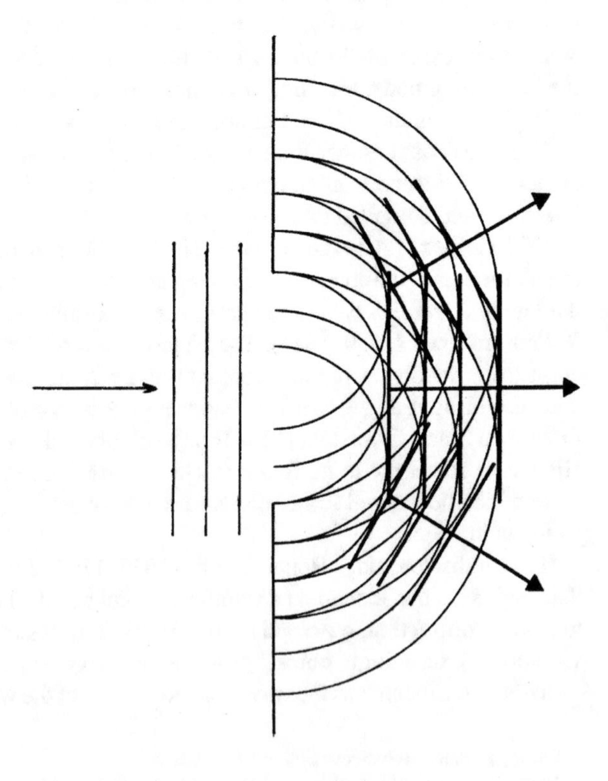

Figure 5.1 Huygens's construction in case of a single aperture. Every point of the aperture may be regarded as a source of spherical waves.

the velocity of light to be slower in the denser medium, the law of refraction. Huygens also discovered the phenomenon of polarization, but confessed his inability to explain it. A first basic fact about the nature of light was established by Isaac Newton (1642–1727) in 1664. He let light enter a dark room through a round hole in a shutter and saw that, when the light emerged from a prism and was collected on a screen, it had been split into the colors of the rainbow. He concluded that each color had its own specific refraction index and gave the name of *spectrum* to this colored image of the sunlight. As regards the mathematical model, Newton opposed the wave theory that in his time meant longitudinal waves — vibrating in the direction of propagation like those of a coiled spring — as inconsistent with experiments of polarization. Though declining to be involved in "troublesome and insignificant disputes," he favored, as more acceptable, a "corpuscular theory" and on the basis of his authority a century went by with only isolated dissenters, like Benjamin Franklin (1706–1790) and Leonhard Euler (1707–1783).

It was the practicing physician Thomas Young (1773–1829) on the staff of St. George's Hospital in London, a man of diverse talents — being also a physicist, a linguist, and an archeologist who participated in the deciphering of the Rosetta stone — who revived the wave theory. After discovering the cause of astigmatism, he turned his attention to light and explained the different spectral colors in terms of different wavelengths that he was able to determine approximately for the first time [1801]. Right afterwards Young [1802] enunciated his *principle of interference* and demonstrated it before the Royal Society of London in 1803. He made two tiny holes close together on a screen and allowed monochromatic light, coming from a distant source, to pass through them to be collected on a second screen. There a pattern of fine bands, alternatively light and dark, appeared (Fig. 2.22, 5.7, 5.8). He explained the phenomenon in terms of the different routes followed by the light emerging from the tiny holes: If the length of such a difference is a multiple of the wavelength the two waves reinforce each other (constructive interference), if an odd multiple of half wavelength they cancel out (destructive interference).

To fit with experiments on the interference of polarized light, Dominique François Arago (1786–1853) pointed out in 1817 that the waves or vibrations of the light had to be assumed transversal, that is at right angles with the direction of propagation. Meanwhile a prize was set up by the Academy of Sciences in Paris on the subject of diffraction. In spite of strong opposition, mainly by Laplace and Biot, the prize was awarded in 1818 to Augustin Jean Fresnel (1788–1827) who put forward a theory based on Young's principle and Huygens's envelope construction. Then Poisson predicted, on the basis of Fresnel's theory, a bright spot in the center of the diffraction pattern of a small circular disk. Impressively enough it was confirmed by experiments made by Arago.

To measure the relative velocities of light in air and water, an experiment was carried out in 1850 by Jean Bernard Léon Foucault (1819–1868) and by Armand Hippolyte Louis Fizeau (1819–1896) who had just determined the velocity of light in air by the first nonastronomical experiment. The result was consistent with the wave theory, namely a lower velocity in the denser medium. It was opposite to that predicted by the corpuscular theory, which came out completely discredited.

The next leap forward came, at about the same time, from another branch of science when James Clerk Maxwell, summing up all previous experiments on electricity and magnetism, began to formulate his mathematical theory of electromagnetism, which culminated in the celebrated equations named after him. They relate the electric and magnetic fields and the spatial and temporal derivatives of the electric charge density and electric current density. In the special case of zero charge and current density in a homogeneous medium, he could solve the equations. The solutions were waves, propagating with finite velocity given by a formula that depended only upon the electric and magnetic constants of the medium. These constants can be deduced from results of electric measurements. When in 1856 such a speed was calculated for the vacuum by Rudolph Kohlrausch (1809–1858) and Wilhelm Weber (1804–1891) it turned out to be close to the speed of light. This led Maxwell to conjecture that light waves are electromagnetic waves.

Starting from 1879 finer measurements conducted by Albert Abraham Michelson (1852–1931), in what turned out to be a passion for life, showed the speed of light to be equal to the value calculated by Kohlrausch and Weber within experimental errors. When speaking of velocity one should say of course what the reference frame is. The existence of an immobile medium, the ether, still believed in at the time, provided a natural reference frame. Michelson set up in 1881 an experiment of interferometry to prove the existence of the supposed "ether drift" caused by the motion of the Earth. The result was negative and the experiment would be repeated several times by Michelson and others as late as 1930, since the existence of this unique intangible medium was believed regardless of the impressive list of properties that went with it.

Only in 1887, eight years after Maxwell's death, was Heinrich Hertz able to generate, transmit, and receive radio waves in his laboratory thus proving experimentally the existence of electromagnetic waves and the validity of Maxwell's theory. His achievement was immediately recognized.

With the main features of the phenomena connected with the propagation of light explained by the electromagnetic theory, the wave model reigned supreme. The field was to rest, but not for long.

5.2 Spectroscopy: a close look at new worlds

Well after the first spectroscopic observation made by Newton in 1664 and right before Young's approximate determination of the wavelengths of the seven colors of Newton's experiment, infrared and ultraviolet rays were discovered by a careful analysis of the spectrum of sunlight.

Infrared radiation, standing between visible light and radio waves, can be detected as warmth by the body. It had been discovered by the German born self-taught astronomer William Herschel (1738–1822). Herschel, who as a boy played as his father did in the band of the Hanoverian Guards, moved to England after the French occupation of Hannover of 1757. He became organist of the chapel in Bath, then turned to the theory of music and, led by his intellectual curiosity and boundless energy, moved to optics, to telescope construction, and finally to night sky observation. Here he found an endless task that led him, through observations stretching over 20 years, to catalog some 2,500 nebulae and star clusters starting from some 100 previously known milky patches. In 1781 during his third survey of the night sky he came across an object that did not appear to be a star. It was a new planet, Uranus, whose discovery made him famous. In 1800, while studying the solar spectrum with the aid of thermometers he found that the greatest effect was taking place beyond the red end.

Spurred by Herschel's discovery of the infrared radiation, the German physicist Johann Wilhelm Ritter (1776–1810) searched beyond the other end of the visible spectrum and one year later, in 1801, found that the chemical effect of solar radiation upon silver salts extended beyond the violet. He had discovered more invisible radiation present in sunlight: the *ultraviolet rays*.

In 1814 Joseph von Fraunhofer (1787–1826) repeated Newton's experiment with improved apparatus. Then he could see that the continuous spectrum of the sun does not vary smoothly from one color to the next but is interrupted by hundreds of dark lines, named after him "Fraunhofer lines." He proceeded to determine the wavelengths of the missing radiation corresponding to many of these. He also observed, as did others, that certain bright lines in the spectra of flames looked as though they coincided with dark lines in the solar spectrum. Nevertheless he failed in tying together the two phenomena.

This was the contribution of Gustav Robert Kirchhoff (1824–1887). By using carefully purified substances he demonstrated in 1859 that each chemical element has a unique characteristic spectrum (*emission spectrum*) and also found that when light passes through a gas, the gas absorbs those wavelengths that it would emit if heated (*absorption spectrum*). He concluded that the Fraunhofer lines are due to the absorption by different chemical elements present in the cooler outer layers of the solar atmosphere. In collaboration with Robert Wilhelm Bunsen, he then proceeded in 1861 to systematically compare the lines in the solar spectrum with those in the flames (and spark spectra) of the purest elements available, thus making the first chemical analysis of the Sun's atmosphere. His discoveries opened up not only the new field of spectrochemical analysis — which in laboratories adds to the other traditional methods of chemical analysis — but also which the new field of astrophysics. Indeed before spectroscopy was introduced there was no way to know anything about the chemical composition of celestial bodies, except for analyzing the occasional meteorites that fell to Earth.

Then, the Swedish physicist Anders Jonas Ångström (1814–1874) measured the wavelength of about a thousand Fraunhofer lines and published in 1868 his great map of the solar spectrum. He used as a unit of length 10^{-10} m, called after him the *angstrom*. By this time, spectroscopy had evolved to such a level that "helium was discovered in the chromosphere of the sun... as a series of unusual spectral lines twenty-three years before it was discovered mixed into uranium ore on earth" (Rhodes [1998]).

Spectroscopy is also used to study properties of atoms and molecules. In fact at the beginning of the 1900s, by the theory of quantum mechanics it could be explained how spectra are formed and why different elements have different spectra. Then spectroscopy also became a tool to obtain clues on the atomic structure of matter.

The methods for dispersing spectra are more than one. Besides refraction through a glass prism, as in Newton's experiment, one can use *diffraction gratings* made of closely packed parallel slits in a number that can vary from a hundred to a few thousand per millimeter. The different wavelengths of the incident light exit the gratings at slightly different directions and can be observed separately.

Other devices for dispersing spectra are called *interferometers* or *Fourier spectrometers*. They divide the incident beam into two beams (by the use of a semitransparent mirror) which are made to traverse different paths before recombining and giving rise to interference fringes. The Fourier transform has to be performed on the data to obtain the spectrum. This method, started around the end of the 1960s in advanced researches, underwent a great diffusion in the 1970s due to the increased capability of powerful computers for the calculations of the FFT (Bell [1972]). More on interferometric methods is found in Sections 9.8 and 9.9.

5.3 The mathematical model for diffraction: the Fourier transform in two dimensions

Properties of light such as reflection, refraction, interference, and diffraction are shared by the whole electromagnetic spectrum. In order to illustrate the general phenomenon of diffraction at infinity, which plays a role in optical instruments such as diffraction gratings (Section 5.2) and telescopes — as well as x-ray diffractometers (Chapter 6) and radiotelescopes (Chapter 9) — we shall define the Fourier transform in two dimensions and work out a couple of examples, relevant for the next section.

The Fourier transform of a function of two variables $f(u, v)$ is defined quite naturally by the very same formula (2.13) suitably interpreted. If $\mathbf{t} = (u, v)$ and $\mathbf{\Omega} = (\xi, \eta)$, then $\mathbf{t} \cdot \mathbf{\Omega} = u\xi + v\eta$ is the scalar product of the two vectors and $d\mathbf{t}$=$du\,dv$ so that the formula reads

$$\hat{f}(\xi, \eta) = \int_{-\infty}^{\infty} \int_{-\infty}^{\infty} f(u, v) e^{-i(u\xi + v\eta)} du\, dv. \tag{5.1}$$

In the special case in which f is the product of two functions, one depending only upon u and the other only upon v, namely $f(u, v) = h(u)g(v)$, then clearly

$$\hat{f}(\xi, \eta) = \int_{-\infty}^{\infty} e^{-iu\xi} h(u) du \int_{-\infty}^{\infty} e^{-iv\eta} g(v) dv = \hat{h}(\xi)\hat{g}(\eta) \, ,$$

that is $\hat{f}$ is the product of the corresponding Fourier transforms of one single variable.

In particular if $f = \chi_Q$ is the characteristic function of the square $Q = [-1, 1] \times [-1, 1]$, and by this we mean the function that takes the value 1 on the square and zero outside, then with the notation of Section 2.7 it is $\chi_Q(u, v) = r(u)r(v)$ and so the Fourier transform $\hat{\chi}_Q(\xi, \eta) = 4(\xi^{-1} \sin \xi)(\eta^{-1} \sin \eta)$. It shows a large central maximum surrounded by a regular pattern in the horizontal and vertical directions of other maxima of decreasing value. Such a decrease is different in different directions. The maxima diminishes rapidly as ξ^{-2} along the bisector $\eta = \xi$ and less rapidly, as $|\eta|^{-1}$, along the η axis and similarly as $|\xi|^{-1}$ along the ξ axis. Consequently, the on-axis maxima are more prominent in pictures (Fig. 5.5).

If a dilation (Section 2.8) is applied in the y variable to obtain for instance the rectangle $R = [-1, 1] \times [-4, 4]$ then the Fourier transform of χ_R is calculated as $(\xi^{-1} \sin \xi)(\eta^{-1} \sin 4\eta)$ by (2.17). This follows from the Fourier transform relative to the square, by a compression of 1/4 in the η variable (Fig. 5.5). The cross nature of the two figures, square and rectangle, is therefore reflected on the Fourier transform side.

Relative to another geometrical property, namely radiality or circular symmetry, it is easily proved that the Fourier transform of a radial function is radial too. In the case of the unit disk D it is therefore natural to use polar coordinates (r, θ) and (R, Ψ) to see calculations drop to one dimension $\hat{\chi}_D(R, \Psi) = \hat{\chi}_D(R, 0)$. Setting $\chi_D = \chi$ for simplicity of notation, from

$$u\xi + v\eta = Rr(\cos\theta \cos\Psi + \sin\theta \sin\Psi) = Rr\cos(\theta - \Psi)$$

and $du\, dv = r dr d\theta$ it follows that

$$\hat{\chi}(R, 0) = \int_0^1 \int_0^{2\pi} e^{-iRr\cos\theta}\, d\theta\, r\, dr = \int_0^1 \int_0^{2\pi} e^{iRr\sin\theta'}\, d\theta' r\, dr$$

by the change of variable $\theta = \theta' + \frac{\pi}{2}$. Therefore,

$$\hat{\chi}(R, 0) = 2\pi \int_0^1 J_0(Rr) r\, dr$$

where J_0 is the Bessel function of order zero. Bessel functions of order $k = 0, 1, 2, \ldots$ defined by

$$J_k(x) = (2\pi)^{-1} \int_0^{2\pi} e^{ix\sin\theta} e^{-ik\theta}\, d\theta$$

make up an infinite family of smooth oscillating functions, largely tabulated and studied due to their usefulness and special properties (Watson [1944], Whittaker and Watson [1946]). For instance

$$\frac{d}{dx}(x J_1(x)) = x J_0(x).$$

Moreover $J_1(x) \cong x/2$ for x tending to zero and $J_1(0) = 0$. Hence by another change of variables $x = Rr$ and by the fundamental theorem of calculus,

$$\hat{\chi}(R,\ 0) = \frac{2\pi}{R^2} \int_0^R J_0(x) x dx = 2\pi \frac{J_1(R)}{R}.$$

From this celebrated formula — first obtained in 1835 by the English astronomer George Biddell Airy (1801–1892) — the properties of $\hat{\chi}(R,\ 0)$ can be derived: It is smooth, it takes the value π at the origin, and it decreases rather quickly being dominated by $R^{-3/2}$ for large values of R. The last property comes from the fact that, for large values of R, all Bessel functions are dominated by $R^{-1/2}$, namely

$$J_1(x) = \sqrt{\frac{2}{\pi x}} \cos(x - \frac{3\pi}{4}) + \text{negligible terms}.$$

The formula shows that the decrease is not monotone but it has an oscillatory nature and so does $\hat{\chi}(R,\ 0)$. The qualitative graph of the positive function $|\hat{\chi}(R,\ 0)|^2$ is in Figure 5.2. By a rotation around the vertical axis, the graph of $|\hat{\chi}(R,\ \Psi)|^2$ can be obtained (Fig. 5.4). The separation between two successive minima, corresponding to the roots of $J_1(R) = 0$, is not exactly π, but rather tends to π as R tends to infinity. Similarly, for the maxima: the principal one at the origin has value π^2 and the next one at $R = 5.136$ has value 0.172 approximately.

As regards dilations and translations, formulas analogous to (2.17), (2.18), and (2.19) hold. Similarly to (2.14), the double Fourier transform has an inverse equal to the Fourier transform itself except for a sign change — denoting rotation through π — and the multiplicative constant $(2\pi)^{-2}$ Hence if the Fourier transform is applied to $f(u,\ v)$ twice in succession, $f(-u,\ -v)$ is recovered except for a multiplicative constant.

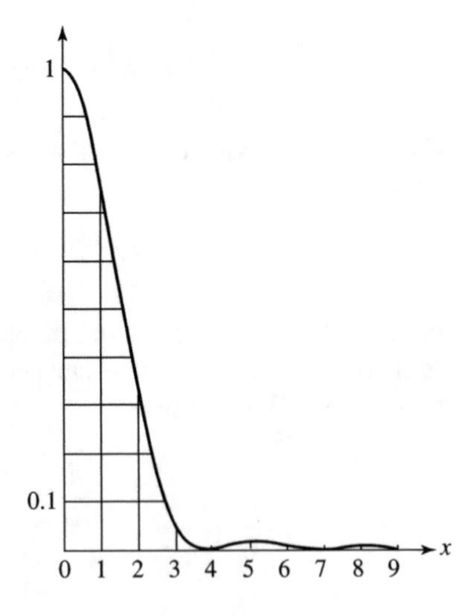

Figure 5.2 The graph of $(2J_1(R)/R)^2$ The principal maximum attained at the origin has value unity, the next one at $R = 5.136$ has value 0.0175 approximately.

5.4 Optical transforms

The phenomenon of *diffraction*, the seeming deviation of light from rectilinear propagation — after passing through a small aperture — produces interference fringes in regions otherwise belonging to an homogeneous shadow. Geometrical optics — which describes light rays by intensity and direction of propagation — does not account for diffraction since it disregards the oscillatory nature of light by making its extremely small wavelength equal to zero.

Diffraction can be made evident in special conditions that become simpler if one employs a parallel and monochromatic light beam, like that obtainable from a laser. By Huygens's envelope construction, every point of an aperture placed on the path of the beam becomes a source of light of the same frequency and emits in all directions. If the emerging rays are collected on a screen, three different situations occur depending upon how far away the screen is set.

If the screen is very close to the aperture, the shape of the aperture is reproduced. As the screen is moved away, a set of fringes appears approximately restricted to the geometrical projection of the aperture (*Fresnel diffraction*). Finally if the screen is moved very far away, relative to the dimensions of the experiment, then an extended system of fringes appears that remains unaltered as the distance is further increased, except for enlargements (*Fraunhofer diffraction* or *diffraction at infinity*).

The mathematical model of this limiting case is somewhat simpler and can be found in Section 9.7. It shows that the electrical field on the screen is proportional to the Fourier transform of the characteristic function χ_A of the aperture A, the wavelength of the light taken as a unit of measure. Thus the intensity of the collected light is proportional to $|\hat{\chi}_A|^2$ and so is the illumination on the screen that gives the picture. Such a picture is called the *optical transform* of the aperture. As we shall see, for the phenomenon to show itself conspicuously, the dimensions of the aperture have to be very small, comparable to the wavelength of the light employed.

If the aperture is a circular one, like the disk D of the previous section, the screen will show a bright central spot surrounded by rings, alternately dark and bright, whose illumination decreases rapidly as in Figure 5.3. This is in accordance with the calculations in Section 5.3 showing $|\hat{\chi}_D(R, \ \Psi)|^2$ to have a principal maximum at the center and secondary maxima of rapidly decreasing value set on concentric circles, called *Airy rings* (Fig. 5.4). The rings are visible under normal circumstances, as when a distant star is observed through a telescope with too small an aperture.

The optical transform of a rectangular aperture is in Figure 5.5. To a slim aperture there corresponds a diffraction pattern with a large central maximum, as previously calculated. The dimensions of the aperture are critical. If much larger than to the wavelength of the light employed, the phenomenon of diffraction becomes imperceptible, for the optical transform undergoes a contraction by (2.17). On the screen only the central maximum shows since the secondary ones, very densely packed together with the minima, are as a matter of fact indistinguishable. On the contrary if the dimensions are much smaller, then again by (2.17), the central maximum becomes very large and at the same time very weak so that the screen appears dark as if it

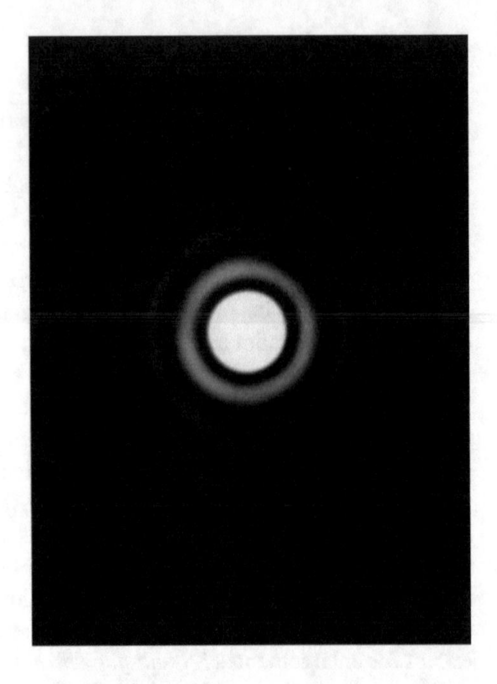

Figure 5.3 Optical transform of a circular aperture. (*Courtesy of S. G. Lipson, Department of Physics, Technion University*)

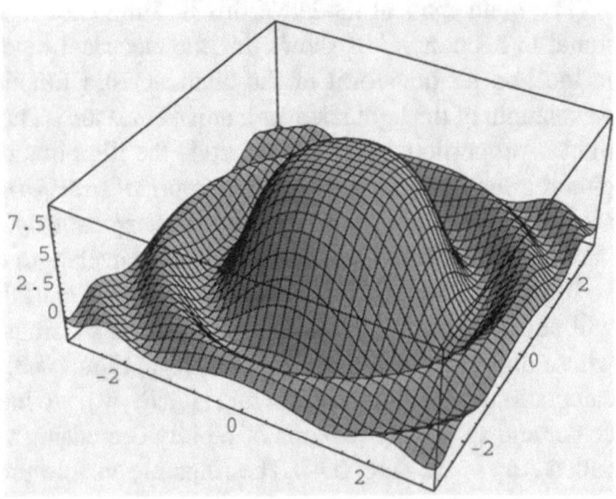

Figure 5.4 Graph of $|\hat{\chi}_D(R, \Psi)|^2$ (see text).

were not illuminated at all. Only in the case where the dimensions are comparable to the wavelength of the light does the phenomenon stand out clearly. An aperture in the shape of a small duck, as in Figure 5.6 a), has its diffraction pattern as shown in Figure 5.6 b).

If one wishes to reconstruct the shape of the aperture from the optical transform — for a purely speculative reason in this case, the aperture being known to start with — it would suffice to collect the light coming from the optical transform on a screen placed on the focal plane of a lens. Indeed lenses have the ability to transform

Figure 5.5 To the right the optical transform of the rectangular aperture shown to the left. (*Courtesy of S.G. Lipson, Department of Physics, Technion University*)

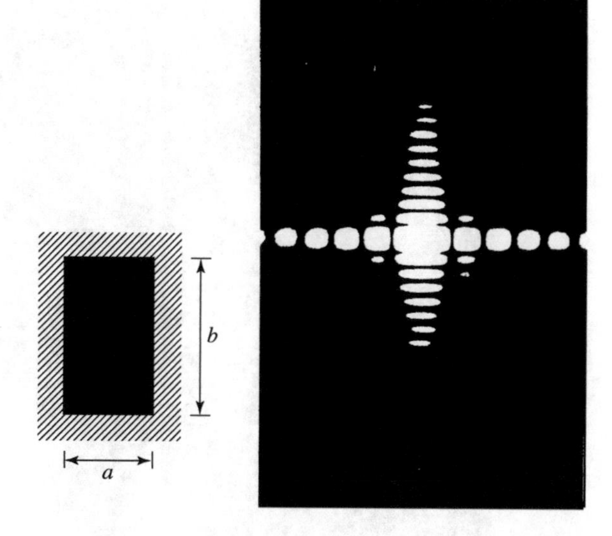

parallel rays — rays converging at infinity — into rays converging to a point of the focal plane. This is equivalent to applying the Fourier transform a second time. Mathematically, with $\chi_A(u, v)$ defining the aperture, $\chi_A(-u, -v)$ is reconstructed and the shape of the aperture is therefore obtained.

Figure 5.6 shows two reconstructions d) and f) of the aperture, each using smaller and smaller pieces c) and e) of the optical transform. A progressive loss of details, or resolving power, is observed.

By this procedure, used in conjunction with x-rays, the atomic structure of crystals could be recovered — the wavelength involved being comparable with the details of interest — if it were not for unfortunate facts. The diffracted x-rays are scattered in all directions, even backward, and moreover x-rays can only be slightly deflected. In other words the diffracted x-rays cannot be focused. Only calculations are left to help. By another unfortunate circumstance the intensity of the light alone can be measured — that is the optical transform — not the full Fourier transform. Indeed the phase is lost in the measurements. As a consequence the reconstruction is not straightforward, as it would otherwise be (Chapter 6). Figure 5.7 shows the optical transform of two horizontally spaced circular apertures centered at $(a, 0)$ and $(-a, 0)$. The Fourier transform in question is equal to $2\cos(a\xi)\hat{\chi}_D(\xi, \eta)$. The transform clearly has extra zeros at the roots of $\cos a\xi = 0$ — namely the vertical lines $\xi = k\pi/2a$ with k any integer (Fig. 5.7)–marking the appearance of interference fringes. In the case of rectangular apertures centered at $(-a, a)$ and $(a, -a)$, the Fourier transform in question is equal to $2\cos(a\xi - a\eta)\hat{\chi}_R(\xi, \eta)$ which has extra zeros at $a\xi - a\eta = k\pi/2$. These are all lines parallel to the bisector $\eta = \xi$ (Fig. 5.8).

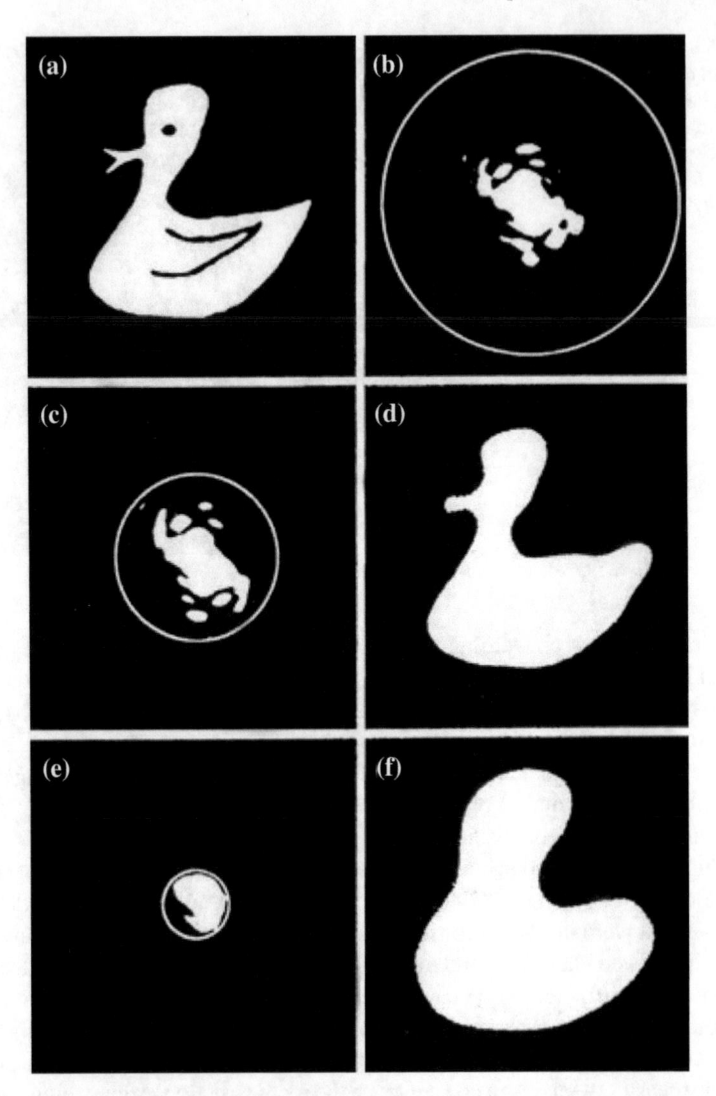

Figure 5.6 a) aperture; b) optical transform; c) and e) smaller and smaller pieces of the optical transform; d) and f) reconstructions of the shape of the aperture. (Taylor C. A., Lipson H., *Optical Transforms*, Bell, [1964].) (*Courtesy of S. G. Lipson, Department of Physics, Technion University*)

Figure 5.7 Optical transform of a system of two circular apertures spaced horizontally. (*Courtesy of S. G. Lipson, Department of Physics, Technion University*)

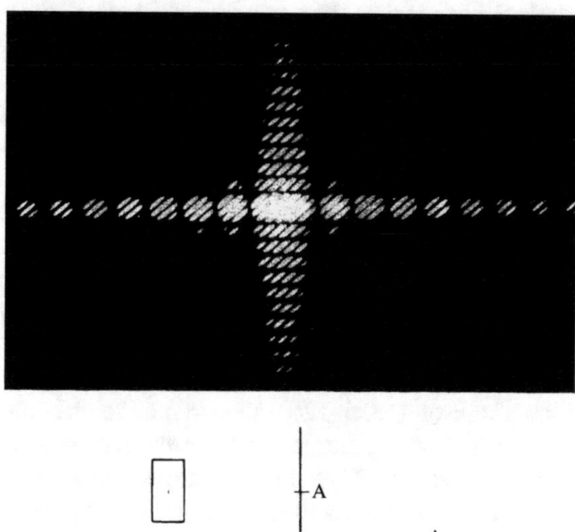

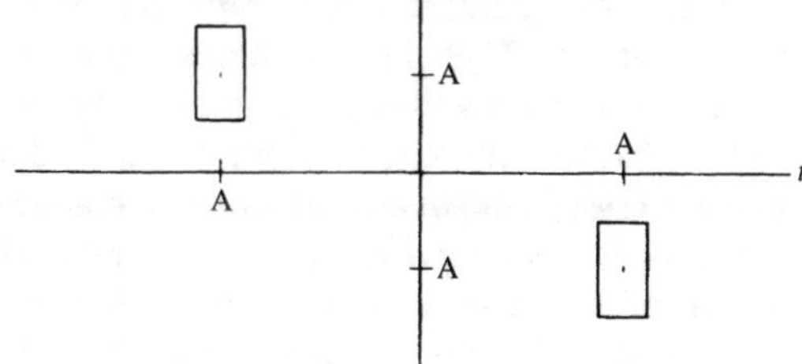

Figure 5.8 Above, the optical transform of the system of two rectangular apertures shown below. (*Courtesy of S. G. Lipson, Department of Physics, Technion University*)

5.5 X-rays and the unfolding of the electromagnetic spectrum

That part of the electromagnetic spectrum that conveys shape and colors to our vision, so wonderfully that it has inspired artists ever since, is a very tiny part of the whole electromagnetic spectrum. It extends less than an octave, between 4×10^{-5} cm (violet) to 7×10^{-5} cm (red). Above and below in a feast of numbers, invisible radiations extend with longer and shorter wavelengths, even thousands of billions of times longer (radio waves) and even a million billion times shorter (gamma rays).

The virgin territory of the electromagnetic spectrum in the direction of the shorter wavelengths, beyond the ultraviolet, was entered at the end of 1895 by the German physicist Wilhelm Conrad Röntgen (1845–1923). He discovered *x-rays* and for that was awarded in 1901 the first Nobel prize for physics.

Röntgen was experimenting with cathode rays when he noticed that light was given off by a fluorescent screen nearby, even when the tube was completely wrapped in black paper. He studied the radiation and even took the first x-ray photograph that showed the bones of his wife's hand. Within a period of months x-rays started revolutionizing medical diagnosis in orthopedics and dental medicine. Röntgen was uncertain about the nature of the radiation, which appeared to lack the basic properties of reflection and refraction (indeed the path of x-ray beams can be deflected only slightly). He called it *x*-radiation, but it also went under the name of Röntgen radiation.

X-rays are produced by energetic electrons as they accelerate or decelerate: In x-ray tubes electrons decelerate suddenly while hitting a metal target; in synchrotrons they accelerate while bending their path. Almost a century later, a more sophisticated use of x-rays will lead to remarkable imaging techniques, such as computerized tomography, microtomography (Chapter 7), and contrast-phase imaging (Section 5.11).

A couple of years after Röntgen's work, in 1897, the English physicist Joseph John Thomson (1856–1940) discovered the electrons ("corpuscules," as he called them) while attempting to establish the nature of cathode rays themselves, until then controversial. They were electrons indeed, for which he provided evidence. Another fifteen years will have to go by to assess the nature of the very same x-rays. The German physicist Max von Laue, thinking that the regular arrangement of atoms in crystals might provide a natural grating to produce interference patterns — if radiation of appropriate wavelength (about 10^{-8} centimeter) were employed — suggested trying with the x-radiation. The experiment, carried out by Walter Friedrich and Paul Knipping, was successful (Friedrich, Knipping, and von Laue [1912]) and made the year of the announcement of their result a most important juncture in physics: The electromagnetic nature of x-rays was established and, at the same time, a way to study matter at the atomic scale was opened up (Chapter 6).

Though harmful in large doses for living creatures, due to their penetrating power and ability to ionize atoms, x-rays provide a unique experimental tool. As Margaritondo [1988] points out, "x-ray experiments have produced more Nobel Prize-winning results in different disciplines than any other kind of experimental techniques."

Following the discovery of radioactivity in 1896 by Henri Becquerel (1852–1908), which earned him in 1903 the Nobel prize (shared with Pierre and Marie Curie), came that of gamma rays. They first evidenced themselves as an even more penetrating radiation emitted in the decay of radioactive substances. Discovered by Ernest Rutherford, their electromagnetic nature was assessed in the same year as Laue's experiment. Very harmful for living creatures, they are used to penetrate metal castings to show defects by means of radiographs and even to sterilize medical supplies.

Gamma rays and x-rays are present in the radiation coming from space but are prevented from reaching the Earth by the atmosphere, which acts as a filter (Fig. 9.1). The first x-rays of solar origin were detected in 1949 by Herbert Friedman and coworkers using a V-2 missile for high altitude research at the White Sands Proving Ground in New Mexico. The V-2 nose cones "often carried at altitudes better than 160 km more than a ton of payload" of scientific instruments (Stuhlinger and Ordway [1994]), so by the time the last V-2 was launched in the fall of 1952 a rich harvest of information on cosmic rays and high altitude atmosphere had been obtained. The first nonsolar x-ray source was discovered in 1962 by the Italian physicist Bruno Rossi, at the time at MIT, in collaboration with Riccardo Giacconi (Morrison [1967]). The best known sources are the Crab Nebula and Sco X-1 in the Scorpius constellation. Gamma ray sources began to appear around 1978 in balloon observations (Bignami and Hermsen (1983)). All recordings, at first, were of low signal-to-noise ratio and controversial.

In Figure 5.9, the electromagnetic spectrum: the relation between the wavelength λ and frequency v is given by $\lambda v = c = 3 \times 10^{10}$cm/s. As a unit of energy the electron volt (eV) is used, which is the energy gained by one electron as it is accelerated in a vacuum through a potential difference of one volt. Its multiples are shown in Table 5.1.

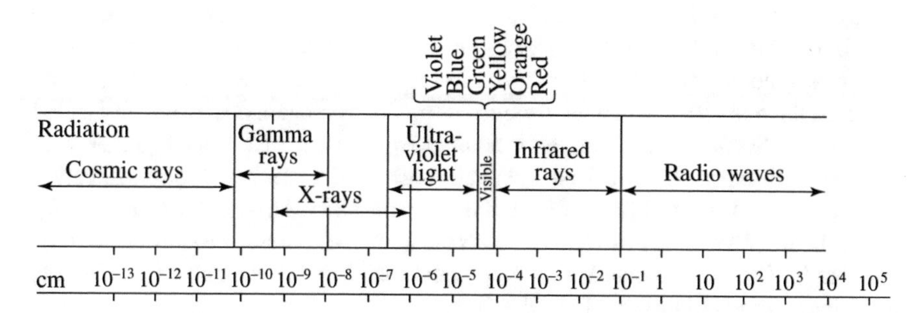

Figure 5.9 Wavelengths of the radiations of the electromagnetic spectrum.

Table 5.1 Multiples of the electron volt (eV).

kiloelectron volt keV = 10^3eV
megaelectron volt MeV = 10^6eV
gigaelectron volt GeV = 10^9eV
teraelectron volt TeV = 10^{12}eV

The ionizing power of x-rays and gamma rays and more generally events of interaction between light and matter occurring within the atom are not explained by the wave theory. Historically the first occurrence, which came to be known as the "photoelectric effect," was already observed by Hertz in his famous experiment of 1877. It consists of the emission of electrons by metals and a variety of other materials, found to be photosensitive, when radiation of sufficiently high frequency — depending upon the material — strikes them. One characteristic of the emission was surprising and could not be explained: the maximum kinetic energy of the emitted electrons was independent of the intensity of the light, which was expected, and depended instead on the frequency.

To explain the photoelectric effect, Albert Einstein published in (1905) a theory of light that revived the corpuscular one under a new form. Going back to a hypothesis formulated by Planck a few years earlier, Einstein assumed light to be composed of quanta, or photons, whose energy is proportional to the frequency of the light. The assumption of an ether was dropped. For this work Einstein received the Nobel prize in 1921.

It was only with quantum mechanics, whose main development took place between 1925 and 1935, that the wave model and the corpuscular model would be encompassed in one single theory that specifies their respective ranges of validity.

5.6 Synchrotron radiation, first seen in the stars

The brightest x-rays available are to be found in the so-called synchrotron radiation produced at the several storage rings in operation worldwide (Fig. 5.10, 5.11). An extremely broad continuous spectrum is among the main characteristics of the radiation, which relies on the theory of special relativity for its mathematical model.

It was synchrotron radiation in the radiofrequencies that was observed by Grote Reber in the late 1930s when he discovered the first radiogalaxies, originally called radio stars (Section 9.10). The enormous energy output that went together with enormous distances remained unexplained until 1950 when synchrotron radiation was advanced as the mechanism of emission by radio stars (Alfvén, Herlofson [1950], Kiepenheuer [1950]). For astronomy it was the first nonthermal emission and marked the decade. The modern detailed development took place in the hands of the Russian Iosif Samvilovich Shklovsky [1960].

Closer to us, Jupiter emits synchrotron radiation due to high-energy electrons trapped in its Van Allen belts and located at a distance between 30 and 100 radii

Figure 5.10 The European Synchrotron Radiation Facility (ESRF) in Grenoble.

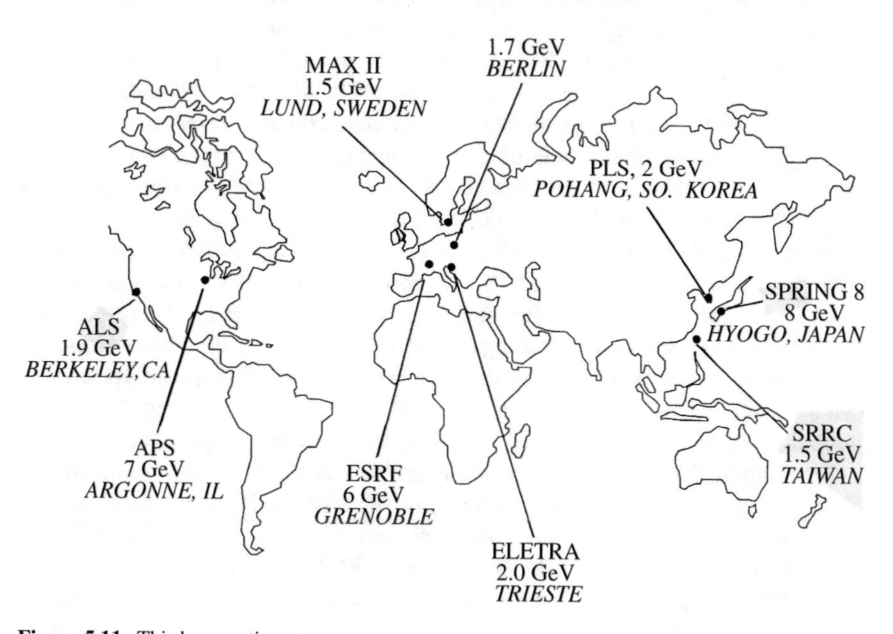

Figure 5.11 Third generation sources.

of Jupiter's surface. The space probe Pioneer 10 encountered Jupiter on December 4, 1973 and measured electron energies of about 5 MeV and a weak magnetic field of about 10^{-4} tesla. Synchrotron radiation is also observed in conjunction with sunspots. In a laboratory, it was first detected in 1946 by J. P. Blewett working at the 100 MeV betatron (beta particles is an old name for high-energy electrons) of the General Electric Research Laboratories in Schenectady, New York. The next year it was observed visually by F. Haber, a technician of the group of H. C. Pollock and collaborators, working in the same laboratories at the 70 MeV synchrotron. (From this comes the name of the radiation.) Their announcement in the *Physical Review* read as follows: "High energy electrons which are subject to large accelerations normal to their velocity should radiate electromagnetic energy. The radiation from electrons in a betatron or synchrotron should be emitted in a narrow cone tangent to the electron orbit, and its spectrum should extend into the visible region. This radiation has now been observed visually in the General Electric 70-Mev synchrotron" (Elder et al. [1947]). The emotion that reverberates in the preceding sentence goes now hidden under the technical description: "The radiation is seen as a small spot of brilliant white light by an observer looking into the vacuum tube tangent to the orbit and toward the approaching electrons. The light is quite bright when the x-ray output of the machine at 70 Mev is 50 roentgen per minute at one meter from the target and can still be observed in daylight at outputs as low as 0.1 roentgen." Their synchrotron had an electron orbit radius of 29.3 cm and a peak magnetic field of 0.8 tesla.

The sources of electromagnetic radiation are oscillating electric charges and currents. For instance radio waves are generated by an alternating current in an antenna.

The phenomenon of emission of electromagnetic radiation by a charged particle whenever accelerated or decelerated was first considered by the Irish physicist Joseph Larmor (1857–1942) who calculated the energy radiated by an accelerated (nonrelativistic) electron while teaching at Cambridge (Larmor [1897]). This early research by Larmor was followed by the work of A. Liénard in 1898 and G. H. Schott in 1907 (Schott [1912]). It was shown that the acceleration was responsible for an electromagnetic field decreasing as $1/R$, where R denotes the distance from the accelerated charge, which for big values of R, is much larger than the field generated both by a static charge and by a charge in uniform motion. The energy lost by the circulating electrons was calculated as well as the angular and spectral distribution of the radiation. The study of the radiation, emitted by electrons moving in circular orbits, was driven by the desire to obtain atomic models. However, precisely the atomic model failed since the electrons, radiating continuously, would lose energy and spiral into the nucleus in no time, such as 10^{-15}s. Soon after in 1913 Niels Bohr (1885–1962), realizing that classical mechanics alone could never explain the atom's stability, proposed his model of quantized orbits and the preceding work was forgotten only to be rediscovered forty years later.

In the early 1940s, what would become known as "synchrotron radiation" gained attention in connection with electron accelerators. There was a loss of energy that went together with a loss of speed, chiefly responsible for limiting the electron energy achievable in an accelerator. This made it difficult and costly to push electrons to higher energies.

From a theoretical point of view the treatment, based on the theory of special relativity due to the relativistic speed of the electrons, was developed by several authors. First were D. Ivanenko and J. Pomeranchuk [1944] in the Soviet Union and J. Schwinger [1946] in the United States. In 1946 Blewett calculated the shrinkage in the electron orbit, in a betatron, due to radiation losses and his measurements were in agreement.

The characteristics of the radiation were studied notably by Pollock and collaborators in the late 1940s and by F. A. Korolev with his group working at the 250 MeV synchrotron of the Lebedev Institute in Moscow in the 1950s. Others followed. These investigations verified the basic theoretical predictions and provided experience in the use of radiation.

A new course, which would start a revolutionary trend, was taken by D. H. Tomboulian and P. L. Hartman [1956] who saw beyond the technical problem for accelerator builders and put the synchrotron radiation to work. They used it in connection with far-ultraviolet spectroscopy while working at the 320 MeV electron synchrotron of Cornell University. A few years later L. G. Paratt [1959] realized its relevance for x-ray experiments and predicted "a boom in many aspects of x-ray physics." Today synchrotron radiation is regarded as a unique scientific resource for basic and applied research and technology.

5.7 Unique features of the synchrotron light

Before entering more technical aspects concerning the synchrotron radiation, it might be worthwhile to highlight the important features that go with it.

In the x-ray region, particularly interesting for applications, synchrotron radiation offers a continuous band of wavelengths — while conventional sources make only few wavelengths available — and also radiation of extremely high intensity and collimation. Some advantages associated with these features can be anticipated, for example, in the field of crystallography (Chapter 6).

To recover by x-ray diffraction methods the spatial atomic structure of crystals, high intensity x-ray beams are needed especially in the case, often encountered, of very small crystals, that is crystals that might have the smallest dimension of the order of 10^{-2} mm. Indeed it is difficult to grow well-ordered protein crystals (Section 6.7) so that the available samples are usually small. Small are also synthetic mineral crystals and again small dimensions are found in mineral crystals extracted from polycrystals. In the diffraction experiment, the intensity of the diffracted beam is proportional to the intensity of the impinging beam and to the volume of the crystal samples, among other parameters. So when samples are small, a high intensity impinging beam is needed to have diffraction peaks noticeable against the background and hence measurable.

Collimation — meaning beams of rays highly parallel or equivalently beams with small divergence — realizes optimal conditions to give rise to sharp diffraction peaks. This feature is especially important when the diffraction peaks are densely packed as in the case of macromolecular crystals (Section 6.7).

Synchrotron radiation has also a good degree of coherence but it does not have complete coherence, which would be the case if all points of the source were to emit radiation in phase. If a coherent source were available, experimental settings could be devised to obtain information on the phase of the diffracted beam, presently lost (Sections 6.4 and 6.7). For instance, after splitting the coherent beam into two beams, one could be used for the diffraction experiment while the other could be kept intact and then made to interfere with the diffracted beam. From the recorded intensities of the interference experiment the needed information on the phase of the diffracted beam could be obtained. Synchrotron builders are indeed improving coherence as well as intensity (Section 5.11).

5.8 Storage rings, among the biggest machines in the world

Electrons moving in a circular path — and more generally changing the direction of motion — experience an acceleration that points transversally in the centripetal direction, if the scalar velocity is constant. (In the case of the radio waves generated by an alternating current in an antenna both the velocity and the acceleration are in the same direction.)

Synchrotrons and storage rings enable an electron beam to move at relativistic speed on a closed circular orbit and in turn to produce extremely intense and short bursts of radiation with a large spectrum of frequencies. Synchrotrons briefly accelerate the beam to speeds very near to the speed of light, while storage rings — which are a specialized form of synchrotrons — keep a relativistic electron beam circulating for many hours at constant speed. They usually work in sequence, with the synchrotron functioning as a booster ring.

In the early years synchrotron radiation was not considered important enough and could only be obtained from accelerators built, operated, and optimized for elementary particle experiments and now called synchrotron radiation sources of the "first generation." In this "parasitic" way pioneering work was carried out in the 1960s at synchrotrons in Frascati (Italy), at the University of Wisconsin, in Hamburg (Germany), Novosibirsk (Siberia) and Tokyo.

The switch to storage rings, whose emission has much more stable characteristics, was initiated in 1961 in Frascati, Italy, with *AdA* (Anello di Accumulazione, the Italian for storage ring) followed by the Tantalus ring at the University of Wisconsin which, originally designed for studies of future high-energy physics machines, got transformed into a storage ring in 1968 and served as a test case for the following one in Brookhaven, New York. The modifications channeled some of the radiation through beamlines to user end stations. In 1970 the first storage ring — 380 MeV SOR ring — intended only as a light source was designed at the University of Tokyo and started operation in the mid-1970s. Subsequently others were built in countries such as Brazil and China besides Russia, the U.S., and Europe. Named storage rings of the "second generation," they usually carry a large number of tangential ports, called

Table 5.2 Main facilities worldwide.

Facility	ESRF European Synchrotron Radiation	APS Advanced Photon Source	SPring-8 Super Photon ring-8 GeV
Energy	6.0 GeV	7.0 GeV	8.0 GeV
Circumference	844 m	1,104 m	1,436 m
Began Operation	1994	1996	1998

beamlines, that channel the radiation to experimental stations (Fig. 5.13) serving some thousands of users per year.

The enormous productivity and significant technical advances (undulators) were a stimulus to build a "third generation" of dedicated storage rings throughout the world. They can be subdivided mainly into two categories: rings 30–60 meters in diameter, designed primarily for ultraviolet and soft x-rays, and rings 300–500 meters in diameter, designed primarily for hard x-rays (Fig. 5.11). The cost is in the order of several hundred million dollars each.

Synchrotrons as well as storage rings consist of a metal vacuum chamber in the shape of a ring or a doughnut, with magnets distributed around it. The diameter of the ring can be up to a few hundred meters and so the circumference can reach the impressive size of more than one kilometer: It is 844 m in case of ESRF (Fig. 5.10) and 1,104 m in case of APS and 1,436 m in case of SPring-8 (Tab. 5.2, Fig. 5.11). In the vacuum chamber of the ring — whose cross sectional area is in the order of tens of square centimeters — the pressure, starting from atmospheric, is brought down to about one thousandth of one billionth of that, comparable to pressure in outer space. Collisions of the circulating electrons with the residual gas molecules, which primarily affect the average lifetime of the beam, are therefore rare, allowing the beam intensity to decrease slowly, with a lifetime of many hours. Unfortunately the synchrotron radiation itself liberates gases from the surface of the vacuum chamber. This is monitored by vacuum pumps that keep the pressure at the indicated value: If pressure rises above that the beam is lost.

Electrons, generated by an electron gun, are injected into the evacuated ring by a suitable injection system based on a combination of accelerating devices, including normally a linear accelerator briefly called *linac*.

The magnets surrounding the doughnut, in an arrangement called a *magnetic lattice*, are the main elements controlling the basic features of the beam. They are essentially of two types: *bending magnets* and *focusing magnets*. The first ones, as the name says, are responsible for bending the path of the electrons which is made by circular arcs — corresponding to the bending magnets — and straight lines in between (Fig. 5.12). If only the bending magnets were used, the electron beam would progressively grow in transverse directions, the same as with a visible light beam. Therefore the need for the so-called focusing magnets, which act as lenses and keep the electrons confined to a tight beam of vertical and horizontal dimensions that can be much less than one millimeter, comparable to the size of a human hair.

The straight sections — some tens in number — are available for insertion devices, such as *undulators*, used to enhance the brightness of the light.

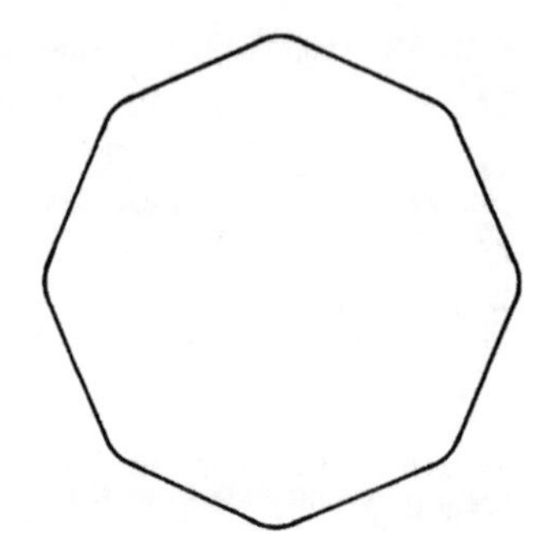

Figure 5.12 Schematic picture of a storage ring. Straight sections alternate with curved sections.

The energy lost in the emitted radiation must be restored. This is obtained by applying, in the so-called *radio frequency cavities*, an oscillating electromagnetic field that imparts longitudinal electrical kicks and makes the electrons go around in bunches, like an enormous rotating pearl necklace. Therefore the radiation is emitted in bursts.

The magnet configuration and the parameters of the radio frequency system determine the length of the bunches, typically of the order of a centimeter. The length of the bunches and their velocity v, essentially equal to the speed of light c, determine the duration of the bursts. For a 3 cm long bunch and $v = c$ it is 10^{-10}s. The number of bunches, usually in the hundreds or a few or just one depending upon the experimental needs, together with the ring perimeter, determines the burst frequency or repetition rate of the order of a hundred million times per second.

As we shall see the most important features of the synchrotron radiation are determined by the electron speed.

5.9 Spectrum of the synchrotron light

The electrons in a storage ring travel from several hours to a few days at a speed v near the speed of light $c = 3 \times 10^{10}$cm/s. To give an example, at the Advanced Light Source facility (ALS) in Berkeley, the ratio $v/c = 0.99999996$. Every time the electron bunches go through a bending magnet they emit radiation that is channeled, via a beamline, to an experimental station (Fig. 5.13). The speed at which electrons travel in a storage ring is a key point, since the angular pattern of the emission at a distance ("far field") strongly depends upon that (Fig. 5.14). If electrons were made to change direction by the bending magnets while moving at a speed much less than the speed of light, they would emit radiation in almost all directions. (The angular distribution of the radiation pattern in three dimensions can be obtained from that in Figure 5.14 (top) by a full rotation around the vertical axis which is the direction of the acceleration (Tomboulian, Hartman [1956]).) The radiation would be very weak and of low frequency.

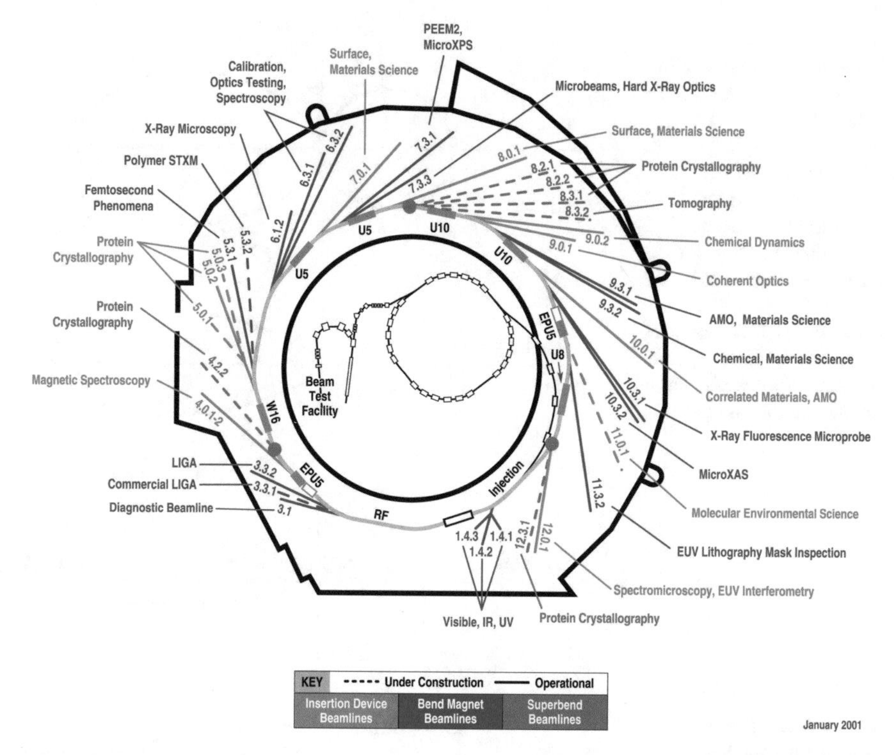

Figure 5.13 Advanced Light Source facility at Lawrence Berkeley National Laboratory, Berkeley, Ca. The ring at the center is the booster synchrotron. The beamlines are updated to January, 2001. (*Courtesy of ALS*)

If instead the electrons are accelerated while moving at a speed in the range of the speed of light — and are observed in a laboratory frame, itself moving with Earth at a small velocity compared to that of light — huge changes take place, in accordance to the theory of special relativity. The radiation gains enormously in intensity and frequency and strikingly in directionality (Fig. 5.14). At half the speed of light the radiation appears already noticeably "pushed forward." In the extreme case of $\beta = v/c$ very close to 1, the radiation turns out to be confined to a narrow cone centered around the direction of motion, or instantaneous velocity vector, so as to look like a well-focused searchlight beam sweeping around, as the electrons describe their circular orbit (Fig. 5.15).

The radiation is made up of nearly parallel rays, since the aperture of the cone might be even less than 0.0001 radians (about 0.006 degree). Given by $\sqrt{1-\beta^2}$, it goes to zero as β tends to 1. The inverse of this important parameter is usually denoted by γ, that is $\gamma = 1/\sqrt{1-\beta^2}$.

The pure number γ turns out to be a ratio involving the rest mass of the electron m_0 and m, the mass of the electron moving at speed v. It is $m = \gamma m_0$, as assigned by the theory of special relativity. Published in 1905 by Einstein, this theory establishes the

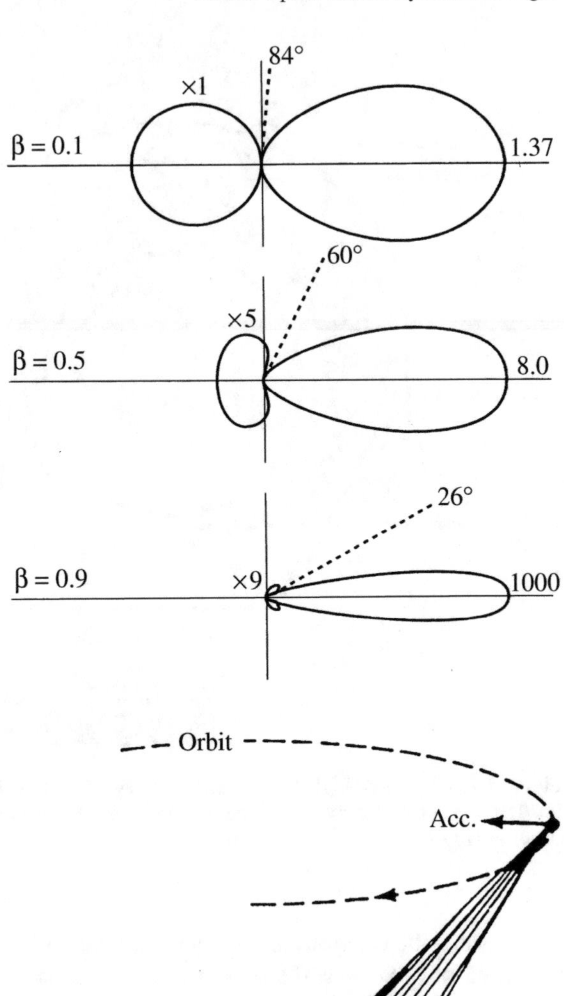

Figure 5.14 Angular radiation pattern drawn in the orbital plane for different values of $\beta = v/c$. In the direction of the dotted line the radiation intensity is zero. To make evident the backward lobes of the radiation their intensities have been multiplied by the indicated factors. (A. Balzarotti [1975], "*Luce di Sincrotrone*," *Enciclopedia delle Scienze Fisiche*, Istituto della Enciclopedia Italiana, pp. 232–240.)

Figure 5.15 The narrow cone represents the angular distribution of the emitted intensity from an electron moving with a velocity close to that of light. (D. H. Tomboulian, P. L. Hartman [1956], "Spectral and angular distribution of ultraviolet radiation from the 300-Mev Cornell Synchrotron,"*Phys. Rev*. 102, pp. 1423–1447, © American Physical Society)

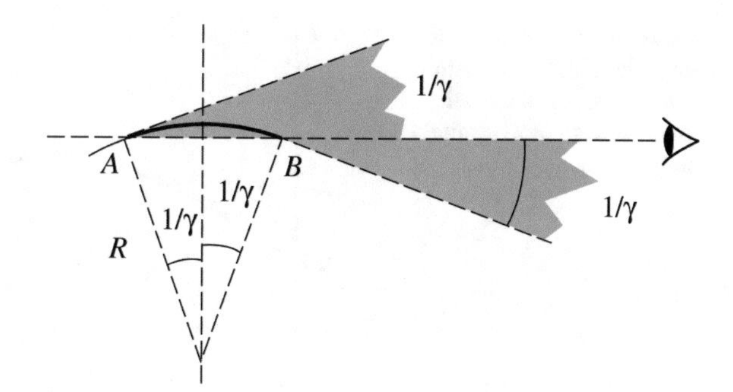

Figure 5.16 The geometry involved in synchrotron radiation emission.

equivalence of mass and energy through the famous equation $E = mc^2$. The electron energy at rest is calculated in $m_0c^2 = 0.5$ MeV. At ALS the energy of every circling electron equals 1.5 GeV, corresponding to a value of about 3000 for γ.

The total power radiated by a single particle, that is the energy it emits per unit time as electromagnetic radiation, increases as the fourth power of E and decreases as the fourth power of its rest mass. This last property explains why electrons are used and protons, which have a much larger mass, are not.

Finally, if the motion of the accelerated electrons were to be rectilinear instead of circular, then the power emitted would become $1/\gamma^2$ smaller. This explains the shape of the machines (Jackson [1975]).

To describe the spectrum of the radiation, we shall assume for simplicity to have only one electron traveling in the ring. The narrow cone of radiation, emitted along the instantaneous velocity vector, implies that the radiation will be visible or detectable only when the particle velocity is directed toward the observer (detector). In Figure 5.16, the trajectory of the emission — as far as the fixed observer is concerned — is the arc AB whose length is $d = 2R/\gamma$ and $2/\gamma$ is the measure in radians of the characteristic angle of emission. R denotes the local radius of curvature.

The electron describes the trajectory AB during an interval of time of length

$$\Delta t = d/v = 2R/\gamma v.$$

The duration of a burst of radiation, or pulse, detected by the observer will now be estimated.

The small arc AB can be approximated by the segment AB. In time Δt the front edge of the pulse travels a distance $D = c\Delta t$, since it is moving at the velocity of light. In the same time Δt the electron, moving at speed v, travels in the same direction the distance d calculated above. Therefore the rear edge of the burst, which ends the burst, is at the distance $D - d$ behind the front edge. This is the pulse length in space, which makes

$$\frac{D-d}{c}$$

the pulse duration in time.

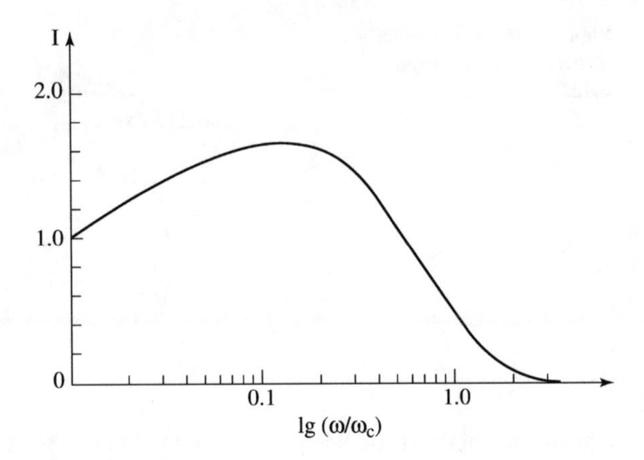

Figure 5.17 Spectrum of a single burst. The ordinate gives the spectral intensity (number of photons emitted per second).

Since $\beta = \sqrt{1 - 1/\gamma^2}$ and $1/\gamma^2$ is close to zero, by series expansion $\beta \cong 1 - \frac{1}{2}\gamma^{-2}$ which makes $1/\beta \cong 1 + \frac{1}{2}\gamma^{-2}$. Thus the pulse duration is

$$\frac{D-d}{c} = \frac{2R}{\gamma c}(\frac{1}{\beta} - 1) \cong \frac{R}{\gamma^3 c}.$$

Such an extremely short-lived pulse has a wide range of frequencies, extending essentially up to the critical frequency $\omega_c \cong \gamma^3 c/R$ (Fig. 5.17) by the indeterminacy principle (2.17). Since the electron keeps circulating, the burst will repeat every $2\pi R/c$ seconds (Fig. 5.18). In other words, the signal detected by the fixed observer is periodic with period $2\pi R/c$. Its Fourier transform is therefore discrete with fundamental frequency $\omega_0 = c/2\pi R$. The spectral intensity is that in Figure 5.17 sampled at $\omega = n\omega_0,\ n \geq 1$. Hence, the number of relevant harmonics is approximately $\omega_c/\omega_0 \cong \gamma^3$.

In the case of the ESRF in Grenoble $2R = 300m \cong 10^4$ cm, making the period of the order of 10^{-6}s, and $\omega_0 \cong 1$ MHz. Moreover $\gamma \cong 10^4$ so that $\gamma^3 \cong 10^{12}$. Hence

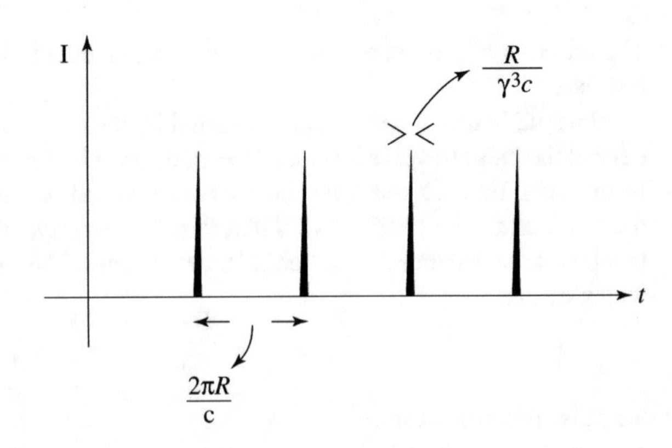

Figure 5.18 The short bursts of detected radiation.

the number of harmonics is enormous, with the critical frequency $\omega_c \cong 10^{12}$ MHz well into the region of x-rays.

Of the large spectrum, which for all practical purposes can be regarded as continuous, only the radiation from the far-ultraviolet to x-rays is normally used since even brighter visible beams can be obtained by the more readily available lasers.

5.10 Brightness and undulators

The brightness of the radiation, an essential parameter for experiments, underwent an exponential increase over the years. Its technical definition makes it proportional to the number of photons emitted per second, in the given bandwidth, and inversely proportional both to the cross sectional area the radiation originates from and to the angle of emission. Thus brightness, measured in photons $s^{-1}mm^{-2}mrad^{-2}Hz^{-1}$, takes into account not only the intensity of the radiation but also its concentration. Indeed it makes a big difference if a certain number of photons (per second) is radiated in all directions, as the sun does, or is concentrated in a small cone of directions as in the case of the synchrotron radiation.

The number of photons being large and the other two quantities of the definition mentioned above being small, the synchrotron radiation has extremely high brightness, millions of times higher than any other available source in the same range. Storage rings of third generation allowed a dramatic increase of brightness throughout the use of insertion devices called undulators.

Made by two rows of permanent magnets, *undulators* create magnetic fields perpendicular to the electron beam and alternating in polarity many times over a length of one or two meters, in the straight sections of the storage ring. There the electron path bends up and down like a sinusoid whose local radius of curvature changes continuously. As the electrons swerve in their path they emit radiation that will be reinforced in case of constructive interference and cancel out otherwise, depending on the wavelength of the radiation. Hence the spectrum reduces to a certain frequency and the first few harmonics. This is the price for enhancing brightness. The point is that the frequency to be enhanced can be chosen, predetermined, for a given undulator and a given energy by adjusting the magnetic field, that is the distance between the upper and lower row of magnets. Then the desired harmonic can be selected by the use of a monochromator. Hence radiation with a high degree of coherence is made available.

Undulators were studied already during the 1950s both in the United States and in the Soviet Union. They were not implemented for a long time because the electromagnets, going with them, could hardly be stacked in the necessarily short spatial period required and moreover would quickly run into coil-heating problems.

In 1979, K. Halbach of the Lawrence Berkeley Laboratory overcame the problem rather simply and economically by using arrays of permanent magnets, made from rare earth elements and cobalt, which became available at about that time.

In 1980, the first undulator to be used as an x-ray source was installed at Stanford and another one at about the same time in Novosibirsk. The brightness produced by an undulator increases as the square of the number of the undulator periods. Thus 100 periods result in a four order of magnitude enhancement in brightness.

5.11 Applications and the futuristic *free electron laser*

Replacing conventional x-ray angiography in the diagnosis of circulatory diseases is the most direct application of synchrotron radiation in medicine. Soft tissues are poor absorbers of x-rays and moreover detailed images of the coronary arteries have to be taken necessarily while the heart is beating. Hence the need for a contrast agent in the region of the heart. In conventional angiography this is achieved by injecting the contrast agent directly into the arteries. The whole procedure is rather risky.

The tunability of synchrotron radiation allows one to choose frequencies just above and just below the given absorption threshold of the contrast agent. The subtraction of the intensities of emerging x-rays increases the contrast and reduces the need of the contrast agent to that obtainable by a peripheral venous injection. In the mid-1980s, tests were performed at Stanford and since then several thousand patients have taken advantage of this technique.

Other x-ray imaging techniques have been advanced and experimented with. Indeed standard radiographs — an amplitude-contrast imaging — even though adequate in many cases, produce images of a rather rough nature since they allow one to distinguish only between tissues of significantly different density, such as bones and soft tissues.

The high degree of coherence of the synchrotron radiation is exploited in the technique called "phase-contrast imaging." This method has produced remarkably fine images independent of density, which might be about uniform and extremely low as in Figure 5.19.

X-rays can only be slightly deflected since their refractive indexes in different media differ by very little and are almost 1. Hence passing through a sample they get out of step, only just a bit, with respect to the undeflected rays passing outside the sample. Nevertheless, the radiation being highly coherent, this small phase difference suffices to produce at a large distance, compared to the dimensions of the experiment, Fraunhofer interference fringes. An appropriate algorithm, implemented on a computer from the data made available by the fringes, reconstructs the image that is the contours of the regions with different refractive indexes. By taking sets of data of the sample exposed from many different angles, it is possible to reconstruct a slice of it as in the clinical scans of computerized tomography.

Of the many other applications in pure and applied research in physics, chemistry, biology, and medicine some will be presented in later chapters, such as an experimental type of tomography called microtomography (Section 7.8). Crystallographic x-ray diffraction too takes advantage of the high brightness and high collimation of the synchrotron radiation — as well as the wide spectrum of frequencies among

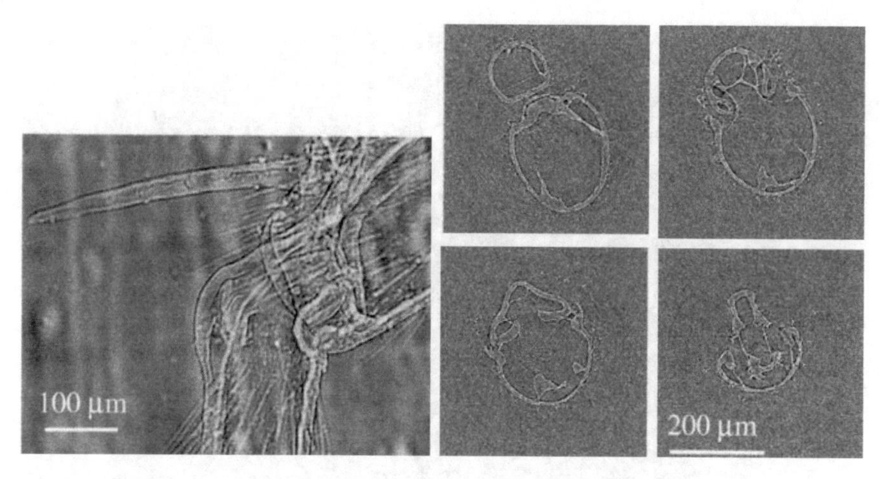

Figure 5.19 A mosquito knee and, to the right, cross sections of it obtained by the technique of phase-contrast imaging. (*Courtesy of A. Snigirev, ESRF*)

which to choose the desired one — to reconstruct the spatial structure at the atomic level of important molecules such as proteins and, following their changes in structure nanosecond by nanosecond, a glimpse of the way they perform their complex biological functions (Section 6.9).

Synchrotron radiation has technological applications too in micromechanics (Schmidt et al. [1996]) and microelectronics, among others. In microfabrication technologies (Schmidt et al. [1996]), synchrotron radiation opens up a completely new range of freedom in terms of shape, material, and production processes by using an advanced type of photolithography called *x-ray lithography* (Fig. 5.20). With miniaturization working for them engineers are already at work building extremely small robots — with all their levers, gears, and correspondingly small electrical motors and electronic controls — that could unclog arteries or fix microcircuitry in spots not otherwise accessible. While users keep profiting from the very special features of this radiation, synchrotron planners are taken by a new goal, the "fourth generation." In fact, present sources are far from fundamental limits. Experiments are in progress to increase coherence and brightness by several orders of magnitude so as to realize x-ray beams comparable in these respects to laser beams, that is complete coherence. Two important projects are taking place, one in Hamburg (Germany) and the other at Stanford (USA). The technical difficulties to be overcome are great but if success greets this project — for the so-called *free electron* laser — completely coherent x-rays will be available and moreover with a peak brilliance exceeding by 10^7 the third generation sources. It will be the x-ray laser which — aside from its military interest — will make possible experiments that presently can only be imagined.

Figure 5.20 Above, a micro gear wheel for a micro gear unit, and below, a planetary gear system. (*Courtesy of the Institut für Mikrotechnik Mainz*)

Chapter 6
X-ray Crystallography: Protein Structure and DNA

6.1 Crystallography from Steno to von Laue

The world contains a large number of chemically distinct substances, some biological such as DNA and enzymes, and others geological such as mica or quartz. There are millions of compounds in organic chemistry and thousands of metal alloys of interest to metallurgists.

Substances occur in either the solid, liquid, or gaseous state. Solids find their representatives in crystals, which have all molecules arranged in structures that are totally ordered as far as spatial distribution, but also include amorphous substances that do not exhibit any order in the spatial arrangement of the molecules and even show intermediate cases such as fibers and polycrystals.

The phenomenon of diffraction of visible light, illustrated by the images of the two-dimensional apertures in Figures 5.3, 5.5, 5.6, 5.7, and 5.8, is used in crystallography as a powerful tool to investigate the three-dimensional structure of matter. In this case, a radiation of much shorter wavelength is needed as well as a model, based on the Fourier transform, to interpret the experimental data.

Since ancient times crystals have been a fascinating enigma. That crystals can grow was noticed and ascribed, by analogy with animals, to food taken in and assimilated to their substance. Only with time and recorded observations has the distinction between growth of crystals and growth of living organisms become established knowledge.

Initially it was attempted to classify crystals according to their shape, which is not a simple matter because even crystals of the same compound can have different shapes depending on the development of their faces (Fig. 6.1). The first basic observation was made by the Danish anatomist and geologist Nicolaus Steno (1638–1686). When cutting quartz crystals, he noticed that it was possible to arrange crystals in space so that, certain (*corresponding*) faces being set parallel, the solid angles between them were constant (Fig. 6.1). The same he observed in hematite crystals and published these observations in 1669 in the treatise *De solido intra solidum naturaliter contento dissertationis prodromus* ("Introduction to a Dissertation Concerning Solids

E. Prestini, *The Evolution of Applied Harmonic Analysis*,
Applied and Numerical Harmonic Analysis,
DOI 10.1007/978-1-4899-7989-6_6

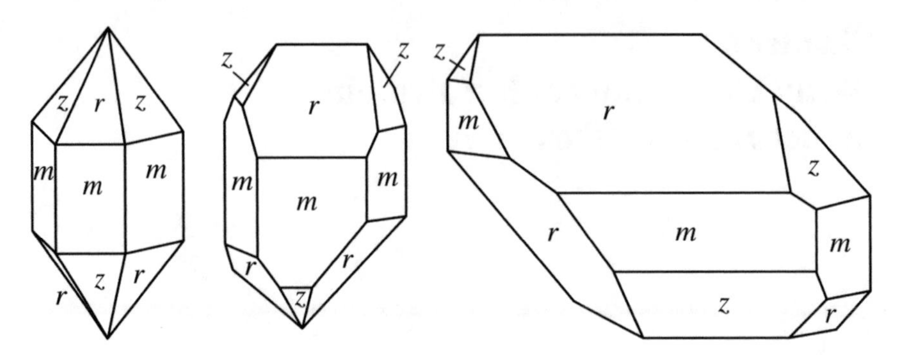

Figure 6.1 Three quartz crystals with corresponding faces developed differently. (B. K. Vainshtein, *Modern Crystallography*, Springer, [1981].)

Naturally Contained Within Solids"), mainly dedicated to his fundamental geological studies. In the treatise, discussions are only outlined, but new insights can be found in almost every paragraph. Guided by his observations—he had even explored alpine grottoes at Lake Garda and Lake Como in Northern Italy—he concluded that fossils were the remains of ancient living organisms and that many rocks were the result of sedimentation, and contrary to the belief that mountains grow like trees, he thought they were due to alterations of the Earth's crust.

In the end, his deeply religious nature and ethical character led him to set aside his revolutionary and controversial ideas and to abandon science for religion. Converted from Lutheranism to Catholicism in 1667, he was made bishop in 1677 and appointed apostolic vicar of northern Germany and Scandinavia. The general validity of the observations, made by Steno for quartz and hematite, was recognized about a century later in 1783 by the French scientist Jean Baptiste Louis Romé de l'Isle (1736–1790) who measured many different crystal species (Romé [1783]). It is the law of *constancy of interfacial angles*. The angles between corresponding faces are indeed characteristic of the substance so that measuring them on a crystal of unknown material might allow identification by comparison with records compiled for known materials.

About the same time the idea took root that the regularity in the exterior appearance of crystals was a manifestation at the macroscopic level of an inner order. Indeed a second fundamental observation was made by the French mineralogist, the Abbé René-Just Haüy (1743–1822), often called the father of crystallography. While teaching in Paris at the Collège Cardinal Lemoine he became interested in mineralogy. By examining the fragments of a piece of calcite accidentally broken, he observed that they were in the shape of rhombohedra and cleaved themselves into smaller and smaller rhombohedra. He concluded that calcite crystals are made by infinitesimal elementary figures of that shape and in his *Essai d'une théorie sur la structure des crystaux* of [1784] he laid the foundation of the theory of crystal structure. Then, in his main work *Traité de minéralogie* of 1801 in four volumes, he applied his crystal theory to mineralogical classification. He also became involved in a controversy with Berthollet who believed compounds to have a variable composition, while Haüy recognized that heterogeneous materials might mix with the compound.

With this new understanding, the picture of crystal growth changed: It was from the outside by deposition of little identical building blocks, side by side and one on top of the other. Only the extremely small dimensions of the blocks—it was thought—was on the way to detect the stair-step nature of the plane face of crystals. The existence of a concrete polyhedron, filling completely the crystal by repetition, nevertheless turned out to be untenable. It made it difficult to explain the phenomena of elastic compression and thermal expansion, and with the advent of atomic and molecular theories it was definitely dropped.

A weaker version of that remained, namely there exists a three-dimensional point system (*crystal lattice*)—identified by the vertices of a unit parallelepiped (*unit cell*) periodically repeated—that, when superimposed on the crystal structure, realizes the periodicity of it. So the unit cell is an abstraction, extremely useful though since it captures the three-dimensional periodic structure of ideal crystals, those extending infinitely in three dimensions, containing no impurities and showing no defects. Knowing a crystal is then reduced to knowing its unit cell.

The unit cell is required to be the *smallest* block containing arrangements of molecules or groups of atoms that, by simple repetition, yield the structure and properties of the crystal. While the crystal lattice is unequivocally identified by the distance between neighboring lattice points along the three lattice axes, there are several candidates for the unit cell, all with the same volume (Fig. 6.2). The choice of the unit cell is made in order to take full advantage of symmetries of the crystal under consideration and to simplify the mathematical calculations. The lengths of the sides of the unit cell are called *periodicities* of the crystal lattice. These are of the order of magnitude of the angstrom (10^{-8} cm) but can be even bigger than 100 angstroms, as in many protein crystals. Within the unit cell the position of each atom is identified by one point. In the simplest crystals—such as copper, silver, and gold—the unit cell contains a single atom, but it might contain up to about one hundred atoms in inorganic crystals and several tens of thousands in protein crystals.

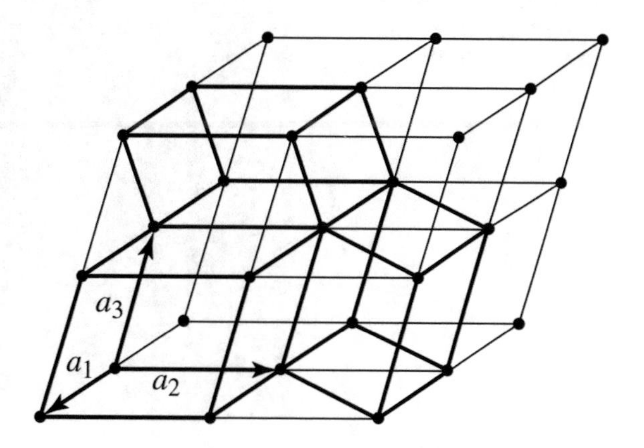

Figure 6.2 Example of a periodic lattice in three dimensions with three choices for the unit cell. (B. K. Vainshtein, *Modern Crystallography*, Springer, [1981].)

A better picture can be obtained by thinking of an electronic cloud, surrounding the nucleus of every atom of the crystal, quantitatively described by an *electron density* function $n(\mathbf{x})$ whose value at every vector position $\mathbf{x}$ is the number of electrons per unit volume and whose maxima correspond to the positions of the atomic nuclei.

The existence of crystal lattices was definitely proved only in the spring of 1912. The German physicist Max von Laue (1879–1960) had a familiarity with and preference for optical problems, manifested also in his doctoral dissertation on the theory of interference in plane parallel plates, directed by Max Planck at the University of Berlin. His optical intuition led him to the crucial idea of sending x-rays through crystals. Laue, then working at the Institute of Theoretical Physics in Munich, argued that if the supposition that the radiation just discovered by Roentgen in 1895 consisted of very short electromagnetic waves, and if the old standing supposition of the regular structure of atoms in a crystal were both true, then x-rays passing through a crystal should cause interference phenomena such as those observed and studied by Fraunhofer using light and diffraction gratings (Fig. 6.3).

A young faculty member in the person of Walter Friedrich—assistant to the Director Arnold Sommerfeld—and Paul Knipping, a doctoral candidate, began experiments in late April. On the second attempt, regularly ordered isolated points appeared on a photographic plate placed behind the crystal (copper sulfate) on the other side with respect to the impinging beam, proving beyond any doubt that x-rays were diffracted out of the primary beam. By May 4 a letter was sent to the Bavarian Academy of Sciences by Laue, Friedrich, and Knipping announcing their result, which proved simultaneously the wave nature of x-rays and the periodic arrangement of atoms

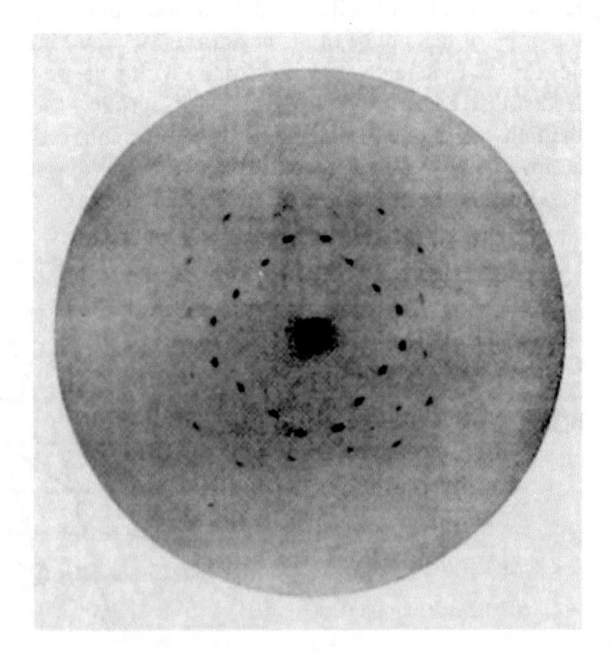

Figure 6.3 One of the first diffraction pictures (zinc blende). The large central spot is due to the incident ray. (W. Friedrich, P. Knipping, and M. Laue, [1913], "Interferenzerscheinungen bei Rontgenstrahlen," *Ann. Physik* **41** (5), pp. 971–988.)

within a crystal. For this work, which opened up in physics and chemistry the new branch of structure of matter, Laue received the Nobel prize for physics in 1914.

Up to then the great part of solid matter was considered amorphous, crystals looking like a rare exception among solids in nature. In the years following Laue's discovery it was found that amorphous solids are few. Materials like rocks, metals, alloys, and cement are crystals or polycrystals. Wood, textiles as well as bones, hair, and muscular fibers have molecules showing preferential orientations. The term "amorphous solid" is therefore reserved for those few substances like certain glasses, resins, and polymers that show no order in their spatial structure. Most solids we deal with show little evidence of their crystalline nature, being composed of an aggregate of minuscule crystals whose reciprocal orientation is random (*polycrystals*).

Before Laue's discovery, crystals were studied almost exclusively with optical instruments such as optical goniometers and microscopes. After 1912, crystals began to be studied at the atomic level with x-rays and later with electronic microscopes. Modern electronic microscopes can penetrate specimens up to 10^{-5} cm thick and achieve enlargements of up to 1 million (Fig. 6.4). Nevertheless, x-ray diffraction is the most commonly used technique. Interacting with the atoms of crystals, x-rays originate a *diffraction spectrum*. This is due to their wavelengths being comparable to the periodicities of the crystal lattice (tenths of an angstrom) and to the distances between chemically bonded atoms (one or two angstroms). Radiation of much longer wavelength, such as solar light (Fig. 6.5), cannot resolve the structural details on either the atomic scale or on the lattice scale. On the other hand, radiation of much smaller wavelengths, such as gamma rays, is diffracted at angles that are too small as codified by Bragg's law. While electron microscopy is the most appropriate tool

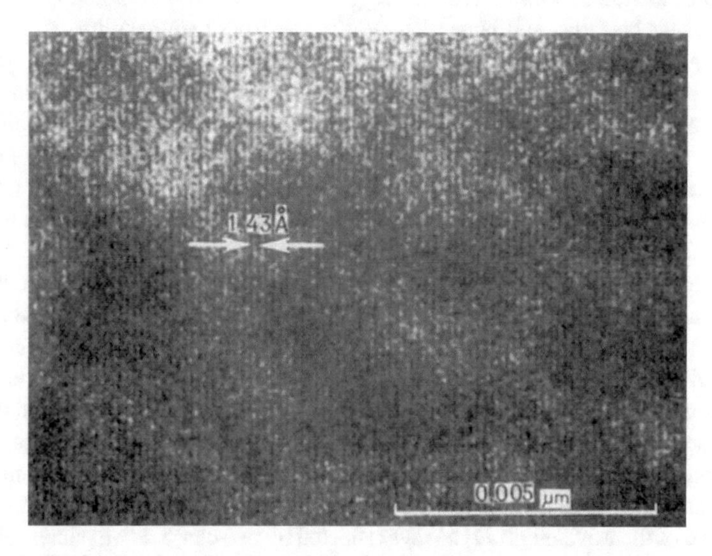

Figure 6.4 The image of a gold crystal obtained by an electron microscope. (B. K. Vainshtein, *Modern Crystallography*, Springer, [1981].)

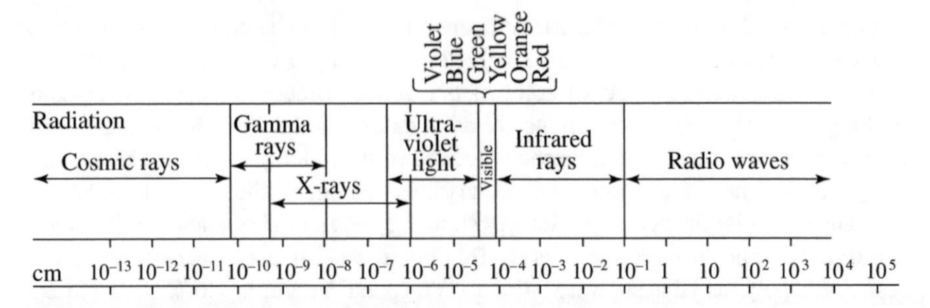

Figure 6.5 The wavelengths of the radiations of the electromagnetic spectrum.

for detecting possible defects in crystals (especially metal alloys), such as incorrect lining up of adjacent grains and missing or extra atoms, x-ray diffraction usually remains the means of choice for the determination of the crystal structure of chemical compounds.

6.2 Bragg's law and the basic role of symmetries

Laue's team experienced some difficulties in the detailed interpretation of the results of their experiments. For example, they could not determine which lattice planes produced which spots on the photographic plate. In the meantime in Cambridge (U.K.) the Australian born Lawrence Bragg (1890–1971) was completing his studies in physics and his father William (1862–1942) was experimenting with gas ionization by means of x-rays. Without any doubt, the favorable scientific situation, combined with the special knowledge in his family, played an important role in the discovery by the younger Bragg. In late 1912, he gave a more detailed quantitative explanation of the phenomenon just observed by Laue and the following year solved the structure of rock salt NaCl as well as diamond, which were the first complete crystal structure determinations ever made (Ewald [1962]). Lawrence Bragg was awarded, together with his father, the Nobel prize for physics in 1915.

The atoms of a crystal, being arranged in a regular periodic pattern, can be thought of as lying on families of parallel and equidistant planes, similar to the numerous lines along which the trees of an ordered orchard appear to be arranged. Every plane of a fixed family reflects a small part of the incident radiation through the electrons of its atoms, while the greater part goes through and gets reflected over and over by more inner planes of the family. The reflected rays show a difference in phase, having gone along different paths (Fig. 6.6). Of all possibilities there are two extreme ones. Constructive interference in case the path difference ABC in Figure 6.6 is an integer multiple n of the wavelength λ: the rays reflected by two consecutive planes reinforce each other. Destructive interference if the path difference is an integer multiple of

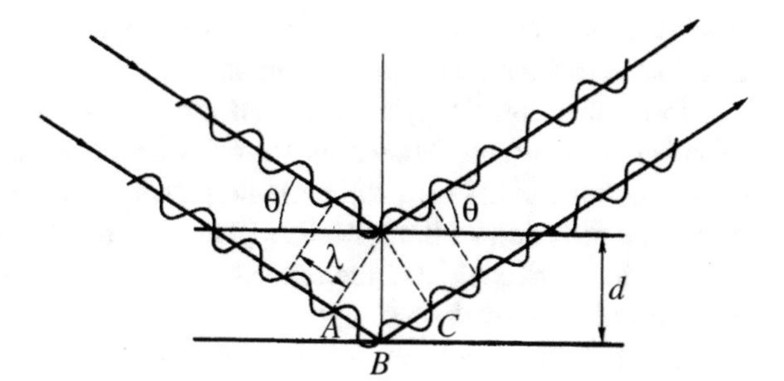

Figure 6.6 Incident rays and rays reflected by parallel atomic planes.

half wavelength: the reflected rays annihilate each other (Fig. 6.7) and no radiation is detected. The remaining directions show intermediate cases.

If d is the interplanar distance and 2θ the angle between the direction of the incident rays and the chosen direction of observation then by simple trigonometry $AB = BC = d \sin \theta$. Hence Bragg's law for constructive interference is

$$2d \sin \theta = n\lambda \tag{6.1}$$

It is right away observed that it requires $\lambda \leq 2d$ for diffraction to take place. Since the distance between adjacent lattice planes in a crystal ranges from a few angstroms

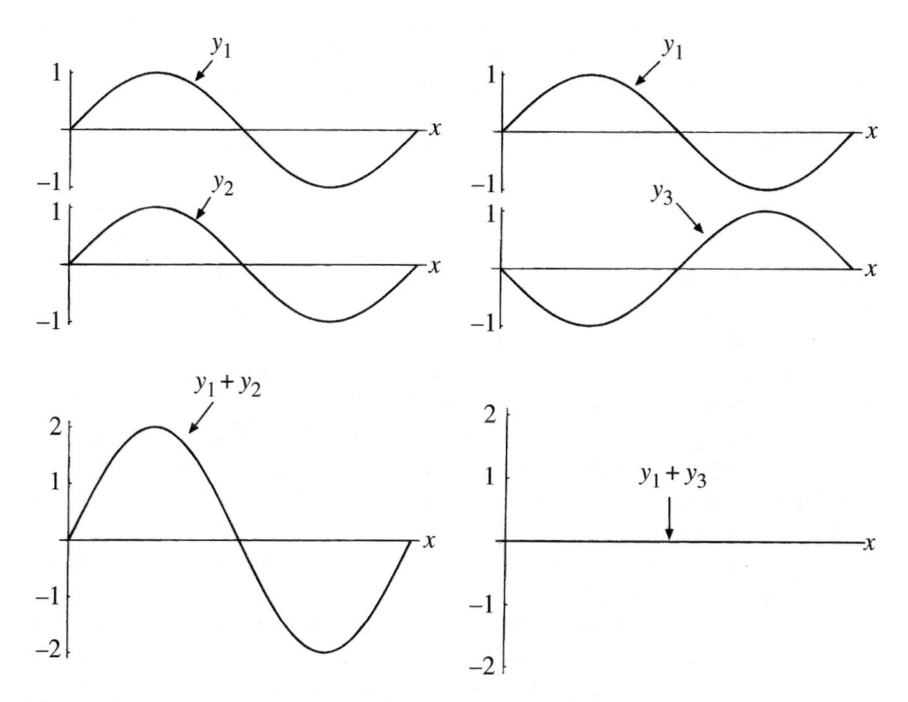

Figure 6.7 Constructive interference (left) and destructive interference (right).

to a few tens of angstroms, this explains why neither visible nor ultraviolet light can give rise to diffraction phenomena in crystals (Fig. 6.5).

Most important, Bragg's law relates the directions of maximum intensity of the reflected waves (made known by the experiment) to the known wavelength of the chosen x-rays and to the (unknown) interplanar distance. One single reflection determines the direction of the reflecting planes—since the normal to them is coplanar with the incident beam and the reflected one—but it does not allow separate determination of d and n. Knowledge of all reflections—in the plane determined by the incident and reflected beam just mentioned—does, since the minimum value of θ, with the exclusion of $\theta = 0$, which is the incident ray itself, corresponds to the first reflection ($n = 1$). In a direction θ satisfying Bragg's law the reflected waves add to each other in a resulting wave that at infinity, that is at a great distance with respect to the dimensions of the crystal, gives rise to one of the spots on the photographic plate, or more modernly to one of the peaks of intensity detected by a counter. These are named *reflections* of first, second, third,... order depending upon $n = 1, 2, 3, \ldots$ the lattice planes, being bisectors of the angle 2θ between the incident rays and the reflected rays (Fig. 6.6), play the role of a reflecting mirror. From this the name of *Bragg reflections* is given to the peaks of intensity.

By changing the direction of the incident ray while maintaining the crystal fixed in the same position, or alternatively by rotating the crystal and maintaining unchanged the direction of the impinging x-rays (*method of the rotating crystal*), it is possible to identify experimentally all the geometrical settings (orientations of the crystal with respect with the incident beam) in which Bragg reflections occur and to measure the angle between the incident and the reflected beams. These data allow the determination of families of lattice planes and of the distance d between them, through Bragg's law.

In the rotating-crystal method, the crystal is mounted with one of its axes—or other important crystallographic direction found by inspection—normal to the monochromatic x-ray beam and with a cylindrical film is placed around it. The axis of the cylinder coincides with the chosen axis of the crystal which will also be the axis of rotation. As the crystal rotates, specific sets of planes will make the correct Bragg angle for reflection, at certain instants. The corresponding spots on the film are distributed with certain regularities and symmetries due to the suitable choice of the rotation axis and to the periodic structure of the crystal itself (Fig. 6.8). The crystal being rotated only around one axis, not every set of parallel planes meets the condition to produce a Bragg reflection. For instance those perpendicular to the rotation axis do not. By rotating the crystal about another axis the complete distribution of reflections is gradually discovered. In place of a film, a counter can be used and the experiment automated to keep a record of all the data involved.

Bragg's law allows the determination of the crystal lattice and unit cell of the crystalline compounds almost instantaneously.

The unit cell, having to fill up by repetition the entire space, leaving no voids and showing no overlaps, cannot have just any shape. Indeed the requirement of periodicity is so strong that there exists only one general type of unit cell, the triclinic, described in Figure 6.9 (the symbol $\neq$ means that equality is not required but it might

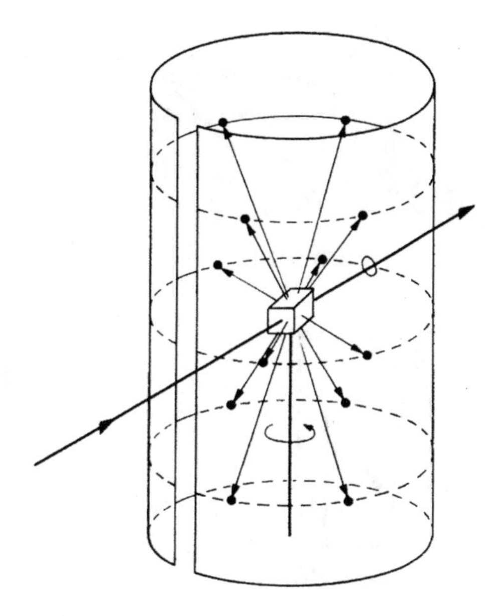

Figure 6.8 Rotating-crystal method. (B. D. Cullity, *Elements* of *XRay Diffraction*, Addison-Wesley, Addison-Wesley, [1978].)

occur). This type specializes into seven different ones. For instance, if all sides and all angles are equal, the unit cell is a cubic. Each type of cell comes with an array of symmetries, for the atomic distribution compatible with the periodicity. The case of no symmetry is included and indeed there are crystals with no symmetries whatsoever, but most have them.

Symmetries play an important role in crystallography and it is easy to understand why. For example, if a body is symmetric with respect to a plane, meaning that the plane cuts the body into two halves such that the reflection of one half produces the other one (mirror), then the whole body is completely determined once half of it is known. Hence knowledge of symmetries reduces the task of the determination of the electron density from the entire unit cell to a specific part of it.

Symmetries, compatible with periodicity of the crystal structures, are *rotation* around one axis of 60°, 90°, 120°, 180°, 360° (the identity)—having *order of rotation n* equal to 6, 4, 3, 2, 1 respectively—*inversion* with respect to a point and *rotoinversion*. Let us observe that reflection with respect to a plane (mirror) is a particular rotoinversion (Fig. 6.10). Each of these symmetries leave at least one point fixed. The set of all symmetries has the mathematical structure of a *group*, since the composition of any two of them, as well as the inverse of any of them, is still a symmetry. By 1890 the German mathematician Arthur Schönflies (1853–1928) had classified all symmetries that three-dimensional periodic structures can have—simultaneously with Evgraph Stepanovich von Fedorov (1853–1919) (Schönflies [1891], Fedorov [1885])—including those with a component of translation. These last kind of symmetries generate *plane glides* (mirror reflection followed by a translation in the plane of the mirror) and *screw axes* (rotation followed by a translation parallel

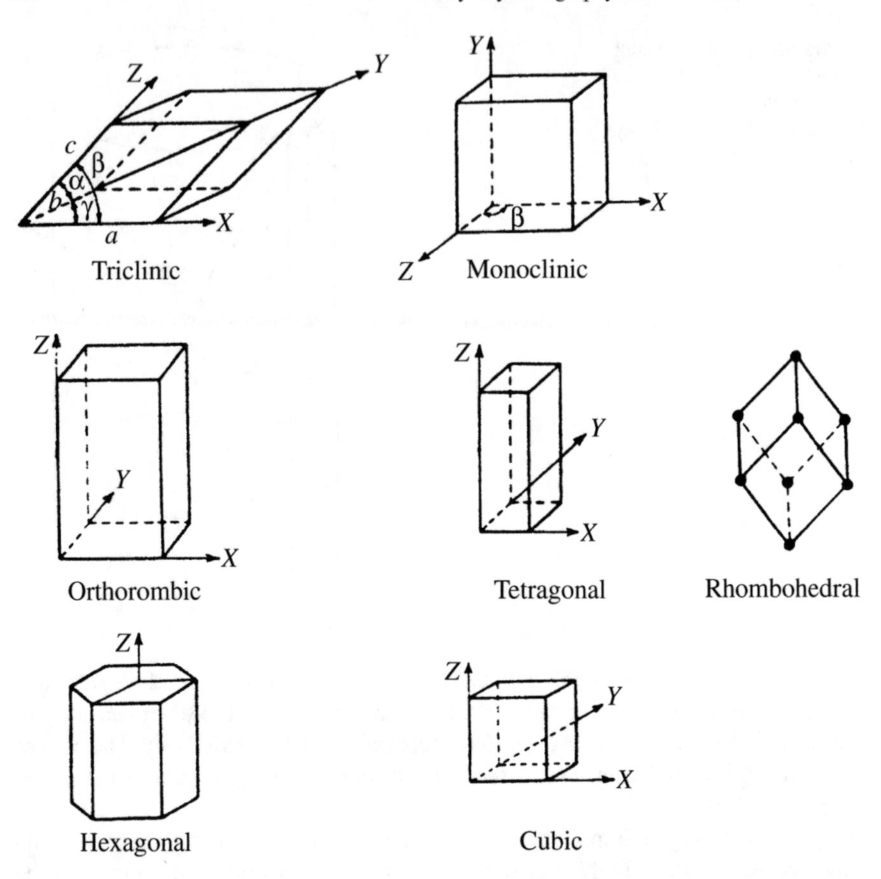

Figure 6.9 Types of unit cells. The sides $a,\ b,\ c$ and the angles $\alpha,\ \beta,\ \gamma$ are as shown. Triclinic: $a \neq b \neq c,\ \alpha \neq \beta \neq \gamma \neq 90^0$; Monoclinic: $a \neq b \neq c,\ \alpha = \gamma = 90^0,\ \beta \neq 90^0$; Orthorhombic: $a \neq b \neq c,\ \alpha = \beta = \gamma = 90^0$; Tetragonal: $a = b \neq c,\ \alpha = \beta = \gamma = 90^0$; Rhombohedral: $a = b = c,\ \alpha = \beta = \gamma \neq 90^0$; Hexagonal: $a = b \neq c,\ \alpha = \beta = 90^0,\ \gamma = 120^0$; Cubic: $a = b = c,\ \alpha = \beta = \gamma = 90^0$.

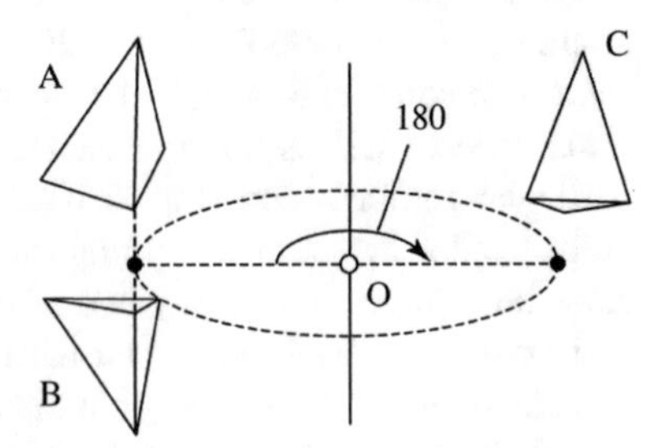

Figure 6.10 Mirror reflection. The asymmetrical tetrahedron B, which is the mirror reflection of A with respect to the horizontal plane, can also be obtained by a 180° rotation of A into C, followed by the inversion with respect to the origin O.

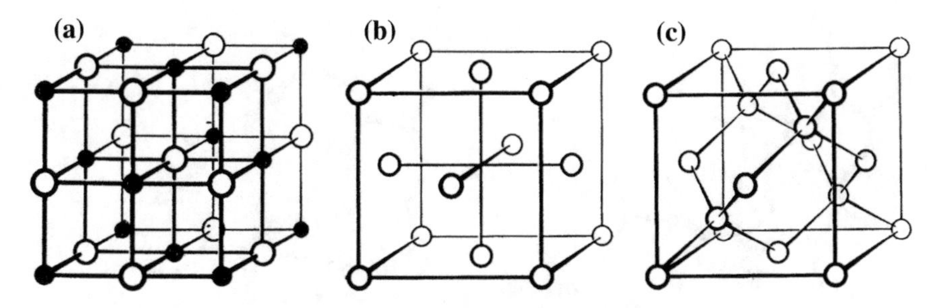

Figure 6.11 a) The simple and highly symmetrical structure of ordinary rock salt NaCl. A sodium atom is at the center of each face and at each vertex, an atom of chlorine is at the center of each side and at the center of the cube; b) structure of copper; c) structure of diamond. (B. K. Vainshtein, *Modern Crystallography*, Springer, [1981].)

to the axis of rotation). In both cases, the translation can only be by an integral fraction k/n of the lattice periodicities, where n is the order of rotation involved.

The work by Schönflies resulted in a list of groups of symmetries made by 230 entries, each of which is called a *space group*. Each type of unit cell is compatible only with a few space groups. There are only two space groups for the triclinic, 13 for the monoclinic, 59 for the orthorhombic, 68 for the tetragonal, 25 for the rhombohedral, 27 for the hexagonal, and 36 for the cubic (International Tables for x-ray Crystallography [1983], McLachlan [1957]). Meanwhile crystallographers found concrete examples of all that. This body of knowledge, in the case of simple structures, allows exclusion of all but a few possible configurations for the distribution of the atoms in the unit cell. By numerical simulation of the diffraction experiment and by comparison with the measured intensities of the Bragg reflections, a unique configuration can be selected. This is feasible for example with sodium chloride NaCl, ordinary rock salt (Fig. 6.11), but also with copper, diamond, fluorite CaF_2, pyrite FeS_2, and calcite $CaCO_3$. As the number of atoms in each unit cell becomes larger and the complexity of the structure increases (Fig. 6.12), this simple procedure becomes unwieldy and does not generate conclusive results. Some experimental work is always required: It is necessary to find the directions in which diffraction takes place and to record the intensity of the diffracted ray, but more powerful procedures must be devised to interpret the experimental data.

6.3 How to see the invisible

To find the electron distribution $n(\mathbf{x})$—whose maxima determine the position of the atomic nuclei—a mathematical model of the diffraction spectrum is needed since the diffracted radiation cannot be focused (but only slightly deviated) either with lenses, as in the optical microscope, or with magnetic fields as in the electron microscope.

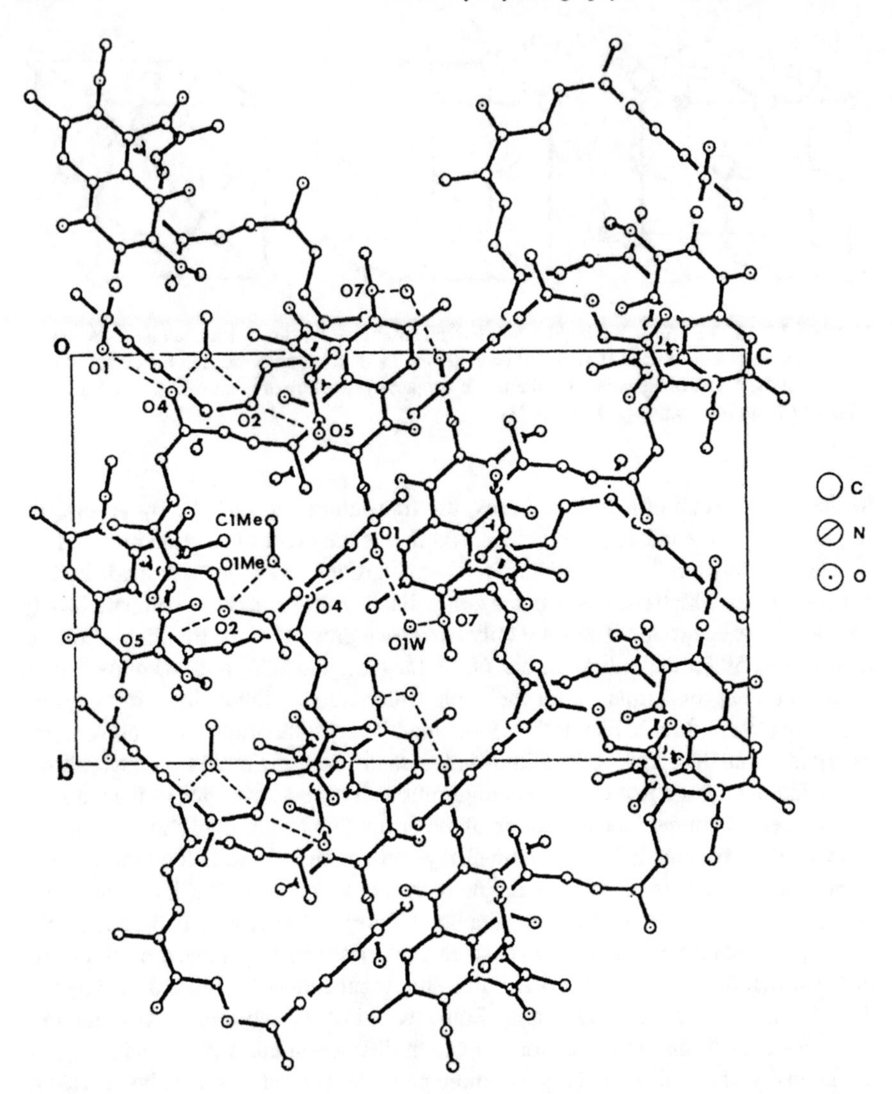

Figure 6.12 Projection on the (b, c) face of the primitive cell of an antibiotic. (Cellai L., Cerrini S., Lamba D. [1995], "30-Dechloro-30-methoxy-25-O-methyl-N-methylnaphthomicin A," *Acta Crystallographica*, Sect. C, pp. 2060–2064.) (*Courtesy of S. Cerrini, Istituto di Strutturistica Chimica, CNR*)

Such a model involves not only the directions of the diffracted rays but also the intensities, which are in an exact relation with the electron distribution.

In a coordinate system with the origin O set in one of the atoms of the crystal, the incident radiation of wavelength λ is expressed by

$$F(\mathbf{x}) = F_0 e^{i\mathbf{k}\cdot\mathbf{x}}$$

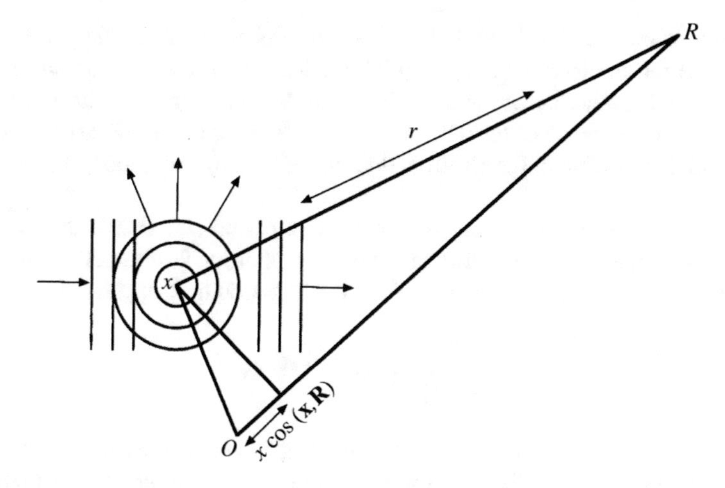

Figure 6.13 Radiation scattering by crystal electrons.

where $\mathbf{x}$ is a point of the crystal and $\mathbf{k}$ is a vector having the direction of the incident radiation and length equal to the *wave number* $k = 2\pi/\lambda$. Since the time evolution is of no interest, the time t has been set equal to zero and so it does not appear. An electron placed at $\mathbf{x}$ scatters some of the incident radiation, as shown in Figure 6.13. At a point $\mathbf{R}$, placed outside the crystal at distance r from $\mathbf{x}$, the scattered radiation is expressed by $F_0 e^{i\mathbf{k}\cdot\mathbf{x}} e^{ikr}/r$. The phase $\mathbf{k}\cdot\mathbf{x} + kr$ is therefore a function of both the phase of the incident radiation and of the length r of the path after the scattering. If $\mathbf{R}$ is at a large distance, compared to the dimensions of the crystal, the approximation

$$r \cong R - x\cos(\mathbf{x},\ \mathbf{R})$$

holds so that

$$kr \cong kR - kx\cos(x,\ \mathbf{R}).$$

Denoting by $\mathbf{k}'$ a vector with the same direction as $\mathbf{R}$ and same length as $\mathbf{k}$, the relation

$$kx\cos(\mathbf{x},\ \mathbf{R}) = \mathbf{k}'\cdot\mathbf{x}$$

holds and consequently

$$(\mathbf{k}\cdot\mathbf{x} + kr) = \ (\mathbf{k} - \mathbf{k}')\cdot\mathbf{x} + kR.$$

The scattering is caused by the electrons, not by the nuclei of the atoms in the crystal. The oscillating electric field of the incident x-rays interacts with all charged particles. In particular it sets into oscillation about its mean position every electron of the crystal. An oscillating electron keeps accelerating and decelerating, hence it emits an electromagnetic wave (Section 5.8) that has the same wavelength and frequency of the incident beam but is changed in phase by $\lambda/2$ (the phase change is the same for all electrons in the crystal and leads to a term of modulus 1 that is dropped in the mathematical model). The beam scattered by an electron is just the beam radiated by the electron. The nucleus too is charged, but its extremely large mass, compared to that of the electron, does not allow any oscillation of appreciable extent.

The scattering by an atom being due only to its electrons, it is reasonable to assume that the amplitude of the scattered radiation is proportional to the electron density $n(\mathbf{x})$. By making the approximation $r \cong R$, the radiation in $\mathbf{R}$ is given by

$$\frac{e^{ikR}}{R} \int_V n(\mathbf{x}) e^{-i(\mathbf{k}'-\mathbf{k})\cdot\mathbf{x}} d\mathbf{x} \tag{6.2}$$

where the integral is taken over the volume V of the crystal. Since $\mathbf{k}$ and $\mathbf{k}'$ have equal length but different direction (of the incident ray and of the scattered radiation respectively), the vector $\mathbf{K} = \mathbf{k}' - \mathbf{k}$ measures the change of direction of the incident ray, due to the blow of the reflection. With this notation and dropping the constant factor e^{ikR}/R, (6.2) becomes

$$\int_V n(\mathbf{x}) e^{-i\mathbf{K}\cdot\mathbf{x}} d\mathbf{x} \tag{6.3}$$

which is recognized to be $\hat{n}(\mathbf{K})$, the Fourier transform of the electron density (Kittel [1971], Stout, Jensen [1968]).

This very same mathematical model works in general, for liquids and gases as well. It is natural to inquire why the Fourier image of a crystal contains hundreds or even thousands of isolated peaks (Bragg reflections), while a liquid has only a few and a monoatomic gas none at all (see the gaussian curve in Figure 6.14). It all depends on the electron density function. In a crystal it is periodic. So its spectrum is discrete and periodic (Section 3.4), itself defining a lattice named a *reciprocal lattice*.

If the dimensions of the unit cell—the so-called *lattice parameters*—are $a,\ b,\ c$, then in the domain of $\hat{n}$, the reciprocal lattice has periodicities given by $2\pi a^{-1}$, $2\pi b^{-1}$, $2\pi c^{-1}$ by the indeterminacy principle (2.17). Therefore a large unit cell generates a diffraction spectrum denser than a small unit cell.

It may be observed that in the above model $\mathbf{K}$ has maximum length $4\pi/\lambda$, while the reciprocal lattice is made by an infinite number of points. Nevertheless it is not really indispensable to know the value of $\hat{n}$ on all of them. Indeed $\hat{n}$ tends to zero at infinity (Section 3.3) and moreover it can be set equal to zero outside a solid sphere of radius equal to the reciprocal of the dimension of the smallest detail to be rendered. This dimension is 1 to 1.7 Å if one wishes to "separate" atoms. Inside such a sphere, $\hat{n}$ is nonzero only at the points of the reciprocal lattice. Hence, to measure nonzero values of it, precise circumstances have to hold. One has to identify an incident beam $\mathbf{k}$ and a corresponding diffracted beam $\mathbf{k}'$ such that the difference $\mathbf{K}$ is a point of the reciprocal lattice. This means that the coordinates of $\mathbf{K}$ must be integer multiples of

Figure 6.14 Fourier images of a crystal, a liquid, and a monoatomic gas.

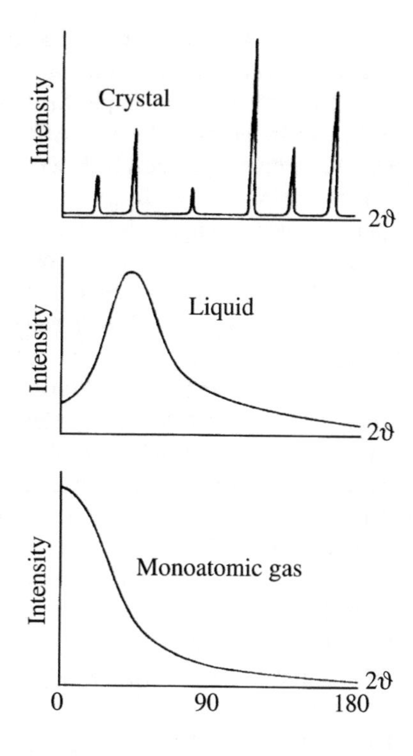

$2\pi a^{-1}, 2\pi b^{-1}, 2\pi c^{-1}$ Since $\mathbf{k}$ and $\mathbf{k}'$ have fixed length $2\pi/\lambda$, only their directions can vary. Usually the direction of the incident radiation $\mathbf{k}$ is varied by keeping fixed the incident beam and rotating the crystal (method of the rotating crystal).

6.4 The phase problem and the *direct methods*

The solution to the problem seems now readily provided by the inversion formula (2.14). However the Fourier transform $\hat{n}(\mathbf{K})$ is in general a complex function, defined by a couple of real-valued functions $|\hat{n}(\mathbf{K})|$ and $\phi(\mathbf{K})$, called modulus and argument (or phase) respectively, as indicated by the formula

$$\hat{n}(\mathbf{K}) = |\hat{n}(\mathbf{K})|e^{i\phi(\mathbf{K})}$$

The intensities of the diffracted x-rays can certainly be measured experimentally, being proportional to the blackening of the photographic plate or to the area beneath a peak of intensity such as measured by a counter. So $|\hat{n}(\mathbf{K})|$ is known. Unfortunately there is no experimental technique to measure the phase. Half of the information needed to reconstruct the crystalline structure is therefore unavailable. In this consists the *phase problem*, of central importance in crystallography.

Ascribing arbitrary values to the phase, one could reconstruct a myriad of functions $n(\mathbf{x})$ by Fourier inversion formula, all consistent with the measured intensities. Note that in the particularly simple case in which the choice is in between the + or − sign, corresponding to the phases 0° and 180°, the number of possible choices is 2^{10}, with ten atoms in the unit cell, and 2^{20}, with 20 atoms. The number of all possible combinations of signs is seen to grow exponentially with the number of atoms, quickly reaching huge numbers.

This seemingly unsolvable problem found its first positive results at the beginnings of the 1950s in the pioneering work, by the chemist Jerome Karle and the mathematician Herbert Hauptman, published in a series of papers in *Acta Crystallographica* in the years 1950, 1952, 1953, and 1954 (see also Karle-Hauptman [1953], McLachlan [1957]). With computers performing the many required calculations, their method could reduce the work of years to the work of days. The method, only gradually accepted by crystallographers because of the unfamiliar mathematical concepts and terminology, over time has been used to determine the structure of the small molecules of many hormones, vitamins, and antibiotics. The two American scientists shared the Nobel prize for chemistry in 1985.

The electronic density is a nonnegative function that takes on large values in the immediate vicinity of the centers of the atoms, while in between atoms it is very small. Moreover, for a given a molecule, the number of atoms (but not their positions) is usually a known quantity. Already from these general considerations it is apparent what severe restrictions the electron density must obey. Taking advantage of that and introducing methods belonging to the theory of probability, it is possible to assign phases probabilistically to the Bragg reflections by successive approximations starting with the reflections of highest intensity (Karl-Hauptman [1953], Lipson-Cochran [1966], Hauptman [1972], Giacovazzo et al. [1992]). These kinds of procedures are known as "direct methods."

The direct methods have made the determination of the structure of molecules with up to a few hundreds of atoms a matter of routine. Occasionally they have been used for molecules of several hundred or even a thousand atoms. However, as the number of atoms per molecule increases, the probability by which the phases are assigned to the Bragg reflections decreases. The problem of strengthening these techniques to solve complex biological structures, such as those of proteins and nucleic acids, is still open. The method of "heavy atoms" (Section 6.7) is an alternative procedure.

6.5 Brief history of genetics

The determination of the structure of genetic material by James Watson and Francis Crick [1953]—working at the Cavendish Laboratory in Cambridge (U.K.), Lawrence Bragg director—is certainly one of the most imaginative discoveries of the twentieth century and one of the salient points in the history of genetics.

The ability of living organisms to pass their qualities to offsprings—for instance hair color, a large nose, big lips—is so obvious that it was quickly noticed by early researchers. Back in human history, the little knowledge of heredity that was acquired led to the improvement of domestic animals and plants by inbreeding and crossbreeding, while legends flourished about bizarre crossbreeds.

The great physician Hippocrates attempted an explanation, but it was the philosopher Aristotle who got closer. Showing deep intuition Aristotle rejected the belief—ingrained in such common expressions as "half blood," "new blood," "blue blood"—of biological heredity as transmission from generation to generation of parts of the body. In his use of the word *dynamis* (power) can be traced the idea that heredity is due to transmission of information leading to the development of the embryo. In the twenty-three centuries that followed, the Aristotelian insight was abandoned without exception and more primitive ideas came back. Indeed heredity as a science did not really begin until the second half of the nineteenth century.

The first to make experiments serving significantly as a basis for a theory was the Abbot Gregor Mendel (1822–1884) of the Augustinian monastery in Brünn, then Austro-Hungarian territory. Mendel began by growing 32 different types of pea plants in the small garden of his monastery and studied them for two years, looking for clearly defined contrasting traits. He selected seven such characteristics: seed form smooth or wrinkled, seed color yellow or green, and so on. Then he proceeded with experimental crossing of species, keeping track of the chosen traits through the first, second, and successive generations of a species. His startling new idea was that exact rules could be found in the reproduction of living organisms.

Relying on a large sample of data, he was led to the conviction that hereditary characters are carried and passed on to subsequent generations by discrete factors, the *genes*, which Mendel called "elements." Furthermore he speculated that genes come in pairs, one inherited from the ovule, the other from the pollen. In Mendel's records was the evidence that inheritance is not a blending of characteristics—as it was thought—but consists in discrete units that are parceled out and reassorted in each generation.

It was 1865 when Mendel presented his conclusions at meetings of the National Science Society in Brünn—in which he used to take an active part—and the following year his article *Versuche über Pflanzenhybriden* (Experiments with Plant Hybrids) appeared in the transactions of the society. At least a hundred and twenty libraries, and surely all major ones, had it on their bookshelves but the article went unnoticed for the next 35 years.

Mendel was fully aware of having broken new ground, for no one had been able "to determine the number of different forms under which the offspring of hybrids appear, or to arrange these forms with certainty according to their separate generations, or definitely to ascertain their statistical relations." Yet the relevance and implications of his discoveries were not grasped. Also his methods were unusual, relying on mathematics. This might help to explain the poor reception.

Mendel's extraordinary discovery escaped even his great contemporary Charles Darwin (1809–1882) in spite of the difficulties met by his theory of evolution in fitting in with the accepted theory of blending inheritance, thus hindering universal acceptance of evolution. Indeed sexual reproduction would tend to rapidly produce uniformity making inexplicable resemblances that could skip one or several generations before suddenly reappearing.

In the years to follow, the studies of life underwent considerable changes that made them more receptive to Mendel's discoveries. Better microscopes and the use of stains to bring out subcellular characteristics led to a central area (the *nucleus*) being distinguished from the surrounding area (the *cytoplasm*) inside the cell, which is the fundamental unit of living organisms. Studies on plants and animal embryos made clear that cells are not formed by assembly of their constituents but by cellular division of the fertilized egg. It was then speculated that the unfertilized egg and the sperm contributed equally to hereditary characters. Since cytological analysis was showing that the egg and the sperm had nuclei of similar dimensions, it was natural to hold the nucleus responsible for heredity. Further studies revealed that inside the nucleus there are rodlike objects, varying with the species under consideration, grouped in homologous pairs (Fig. 6.15).

It was then thought that along these rods, called *chromosomes*, the hereditary units were lined up and it was observed that at the formation of egg and sperm the number of chromosomes splits in two, to recombine in the fertilized egg with half of the material coming from the maternal side and half from the paternal side. These experimental findings were pointing in the same direction as shown by Mendel. At

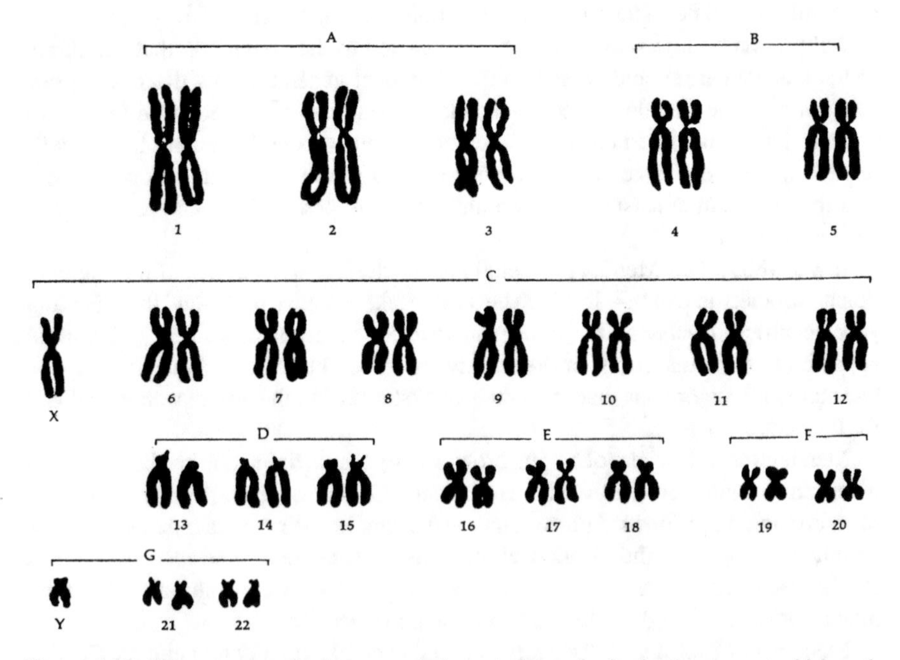

Figure 6.15 Chromosomes of a human male in metaphase of mitosis, the stage at which each chromosome consists of two identical chromatids that will separate in later phases. Morphologically similar chromosomes are grouped together, as shown. (*Pediatric Research Unit, Guy's Hospital Medical School, London*)

the beginning of 1900, three European botanists, writing independently about some experiments performed on peas or flowers similar to the ones carried out by Mendel, found Mendel's article while searching for the literature on the subject and gave this author the deserved credit. The theory of genes was later recognized to apply not only to plants but also to animals including man.

Genetic studies were further accelerated in 1910 by the experiments carried out by the American embryologist Thomas Morgan (1866–1945) on the drosophila. This small fruit fly had many advantages to offer being small, easy to raise, reaching maturity in only two weeks, and finally having only four pairs of chromosomes, compared to seven in the case of peas and 23 in man. Morgan's studies identified the chromosomes linked to sex—XX in the female and XY in the male—and backed the abstract entities in Mendel's work by concrete material visible under the microscope. Morgan showed that each gene could be assigned a precise chromosome region and therefore had a physical basis on the chromosome structure. For his discovery of the hereditary transmission mechanism in drosophila, he was awarded the Nobel prize for medicine in 1933.

It is worth pointing out that already from these initial results great advances followed in agriculture—allowing hybridization to be performed systematically—and in medicine where the genetic cause of some diseases started to be recognized. However the concept of the gene—introduced in 1909 by the Danish botanist Wilhelm Johannsen—remained an abstraction until the 1950s; nor did anyone understand the mechanisms through which genes control the physiology of the cells or through which they replicate themselves during cellular reproduction.

It was necessary to start again from the chromosomes, investigate their chemical composition, and show them to be composed of genetic material. Already in 1869, the same decade of Mendel's contribution, at the University of Tübingen the chemist of Swiss origin Friedrich Miescher had found in the nuclei of white blood cells a substance showing unique characteristics. It was called *nucleic acid* in 1874 when the same Miescher separated it into a protein and an acid molecule; it was later named deoxyribonucleic acid (DNA) to distinguish it from the closely related ribonucleic acid also isolated from cells. Then the DNA was shown to be present in all cells and located precisely in the chromosomes.

After being separated from the proteins, nucleic acids can be broken down into smaller units, the *nucleotides*. Each nucleotide in turn is composed by a sugar, called deoxyribose, a phosphate group attached to the sugar at position 5, and nitrogen groups attached to each sugar molecule at position 1 (Fig. 6.16). The nitrogen groups, called *bases*, are of four different types: thymine, cytosine, adenine, and guanine. Since each base characterizes the nucleotide that carries it, there are four different deoxyribose nucleotides in the DNA. Such a small number led to the assumption that the DNA molecule was a simple one, unable to sustain the complexity of biological heredity. Consequently DNA received little attention until 1940, when some biochemists pointed out that DNA is not composed by a sequence of four nucleotides that repeats itself unchanged, but it is a very long chain made by hundreds or even thousands of copies of the four nucleotides, arranged in sequences that differ according to the number of the nucleotides and the order they appear in

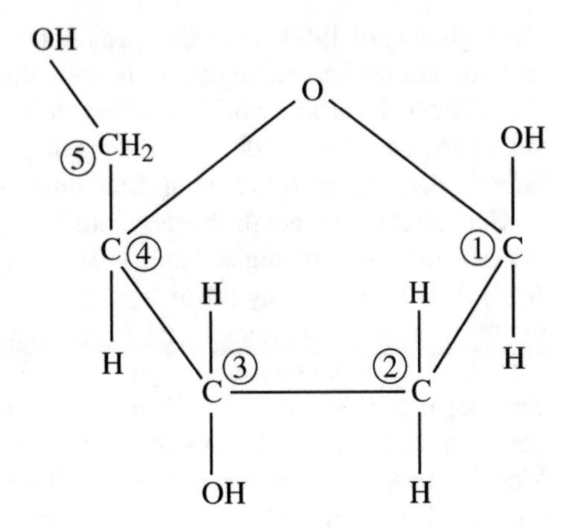

Figure 6.16 Chemical structure of deoxyribose. Nucleotides consist of deoxyribose, a phosphate group attached to the carbon atom at position 5, and a nitrogen bearing group attached at position 1 (single hydrogen atoms have been omitted).

depending on the species. The length and possible variations of DNA could indeed account for the great biological diversity. To give an example, in human beings the total number of nucleotides of the DNA of a single cell is more than three billion.

6.6 DNA through x-rays

Meanwhile some of Lawrence Bragg's students had perfected the crystallographic methods spearheaded by their teacher and were applying them already in the 1930s to determine the spatial structure of organic molecules of increasing complexity.

The first big success in this field was achieved in 1951 by Linus Pauling, who discovered the helical structure of an important protein (myoglobin). His work was based on the construction of models and on his formidable intuition more than on the analytical methods of crystallographers. At that stage it became reasonable to wonder about the spatial structure of the DNA molecule.

Three teams began working on the project: in Cambridge the 23-year-old James Watson, just arrived from the US—from the rigorous teaching of Salvador Luria—to specialize in biochemistry and the physicist Francis Crick always in ebullience but so far inconclusive; in London the experimentalist Maurice Wilkins, physicist, and Rosalind Franklin, crystallographer; at Pasadena, in California, the very same Pauling. Knowing that the genial Linus—"unquestionably the world's most astute chemist," in Watson's own words—was also working at the problem caused apprehension in the competing teams.

Pauling asked Wilkins to see the x-ray pictures of DNA that Franklin was obtaining in his laboratory. Wilkins took his time. Even so Pauling was first in writing up a paper for publication. His model was a three-chain helix with the sugar phosphate backbone in the center and the bases on the outside. And the unforeseeable happened, for the chemistry of the model was erroneous. Watson was to realize something was wrong with it as soon as he had the manuscript, for the location of the essential atoms carried a superficial resemblance to an aborted model he and Crick had pondered about fifteen months earlier. Watson and Crick were relieved and estimated it would take anywhere up to six weeks before Linus was back in full-time pursuit of DNA, calculating that as soon as the paper appeared the flaw in the chemistry of the model would be pointed out.

Watson went to London to inform his contacts there. Franklin had never believed in helices; in her mind not a shred of evidence permitted Linus, or anyone else, to postulate a helical structure. Then Watson talked to Wilkins and learned that Rosalind was about to leave the lab, that together with his assistant Wilkins was quietly duplicating her x-ray work, and most interesting that she had evidence of a new three-dimensional form of DNA, arising from an x-ray pattern she had recently obtained from a highly oriented DNA fiber. When Watson asked what the pattern was like, Wilkins went into an adjacent room to pick up a print of the new form (Fig. 6.17). "The instant I saw the picture my pulse began to race. The pattern was unbelievably simpler than those obtained previously. Moreover, the black cross of reflections that dominated the picture could arise only from a helical structure" and "conceivably, after only a few minutes' calculations, the number of chains in the molecule could be fixed" (Watson [1968]). Pressing Wilkins further Watson learned that together with a colleague he had done some work with three-chain models, but "nothing exciting had come up."

On the one hand Watson became convinced that the two-chain rejection of Wilkins' group was not a full proof. On the other hand he was convinced that "important biological objects come in pairs." He returned to Cambridge to tell that to Crick. Indeed it was not long before they worked out the structure: The helices were two in number, the backbone on the outside and the bases inside (Fig. 6.18). The two

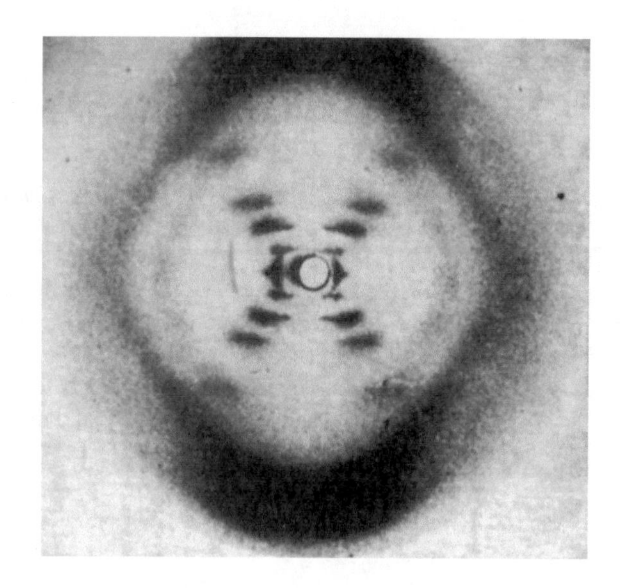

Figure 6.17 The diffraction spectrum of a highly oriented DNA fiber obtained by Rosalind Franklin in 1952. The fiber axis is vertical. (Reprinted with permission from Franklin R. E., Gosling R. G. [1953], "Molecular configuration in sodium thymonucleate," *Nature* 171, pp. 740-741, © Macmillan Magazines Limited)

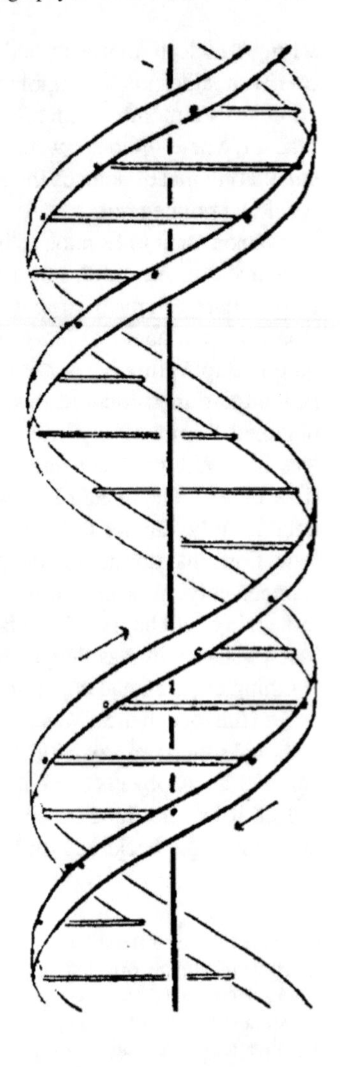

Figure 6.18 The double-stranded helix structure of DNA. (Reprinted with permission from Watson J. D., Crick F. H. C. [1953], "Molecular structure of nucleic acids," *Nature* 171, pp. 737-738, © Macmillan Magazine Limited)

ribbons in the picture symbolize the two phosphate sugar chains and the horizontal rungs the pair of bases holding the chains together. The vertical line stands for the fiber axis. The DNA molecules tightly packed inside the cells, once extracted, are pulled and take on the shape of a fiber. One might think of a long and thin rod made by DNA molecules, rotated and offset one with respect to the other. Then the diffraction experiment can take place. The structure is not periodic in the three dimensions, it is not a crystal and consequently the spectrum is not discrete (Section 3.4). The phase problem becomes unsurpassable for the phase ought to be assigned on a continuum, like the extended dark area surrounding the central "cross" in Figure 6.17. The method then consists in making an "educated guess" for a hypothetical structure, relative to which the Fourier transform is computed. Finally the spectrum so obtained has to be matched against the experimental data. This procedure is known as "trial and error."

Already in Cochran, Crick, and Vand [1952] the intensity distribution in the diffraction pattern of a series of point masses, equally spaced along a helix, had been computed and turned out to involve squares of Bessel functions (see also Wilkins, Stokes, and Wilson [1953]). As Watson [1968] reports, of the two it was Crick who knew the mathematics it takes to interpret diffraction spectra and even considered writing up its simple rules under the curious title "Fourier Transforms for the Birdwatcher."

So from Franklin's experimental data a few important numbers were drawn. If one pictures the double helix as a spiral staircase, the diameter is about 20 Å and the rungs are 3.4 Å apart, each one rotated by 36° with respect to the previous one. Hence ten steps mark one complete turn. The structure is stiffened by an enormous number of weak hydrogen bonds, each consisting of a positively charged hydrogen atom that acts as a bridge between two negatively charged atoms, the first an oxygen atom in a base and the second a nitrogen atom in the associated base.

In Watson and Crick [1953b], the authors cautiously write about their model, "So far as we can tell, it is roughly compatible with the experimental data, but it must be regarded as unproved until it has been checked against more exact results," as it was. Watson and Crick became famous all over the world and in 1962 were awarded, together with Wilkins, the Nobel prize for medicine.

The DNA structure immediately suggested "a possible copying mechanism for the genetic material" (Watson, Crick [1953b]). The DNA helix unzips down the middle and divides into two separate helices, as the bases break apart at the hydrogen bonds. Since bases bind in pairs only selectively—adenine goes only with thymine and guanine only with cytosine—each filament acts as a template for an exact copy of the original double helix, using the raw material (nucleotides) in the cell and through the reforming of the hydrogen bonds.

It is truly a wonder to think that the DNA spirals, approximately one meter long, are so thin and tightly packed to fit into a cell whose transverse dimension can be even only a few microns and that the genetic material transmitted from generation to generation consists of a few thousandths of a gram, to grow several billion times in the adult organism. Indeed each cell has a complete copy of the genetic instructions of the organism, even though different cells express different parts of the genome. This means that a cell of the skin, of the bones, or of the nerves, even though endowed with about the same genes, have different functions as a result of a selective expression or repression of certain genes.

This body of knowledge has been fundamental for the development of genetic therapy. The first patients on which they were attempted—at the National Institutes of Health in Bethesda—two girls were Ashanti De Silva and Cynthia Cutshall, who were vulnerable to every type of infection because of a severe immunodeficiency of genetic origin (*Time International*, June 7, 1993). Even though vital scientific issues are still to be resolved—like possible immune reactions to viruses, used in trials to convey genes into the patient's cells—much hope for the treatment of genetic disorders is tied to this therapy.

6.7 Protein structure: work for a century

The *human genome* has been essentially deciphered. It is a recent achievement well covered by the press. Next in *line* are proteins which are assembled according to instructions set by the genes. All human proteins, expected to be over a million in number, need to be identified and their spatial structure needs to be determined to understand how they work. This has been achieved for only a small percentage of the known proteins, so that the work that remains to be done is enormous. It might easily take a century. As results are obtained, new drugs could be devised, ultimately on an individual basis when needed.

Proteins are macromolecules, in the shape of a long chain or multiple chains, composed by repeating blocks named *aminoacids*, which come in 21 most common kinds. The total number of amino acids in a chain runs into the hundreds if not thousands. Two proteins can be made by the very same amino acids and yet be different—having different tasks—due to the different order the amino acids appear in the chain. The amino acids sequence is being genetically determined. For instance sickle-cell anemia is an inherited disease caused by the replacement of just two amino acids among the about 600 that make up the four chains (*subunits*) of the hemoglobin molecule.

Proteins regulate the metabolic processes in living organisms and for a long time they were thought to be repositories of hereditary characters until experimental evidence pointed in a different direction. Examples of proteins are *collagen*, protein of the bones, tendons, ligaments, and skin; *keratin*, protein of the hair and nails; *hemoglobin*, the oxygen-bearing protein of the blood; *enzymes*, which catalyze chemical reactions in living organisms; *hormones*, true chemical messengers between parts of the body; *antibodies* or *immunoglobulins*.

The protein diffraction spectrum is of fantastic complexity (Fig. 6.19). Remarkable is the task of determining the spatial distribution of the thousands of their atoms, actually impossible until quite recently. Take for example hemoglobin, the most important protein of the red cells, each containing about 280 million molecules of it. The single molecule is large even for today's standards, being composed of 574 amino acids spatially arranged along four chains intertwining in a complicated way, as M. F. Perutz discovered in 1960 (Perutz [1964]). Perutz solved the structure even though not as finely as to indicate the positions of the approximately 10,000 atoms of hydrogen, carbon, nitrogen, oxygen, and sulfur in addition to four most important atoms of iron. In particular he could not locate the about 4,000 hydrogen atoms. Their scattering, due to one electron only, is weak. Each chain enfolds a heme group—the structure that binds oxygen to the molecule—carrying at its center one of the four atoms of iron. The method of "heavy atoms"—now commonly employed when the direct methods are inadequate—was developed by crystallographers to solve the phase problem in relatively simple structures. Perutz, working in Cambridge (U.K.), discovered it could also be applied to proteins. The method consists in modifying the molecule of the substance under investigation by attaching atoms of heavy metals—such as mercury, gold, or platinum—to definite positions

Figure 6.19 A recent image of the diffraction spectrum of a hemoglobin crystal. (*Courtesy of M. F. Perutz, MRC Laboratory of Molecular Biology, Cambridge*)

in the structure. On this occasion Perutz used mercury. The presence of a few atoms of mercury in various positions did not modify appreciably the way the molecule organized itself in a crystal, precisely because of the molecule's big size. So when on his first x-ray photograph the diffraction spots appeared in the same position as in the mercury-free protein but with intensities significantly altered—thus providing information on the phase—Perutz in excitement rushed to Lawrence Bragg's office to tell him that the phase problem was solved and thus the structural problem. True, but only in principle, for an unexpected further five years of work were ahead.

The diffraction spectrum of a protein crystal is made by tens of thousands of reflections. Their intensities have to be measured from a crystal of the pure protein and then again in the patterns of one or more compounds of the protein—properly called *derivatives*—each with heavy atoms attached to different positions in the molecule. In the end tens of millions of numbers must be handled. This would have surely been an impossible task if computers—employed in crystallography from the end of the 1950s—did not have adequate speed.

So, how much time did Perutz devote to the purpose? As he himself writes, "In 1937, a year after I entered the University of Cambridge as a graduate student, I chose the x-ray analysis of hemoglobin, as the subject of my research." At the time the most complex organic substance whose structure had yet been determined by x-ray analysis was a molecule containing 58 atoms. Then Perutz adds, “Fortunately the examiners of my doctoral thesis did not insist on a determination of the structure, otherwise I should have had to remain a graduate student for 23 years.” He was awarded the Nobel prize for chemistry in 1962.

Since then enormous advances have been made in automating the experiment and in the data processing by computers. Software packages make extensive use of the FFT, and specially developed graphics routines allow accurate images of the molecules, to be displayed on the screen directly derived from computations. It is also possible to rotate the images to observe the molecules from different viewpoints, as in the usual visual exploration, and to enlarge details.

Going back to the structure determination it must be added that most of the time is presently taken up by growing the crystal. Indeed one can claim that the true problem of contemporary crystallography is the availability of crystals (McPherson [1989]). Only a small fraction of the thousands of known proteins have been crystallized. This process is not as simple a matter as in the case of common salt for which it suffices to let sea water evaporate. Protein crystals do contain a large amount of solvent, ranging from 30 to 80%. As a result they disintegrate in the absence of mother liquor, they are soft and crush easily making them very difficult to handle. In addition protein crystals exhibit weak optical properties and usually they diffract the x-radiation poorly. Therefore the growth of high quality protein crystals, i. e., suitable for an x-ray analysis, is a time-consuming step. Protein crystals are grown from water solutions in strictly controlled conditions (temperature, acid–base balance (pH), etc.) because proteins, as well as nucleic acids, tend to lose their structure in moderately hostile conditions. For this reason it may happen that the crystals turn out to be not sufficiently “well ordered” because they include heterogeneous forms of the protein. The diffraction pattern from such crystals usually extends to low-medium resolution (2.0–3.0 Å). There are few experts in this field and the methods are essentially empirical.

High quality crystals might require microgravity conditions for growth. This is the case of the crystals of the collagen-like polypeptide that were grown on board the Shuttle *Discovery* during the eight days space mission STS-95, which received much public attention because the 77-year-old astronaut John Glenn was among the crew members. This crystallization experiment was very successful and led to well-diffracting space-grown crystals (Fig. 6.20) (Berisio et al. [2000]). The high resolution (1.3Å) x-ray data collected from these crystals allowed the determination of subtle structural details of the triple helix that are peculiar of collagen and other fibrous proteins. This deeper insight contributed to the clarification of the reasons collagen is so stable (Berisio et al. [2001]).

The understanding of the structure and interaction mechanisms of natural macromolecules is of fundamental importance for biotechnological industries aiming to produce more effective drugs and to develop cereals more resistant to pest or marginal weather conditions. Other techniques come to the rescue, such as nuclear magnetic spectroscopy (Chapter 8).

Figure 6.20 Crystals grown on board the Shuttle *Discovery* during the space mission STS-95 of October 1998. The linear dimensions are of the order of a millimeter. The picture was taken at the Kennedy Space Center the day after landing. (*Courtesy of A. Zagari, Dipartimento di Chimica, Università di Napoli*)

6.8 Seeing the hydrogen atom

In recent last years, thanks to the use of synchrotron beamlines (Chapter 5), to fast and accurate detectors, and improved crystallization procedures (Chayen [1997]), protein structures have been reconstructed at extremely high resolution as well as time-resolved, revealing features that can provide clues to protein function.

In Berisio et al. [1999] the process of deprotonation in a crystalline environment is captured by a sequence of snapshots (Fig. 6.21). As the authors write, "Until quite recently, there was a common belief that in protein x-ray structure determinations hydrogen atoms could not be located and their positions had to be calculated from the known coordinates of C, N and O atoms (Jeffery, Saenger [1981])." The images show a significant effect of pH* on a protein structure (bovine pancreatic Ribonuclease A) in the crystalline state. The hydrogen linked to the $N^{\delta 1}$ atom is present over the whole pH* range, from 5.2 to 8.8, while the hydrogen linked to the $N^{\delta 2}$ atom is released at high pH* values. (The above crystals have been grown and manipulated in a water–alcohol solution. While the commonly known pH refers to water systems, pH* refers to water–alcohol systems. Yet pH and pH* measure essentially the same physical entity, i.e., the level of acidity.)

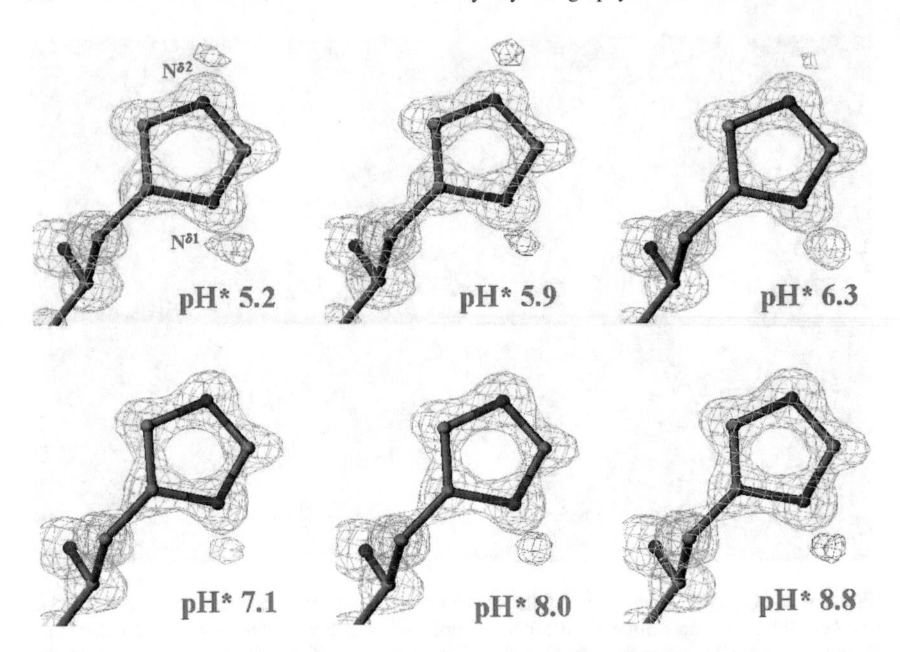

Figure 6.21 Structural features of an enzyme. The balls indicate atoms, the sticks indicate atomic bonds, and the surrounding web are the electronic density's level sets. The sequence of images shows the release of the proton linked to the $N^{\delta 2}$ atom as the pH* increases; in contrast, the hydrogen linked to the $N^{\delta 1}$ atom is present over the whole range. The resolution of the six structures falls in the interval 1.05–1.15 Å. (Reprinted with permission from Berisio et al. [1999] "Protein titration in the crystal state," *J. Mol. Biol.* 292, pp. 845–854.) (*Courtesy of R. Berisio, Centro di Biocristallografia, CNR*)

Approximately 1.0 mm ×0.5 mm ×0.2 mm are the dimensions of the crystals used. The protein itself is medium size, being made of 124 amino acids. The unit cell is 30.4 Å × 38.4 Å × 53.3 Å with angles $\alpha = 90°$, $\beta = 105.7°$, and $\gamma = 90°$. The diffraction data were collected at the storage ring DESY in Hamburg (Germany).

6.9 Vision at the nanosecond: a time-resolved structure

The biological activity of macromolecules is accompanied by rapid structural changes. From a femtosecond to a millisecond or longer is the lifetime of an intermediate. To elucidate the bases of biochemical mechanisms, structural studies of intermediates are carried out. This may be done by artificially prolonging their lifetime or by applying very fast x-ray crystallographic techniques to authentic short-lived intermediates.

Table 6.1 Submultiples of the second (s).

millisecond ms = 10^{-3}s
microsecond μs = 10^{-6}s
nanosecond ns = 10^{-9}s
picosecond ps = 10^{-12}s
femtosecond fs = 10^{-15}s
attosecond as = 10^{-18}s

Time-resolved studies using x-rays, yet at the beginning, take advantage of an intrinsic property of synchrotrons: generation of x-ray pulses 50–150 ps long (Chapter 5). This feature could not be utilized before the synchrotron's third generation, for single pulses were not sufficiently intense then. In Srajer et al. [1996] structural changes that occur in the carbon monoxide complex of myoglobin (MbCO) at room temperature on CO photodissociation have been recorded with nanosecond time resolution. Photolysis was initiated by a 7.5 ns laser pulse and monitored by a 150 ps x-ray exposure.

From x-ray diffraction data collected at the European Synchrotron Radiation Facility in Grenoble, the pictures in Figure 6.22 have been obtained. They feature 1.8 Å resolution and six time delays between laser and x-ray pulses: 4 ns (150 ps x-ray exposure), 1 μs, 7. 5μs, 50.5μ s, 350μ s, and 1.9 ms. The pictures show difference maps between the average structure, at the listed time delays, and the stable MbCO structure. The prominent (negative) feature labeled CO corresponds to loss of CO upon photolysis. The magnitude of this feature declines with time as dissociated CO recombines.

The typical dimensions of the crystals were 0.4 mm $\times$0.3 mm $\times$0.07 mm. The unit cell is monoclinic with dimensions $a = 64.5$Å, $b = 30.8$Å, and $c = 34.8$Å, and angles $\alpha = 90°$, $\beta = 105.8°$, and $\gamma = 90°$.

Fast laser pulses are an essential tool for understanding biochemical mechanisms. As pulses in the femtosecond range were produced, femtochemistry—for which the Nobel prize in chemistry was awarded in 1999 to Ahmed Zewail—could develop. Attosecond pulses were obtained recently, as announced in Corkum [2000] (Table 6.1).

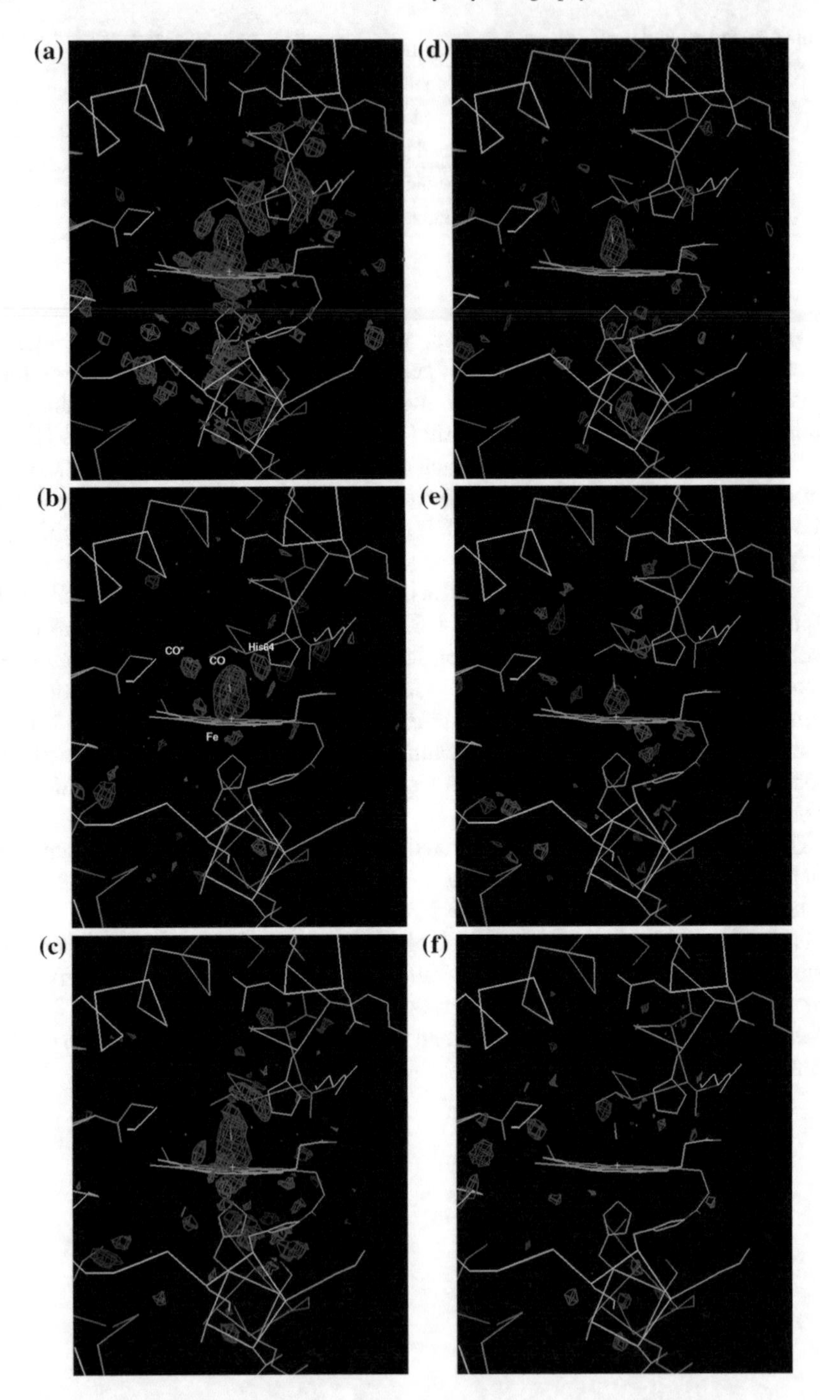

Figure 6.22 a) Reference difference map calculated from deoxy Mb and MbCO; b) 4 nanosecond time delay after CO photodissociation; c) 1 μs (labeled); d) 7.5 μs; e) 50. 5μs; f) 350μ s. (*Courtesy of V. Srajer, Department of Biochemistry and Molecular Biology, University of Chicago*)

Chapter 7
The Radon Transform and Computerized Tomography

7.1 Computerized axial tomography, a technique from the 1970s

The word tomography comes from the Greek "tomos" meaning cut or slice. Computerized axial tomography (CAT), also referred to simply as computed tomography (CT), has been a major event in diagnostic medicine and brought a revolution in radiology together with MRI, which followed later.

It is based on a mathematical relation, called the *Radon transform*, advanced in 1917 by the Bohemian mathematician Johann Radon (1887–1956) in a paper titled "On the determination of functions from their integrals along certain manifolds." Radon was a well-known and well-read mathematician, yet the important applications that were to come for the Radon transform were hardly foreseen at the time.

The first application due to Ronald Bracewell [1956], who further pursued the subject together with A. Riddle [1967], took place in radioastronomy. Bracewell obtained images of the Sun in the microwave region although the available antennas could only measure the total intensity along narrow strips. Using several measurements, or "strip sums," in different directions he produced an image of the sun determining those regions that emit microwave radiation.

In medicine, early work on image reconstruction was done by William Oldendorf [1961] and jointly by D. Kuhl and R. Edwards [1963]. In particular Oldendorf showed experimentally that smaller variations in x-ray absorption could be detected if differences in numbers of photons rather than in film density were recorded.

The mathematical and physical foundations of CT scans were established by the South African born physicist Allan MacLeod Cormack [1963, 1964] while at Tufts University, Massachusetts. His main field of research was subatomic particles. A part-time position as a physicist in a hospital radiology department was the occasion that aroused his interest in the problem of x-ray imaging of tissues of different density.

E. Prestini, *The Evolution of Applied Harmonic Analysis*,
Applied and Numerical Harmonic Analysis,
DOI 10.1007/978-1-4899-7989-6_7

In 1970, the English electrical engineer Godfrey Newbold Hounsfield, member of the Royal Air Force during World War II and expert in electronics and radar, developed a head scanner at the Center Research Laboratories of EMI Ltd., following an idea originating in his work on pattern recognition. The scanner was later installed at Atkinson Morley's Hospital in Wimbledon, England, where in 1972 the first clinical test of computerized tomography was performed successfully. Also in 1972 British patent No. 1283915 was issued in London for: "A method of and apparatus for examination of a body by radiation such as x-ray or gamma radiation." The Hounsfield's scanner by 1973 was the first tomographic equipment to give images sufficiently accurate to be clinically useful and was perhaps the greatest single advance in the field of radiology after the discovery of x-rays in 1895 (Section 5.5).

Since then different companies have developed head and whole body x-ray scanners that have brought dramatic changes in the field of medical diagnosis radiology by making available images of previously inaccessible parts of the body, particularly the brain. Prior to the introduction of this technique, standard x-ray pictures of the brain were generally of little use, being blurred beyond recognition by the skull. A useful x-ray image of the brain's vascular system required procedures involving a few days of hospitalization. Such procedures included injection of a dye, or an "air scan," during which the lower spine was injected with gas that, flowing upward, would fill, and outline the brain cavities. CT made available detailed cross sections of the brain that eliminated the need of "air scans" and greatly reduced the use of dyes. At the same time the amount of x-rays absorbed by the patient was reduced to a fraction.

Cormack and Hounsfield shared the Nobel prize for medicine in 1979 for their contributions to the development of this diagnostic technique.

In suitable units, the density of head tissues varies from 1 to 1.5 with the exception of bone tissue which has density 2. Some features of medical interest are associated with density changes that might be as small as 0.005. Therefore the problem is to reconstruct the tissue density at a sufficient number of points with adequate accuracy.

For the reconstruction problem different approaches have been taken. Iterative algorithms, as in Gordon et al. [1970], Herman, Rowland [1973], and Crowther, Klug [1974], work without the Radon transform being introduced. The original EMI machine used a type of iterative algorithm. To determine the density function, which is assumed to be piecewise constant on the squares of a grid, a system of equations is solved in the unknown values of the density, using the data (*projections*) obtained from the x-ray experiment.

Bracewell's original technique used the method of *back projections* and inverted the Radon transform by convolution and Fourier transformation as most modern machines do (Natterer [1986], Engl et al. [1996], Macovski [1983], Helgason [1980]). This kind of method not only is faster but generally gives reconstructions with better accuracy and spatial resolution. Besides these two major types of algorithms other methods have been developed to address the reconstruction problem. An overview can be found in Herman and Lewitt [1979].

7.2 Beyond classical radiology

The essence of classical radiology consists of taking advantage of the power of x-rays to penetrate tissues. Depending on the density of the tissue, the radiation is absorbed differently or equivalently transmitted differently. So when it emerges from the body it produces on a photographic plate or film an image, a "shadow" of the structures of different densities it went through.

The capability to distinguish between structures is limited by at least two factors. First the x-ray absorption of adjacent structures or organs has to be significantly different to produce a notable contrast in the image. Conventional radiology can reliably distinguish only between the absorptions of air, "soft tissue" and "bone." Bones, absorbing much more than other tissues, are easily recognized; similarly normal and abnormal structures in the lungs are outlined by the surrounding air. Regions with similar x-ray transmittance, such as muscle, blood, organs, and generally most of the "soft tissue" of the body, cannot be easily distinguished from one another. Second the resulting x-ray image, being a projection of all the overlying structures superimposed on one another, turns out to be a kind of "multiple exposure" made by many hypothetical pictures, each of a thin slice of the body (Fig. 7.1). As a consequence, for instance, every detail lined up with the denser bone tissue tends to be excluded. To overcome the limitation mentioned first, that is poor resolution in case of insufficiently different electronic densities, various "contrast agents" are employed. These substances, like barium salts and iodinated compounds, are apt to enhance contrast, having high electronic density. Chiefly used to evidence the hollow lumen of the alimentary tract, or of blood vessels and arteries, they produce images that, in certain cases, might still be of difficult interpretation. Instead, generally speaking, the brain and other solid organs like the pancreas and liver, together with surrounding regions, are particularly difficult to study radiologically.

The second limitation, due to the overlying structures, is addressed by the technique of *stratigraphy*. During prolonged exposures, the x-ray source is moved above the patient along a straight line while it continues to emit radiation. At the same time the photographic plate, below the patient, is moved in the opposite direction. Most structures in the part of the body under examination are blurred in the image, by motion. Only the shadow cast by a layer, containing the pivot points of the x-ray

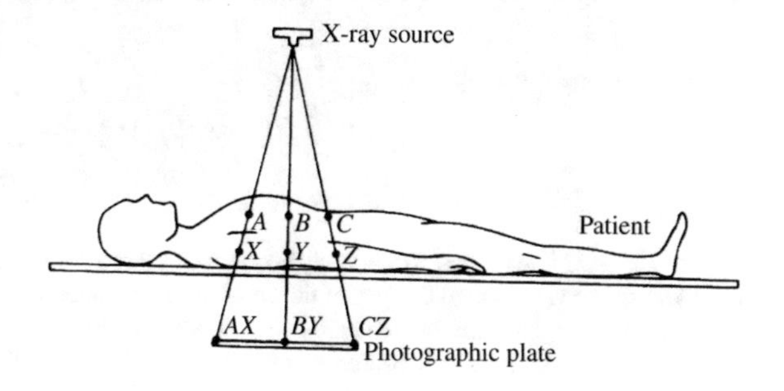

Figure 7.1 In conventional radiography all points lined up on an x-ray path are projected on the same point of the photographic plate.

path between the tube and the film, remains stationary and hence in sharp focus on the plate.

Computerized tomography offers major advantages since the ability to distinguish between regions with small differences in x-ray transmittance and the ability to separate overlying structures are both greatly improved. By a completely novel way of processing x-ray absorption data (projections) obtained from many different angles, CT scanners provide accurate images of the interior of the human body allowing a convenient and quantitative localization of anatomical structures, lesions, or other entities that would be difficult or impossible to determine otherwise.

A CT procedure requires the x-ray source to rotate a full circle around the slice to be imaged. Thus the slice is usually chosen transversely, that is on a plane perpendicular to the long axis of the body. By a set of contiguous transverse images it is possible to create two-dimensional images oriented in different directions as well as three-dimensional images.

A typical CT scanner resembles a large doughnut (Fig. 7.2). The patient lies on a table and the doughnut surrounds the part of the body to be examined. Mounted on a gantry and housing the x-ray tube and detector, the doughnut rotates around the patient. It describes a circle on a plane, at right angles with the main axis of the body, that ideally cuts a slice out of the body. It is the slice that will be irradiated by the x-ray tube in its circular motion, for the purpose of being imaged. The detector, opposite the x-ray tube, records the intensities of the x-rays exiting the body that provide the data for the reconstruction. The basic method, such as that embodied in the first generation machines, employs an x-ray tube and opposing x-ray detector

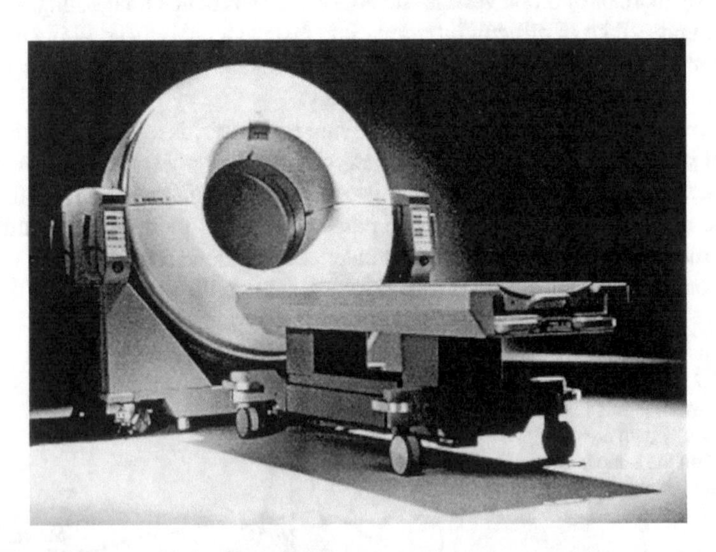

Figure 7.2 A CT scanner schematically consists of a large doughnut, containing an x-ray tube and detector, that rotates around the part of the body to be examined in order to reconstruct the image of a slice at right angles with the main axis of the body. (M. Bertero [1999], Matematica e immagini: alcuni esempi di applicazioni, *Bollettino U.M.I.*, (8) 2-A, pp. 47-67.)

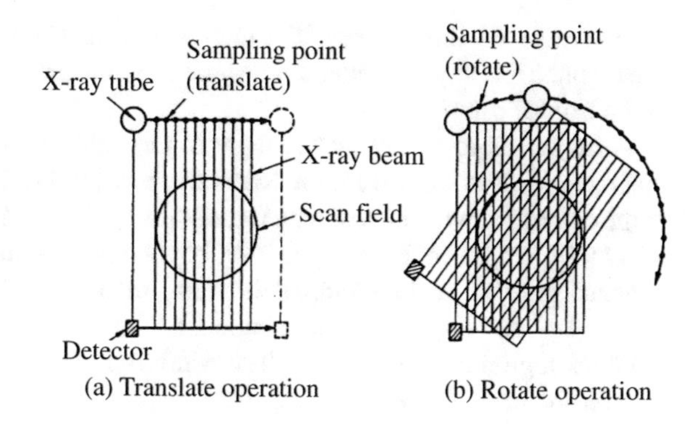

Figure 7.3 Principles of first generation CT scanners. (K. Kimura, S. Koga (Eds.), *Basic Principles and Clinical Applications of Helical Scan*, Iryokagakusha, [1993].)

that translate linearly on a track. Hundreds of narrow beams of parallel x-rays (*pencil beams*) cross the chosen section to have their attenuation recorded. The end of the track having been reached, the x-ray tube and detector rotate to a new angle and the linear motion takes place again (Fig. 7.3). Radon's mathematics showed that if the attenuation in all directions was measured, then the entire two-dimensional density distribution of the section could be reconstructed unless the attenuation is complete. Obviously in practice only a finite number of directions is tested. The x-ray tube makes some 500–1500 equidistant stops on a full circle around the patient to detect at each stop the intensities of some 700–1500 beams. In a standard procedure 10 to 20 sections are reconstructed so that the number of data to be processed per patient might vary from 3.5 to 45 million. With this procedure the scanning time can be as long as four or five minutes per slice, so this method may be conveniently applied only to regions, such as the head, that can be kept motionless for that long time. In place of the pencil beam the second generation uses a *fan beam* of slightly diverging rays, from 3° to 15°, and as much as 60 detectors to match the shape of the beam. In this way projection data for several directions are obtained during a single linear movement. The scanning time is reduced to 10–20 seconds.

The third generation uses a wider fan beam, from 30° to 60°, to cover the entire scanning field so that data acquisition is completed by rotation only. The scanning time is brought down to 1–2 seconds.

In the fourth generation the detector array is set all around a circle and this remains stationary. The x-ray tube might rotate inside it, as in the previous generation, or else outside it allowing a higher spatial resolution due to the reduced diameter of the detector array. In this last case the array must nutate to avoid blocking the fan beam. The scanning time is one second.

Methods of avoiding the mechanical rotation of the x-ray source have been devised (Kimura, Koga [1993]) resulting in high speed machines. The scanning time, reduced to tens of milliseconds, makes it possible to scan the heart and other organs that undergo rapid movements. Meanwhile the computer processing speed has increased to the point that images can be displayed as soon as scanning is completed.

As described, the history of CT has been a constant striving for shorter scanning times, since this implies better image quality, in the presence of sufficient x-ray doses and adequate sampling.

A major improvement was brought to the field by the *helical scan*, a technique that dates back to the late 1980s (Katakura et al. [1989], Kalender et al. [1990]). The name comes from the helical path traced around the patient by the x-ray beam. If the x-ray tube keeps rotating continuously beyond 360° while the couchtop is moved at constant speed in the longitudinal direction, then rather than working on one slice of a certain thickness at a time, in a single 20–30 second operation, the data for a 30–70 cm thick region can be acquired. The width of the reconstructed slices is about two mm for the head and ten mm in general.

A number of reconstruction problems were posed by the helically scanned projection data, like strong motion artifacts, increased slice thickness, position of the slice being imaged not precisely known. Once solved by the development of dedicated processing techniques and interpolated reconstructions, helical scans have provided significantly shorter examination times and larger numbers of images, and moreover they offered the choice of any desired sectional plane which is difficult to do with conventional CT. Also the helical scan together with the development of massive storage memory and dedicated hardware allows high quality three-dimensional images to be obtained during a single breath hold (Kimura, Koga [1993]).

The detection method is electronic in all cases so the analog signal has to be converted to digital to be fed to the computer that reconstructs images by elaborate calculations whose mathematical foundations are described in Section 7.4.

CT images are digital in nature and can be viewed only on a computer monitor or on photographs made from it. The mathematical algorithm produces a numerical description of the density as a function of position. The "computed" image is an array of "CT numbers," relative to a grid of discrete locations, scaled so that water has zero CT number, less absorbing substances such as fat and air have negative numbers (commonly air is set at–1000), and more absorbing substances have positive numbers (commonly bone is set at 1000). In particular the numerical data of the reconstruction overcome the capability to render them by various scales of gray (and also the human visual perception) so the best "window" can be chosen to evidence the structure under examination.

7.3 CT versus MRI

While the machines appear externally similar, CT and MRI are entirely different. The first one employs x-rays and a system that rotates mechanically around the patient, making the scanning time inherently rather long. MRI instead is entirely electronic, thus faster. In addition it does not use ionizing radiation and the magnetic fields employed in clinical tests do not seem to have harmful effects.

Moreover CT is essentially a two-dimensional technique while MRI is essentially three-dimensional, with a reduction in the reconstruction time in case only two-dimensional images are required.

Also while in CT scans the input signal is always the same x-ray beam, in MRI the entering signal can be chosen depending on the object under examination. Similarly the parameter on which to base the reconstruction is always the x-ray absorption with CT, while it can be either one of the two relaxation times or the resonance signal with MRI.

On the other hand, CT has an outstanding density resolution and offers detailed anatomical information. It is superior in obtaining clear images of bony structures, in particular fractures and tissue ossification, and generally speaking, in the examination of the chest and solid organs such as liver and pancreas.

7.4 Projections

We shall consider simple density distributions to illustrate what their projections are and what information they carry. Consider a distribution $f(x, y)$ in the plane given by a small mass centered at A, fully surrounded by a uniformly dense circular shell as in Figure 7.4. At a fixed x_0 along the x-axis, the *projection* P of the distribution along the vertical is given by the integral of f along the vertical line $x = x_0$ as in the formula

$$Pf(x_0) = \int_{-\infty}^{\infty} f(x_0, y)dy.$$

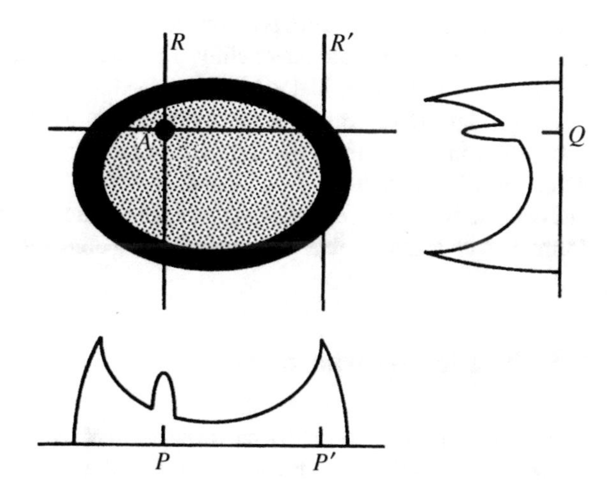

Figure 7.4 Projections along the vertical and horizontal direction of a density distribution. (R. N. Bracewell, *Two-Dimensional Imaging*, Prentice-Hall, [1995].) (*Courtesy of R. N. Bracewell*)

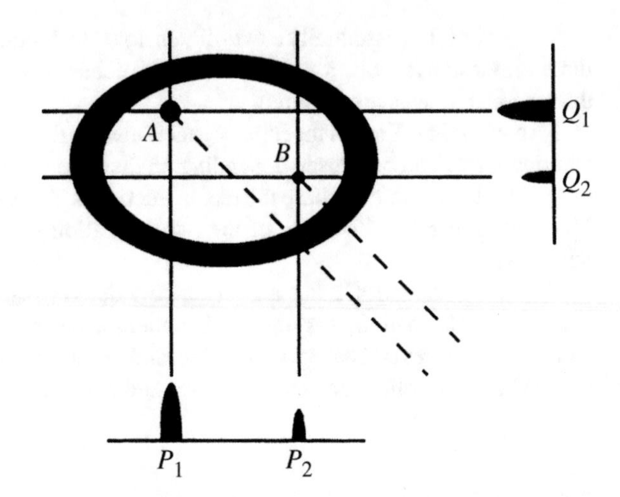

Figure 7.5 Projections of a density distribution containing two compact objects. (R. N. Bracewell, *Two-Dimensional Imaging*, Prentice-Hall, [1995].) (*Courtesy of R. N. Bracewell*)

As x_0 varies along the x-axis, a function is obtained. It is the projection $Pf(x)$. Similarly $f(x, y)$ can be projected along the horizontal direction resulting in another function $Qf(y)$ (Fig. 7.4). The same can be done for any direction in the plane. By assumption a single small mass, ideally a point mass, is known to be inside the shell and the problem is to locate it. Clearly $Pf(x)$ signals the presence of a mass along the line RP but does not give any information whatsoever on the height. Nevertheless if another projection, like $Qf(y)$, is known, then the height too can be uniquely determined.

If there were two masses, at A and B, then two projections would be insufficient to the same purpose if the masses are comparable. For instance the projections Pf and Qf would be the same if A and B were at the marked vertices or at the unmarked vertices of the rectangle in Figure 7.5. A third projection along any other direction is needed to overcome the uncertainty. The question could be slightly more refined if the masses at A and B had to be determined in addition to their location.

Radon, apparently carrying out mathematics for its own sake, established the remarkable fact that the set of all projections along any line L of the plane uniquely determines the distribution f under quite general assumptions, such as f continuous and compactly supported, that always hold in practice. In addition he found an explicit formula to calculate f, thus paving the way to computed tomography.

7.5 The Radon transform

The absorption and scattering of a narrow beam of x-rays directed into the body along a direction L results in an attenuation ΔI of the beam intensity, which is assumed to be proportional to the intensity I itself (measured in photon/s), to the length t the beam went through the body, and to the density distribution $f(x, y)$ as in formula

$$\Delta I = f\ I\ \Delta t$$

In particular the denser the object the higher the attenuation is.

Then along L the following differential equation holds:

$$\frac{dI}{I} = f dt.$$

Denoting by I_0 the input intensity of the beam and by I the output, the equation is solved by

$$\log(I/I_0) = \int_L f(x,\ y)dt, \tag{7.1}$$

where t indicates length along L. The line integral $\int_L f(x,\ y)dt$, or *projection* or *radiograph*, can be calculated since I_0 is known and I can be measured. As L translates parallel to itself the projection $P_L f(s)$ is defined as a function on the line passing through the origin and orthogonal to L.

In the simple case in which $f(x,\ y) = \alpha$ is a constant, which corresponds to a uniform distribution, (7.1) gives

$$I = I_0 e^{-\alpha t}$$

and the attenuation is said to decay exponentially.

Clearly (7.1) is only an approximation since it assumes the beam to be infinitely thin and moreover strictly monochromatic, thus overlooking the important effect of "beam hardening." Low energy x-rays do not penetrate materials as well as high energy ones and are considered "softer." As the beam proceeds through the body, the low energy rays are blocked and the fraction of high energy rays becomes larger and larger with a net result of a more complicated signal. This can be avoided by working with the x-rays obtained at synchrotrons or by passing the beam through a filter to preharden it. The effects of the above simplifying assumptions — a slight loss of resolution is a typical one — have been studied both theoretically and by simulations and at least in principle they can be made as small as desired (Shepp, Kruskal [1978]).

The next step is to invert (7.1) and to find a formula for the density $f(x,\ y)$ starting from the known values of the radiographs.

The transformation that associates the line integrals $P_L f = \int_L f\ dt$ with a function f is known as the *Radon transform*. Radon [1917] proved, in great generality since he worked with hyperplanes in n-dimensional Euclidean spaces, $n \geq 2$, that any function f continuous and compactly supported is uniquely determined by its radiographs and gave a fairly simple inversion formula. To compute the value of f at a point Q in the plane one has to consider the circle C_q centered at Q of radius q (Fig. 7.6). At any point on the circumference there is a tangent line L, and relative to that,

Figure 7.6 Geometry involved in computing the inverse Radon transform.

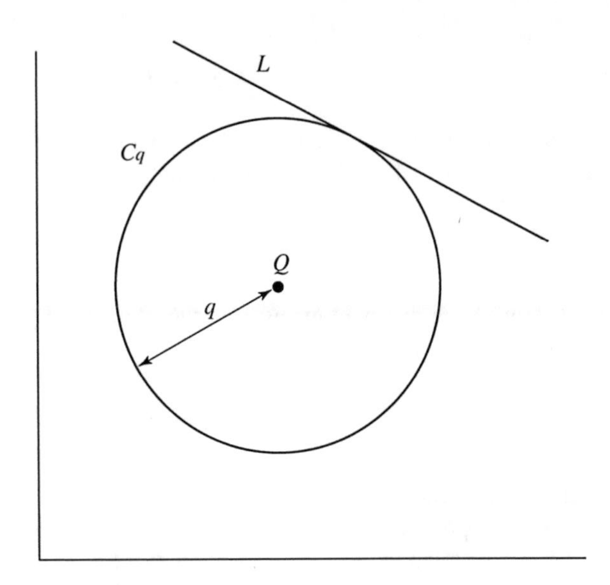

the projection $P_L f$ has a known value. First the average of $P_L f$ on the circumference has to be computed as in

$$F(q) = \frac{1}{2\pi q} \int_{C_q} Pf(L)ds$$

where s is the arc length. Thus, as $0 < q < \infty$ varies, a function is defined that also depends on the fixed Q. Its differential is required by the inversion formula

$$f(Q) = -\frac{1}{\pi} \int_0^\infty \frac{dF(q)}{q}. \tag{7.2}$$

If the value of f at another point $\bar{Q}$ is required, then all circles centered in $\bar{Q}$ of radius q have to be considered, the corresponding $F(q)$ computed, and (7.2) run again.

Radon went so far. His formula brilliantly solved the problem. For those who need to perform the actual calculations a whole new list of relevant questions opens up. Evidently in practice, projections can be calculated only for a finite number of directions. Then uniqueness fails. There are examples (Cormack [1963], Logan [1975], Louis [1984]) of nonzero functions that nevertheless have zero projections in a finite number of directions, however chosen (Fig. 7.7). These are functions that do oscillate and have high frequencies. The problem they pose can be overcome and arbitrarily small details can be rendered if the number of directions is correspondingly high and well distributed (Faridani [2003]). Then there is the problem of the measurements being made on strips rather than on lines. Also (7.2) shows an apparent singularity at $q = 0$ (apparent since when the denominator is zero the numerator is too) and

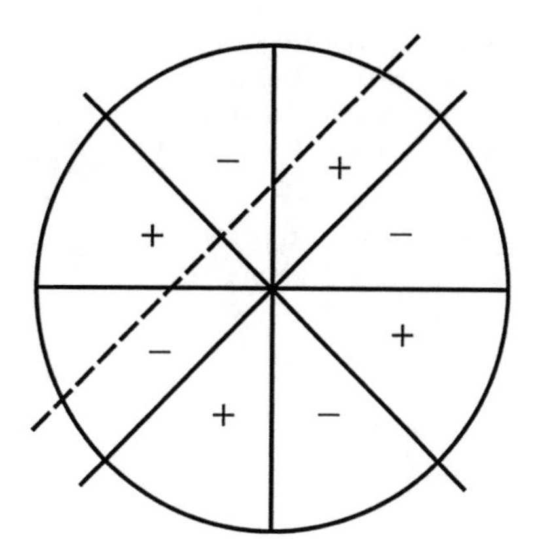

Figure 7.7 An example of a nonzero function having zero projections in each of the four directions horizontal, vertical, 45°, and 135°. The function is constant in absolute value, only the sign changes as drawn.

moreover the problem of the stability of (7.2) under imprecise and rounded off values of the measurements have to be addressed.

Finally the computations required by (7.2) are based on a "circular geometry" not well fitted to the "parallel geometry" of the x-ray device. Indeed all modern machines work with another inversion formula based on the Fourier transform.

7.6 Inverting the Radon transform by convolution and back projections

The reconstruction problem will be solved for the parallel beams of the first generation machines. This is a basic result. A modification works for the fan beams (Scudder [1978]).

The idea that the Fourier transform might play a role comes from the relation

$$\hat{g}(0) = \int_{-\infty}^{\infty} g(x)\, dx, \tag{7.3}$$

which is elementarily obtained from the definition of the one-dimensional Fourier transform (2.13) by substituting $\xi = 0$. To make use of this simple relation in the two-dimensional setting of the Radon transform one is led to choose a suitable reference frame for the projections and related computations (Fig. 7.9). If L is the direction of the x-ray beam, emitted by the tube that translates along S to irradiate the section under consideration (the circle in Figure 7.8), then let $(\tilde{x}, \tilde{y})$ be the new frame, where the $\tilde{x}$-axis is parallel to S and the $\tilde{y}$-axis is parallel to L. Let θ denote the angle of rotation with respect to the original frame (x, y).

In the new frame, the projection $P_L f(s)$ is a function of $\tilde{x}$. As such it shall be denoted by $P_\theta(\tilde{x})$. Similarly the density $f(x, y)$ as a function of $\tilde{x}$ and $\tilde{y}$ shall be denoted by f_θ. Formula (7.3) suggests that we compute the two-dimensional Fourier

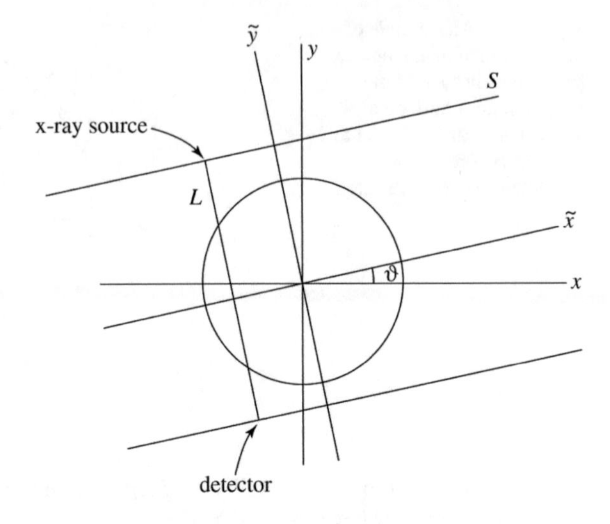

Figure 7.8 The x-ray source translates along the line S and emits radiation in the direction L.

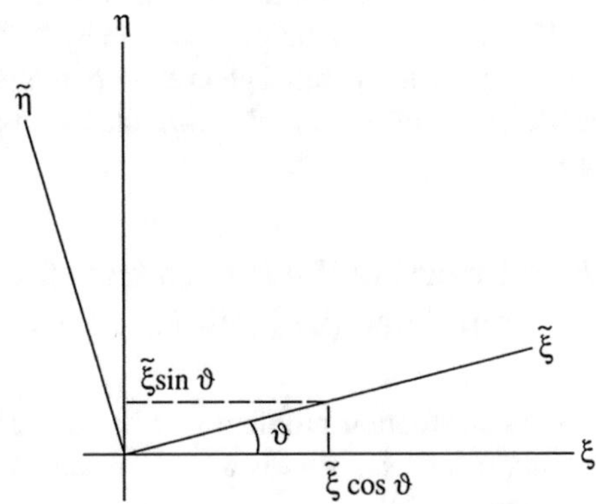

Figure 7.9 The frames $(\xi,\ \eta)$ and $(\tilde{\xi}, \tilde{\eta})$ in the frequency domain.

transform of f_θ for $\tilde{\eta} = 0$, namely on the $\tilde{\xi}$-axis. By the definition (5.1)

$$\begin{aligned} \hat{f}_\theta(\tilde{\xi},\ 0) &= \int_{-\infty}^{\infty} \int_{-\infty}^{\infty} e^{i\tilde{\xi}\tilde{x}} f_\theta(\tilde{x}, \tilde{y}) d\tilde{x} d\tilde{y} \\ &= \int_{-\infty}^{\infty} \left[\int_{-\infty}^{\infty} f_\theta(\tilde{x}, \tilde{y}) d\tilde{y} \right] e^{i\tilde{\xi}\tilde{x}} d\tilde{x} \\ &= \int_{-\infty}^{\infty} e^{i\tilde{\xi}\tilde{x}} P_\theta(\tilde{x}) d\tilde{x} = \hat{P}_\theta(\tilde{\xi}). \end{aligned} \tag{7.4}$$

The Fourier transform is invariant under rotation, as expressed by $\hat{f}_\theta(\tilde{\xi}, \tilde{\eta}) = \hat{f}(\xi,\ \eta)$, where the $(\tilde{\xi}, \tilde{\eta})$ frame is rotated by the angle θ with respect to $(\xi,\ \eta)$. In this last frame $\hat{f}_\theta$ is $\hat{f}$ by the preceding equality and the points $(\tilde{\xi},\ 0)$ have coordinates

$$\xi = \tilde{\xi} \cos\theta \quad \eta = \tilde{\xi} \sin\theta$$

as Figure 7.9 clearly shows. Hence (7.4) becomes the key formula

$$\hat{P}_\theta(\tilde{\xi}) = \hat{f}_\theta(\tilde{\xi},\ 0) = \hat{f}(\tilde{\xi}\cos\theta, \tilde{\xi}\sin\theta). \tag{7.5}$$

What is needed now is the Fourier inversion formula

$$f(x,\ y) = \frac{1}{(2\pi)^2}\int_{-\infty}^{\infty}\int_{-\infty}^{\infty} \hat{f}(\xi,\ \eta)e^{i(x\xi+y\eta)}d\xi d\eta$$

in polar coordinates $(r,\ \phi)$ in space and $(R,\ \theta)$ in the frequency domain, where

$$\begin{aligned} x &= r\cos\phi, \\ y &= r\sin\phi, \\ \xi &= R\cos\theta, \\ \eta &= R\sin\theta, \end{aligned}$$

with $0 < r < \infty$, $-\infty < R < \infty$ and $0 < \phi < 2\pi$, $0 < \theta < \pi$. Since $d\xi d\eta = |R|dRd\theta$ is the relation between the infinitesimal areas, the change of coordinates leads to the known formula

$$f(x,\ y) = \frac{1}{(2\pi)^2}\int_0^{\pi}\int_{-\infty}^{\infty} \hat{f}(R\cos\theta,\ R\sin\theta)\, e^{irR\cos(\theta-\phi)}|R|d\,Rd\theta.$$

This formula becomes

$$f(x,\ y) = \frac{1}{(2\pi)^2}\int_0^{\pi} d\theta \int_{-\infty}^{\infty} \hat{P}_\theta\,(R)\, e^{irR\cos(\theta-\phi)}|R|d\,R \tag{7.6}$$

by (7.5), with $\tilde{\xi}$ replaced by R as far as notation. This is the new inversion formula of the Radon transform. It recovers the density distribution f starting from the Fourier transform of the known projections P_θ.

Most commercial machines work in the spatial domain, to which (7.6) shall now be transferred. The inner integral of (7.6)

$$\int_{-\infty}^{\infty} \hat{P}_\theta\,(R)\, e^{irR\cos(\theta-\phi)}|R|d\,R$$

is the inverse Fourier transform, evaluated at $r\cos(\theta-\phi)$, of the product of the functions $\hat{P}_\theta(R)$ and $|R|$. As described in Section 3.5, it might be equivalently expressed as the convolution

$$(P_\theta * K)(s) = \int_{-\infty}^{\infty} P_\theta(s')\,K(s-s')ds', \tag{7.7}$$

where $s = r\cos(\theta - \phi)$ and K denotes the inverse Fourier transform of $|R|$. Note that K is not a function but rather a distribution, seeing that the transform of $|R|$ is too big at infinity to be integrable. This means that the computation of K is more involved and requires a limiting process as described in Section 3.5.

Now the outer integral in (7.6), known as *backprojection*, gives the density distribution

$$f(x, y) = \frac{1}{(2\pi)^2} \int_0^{\pi} (P_\theta * K)(r\cos(\theta - \phi))d\theta. \tag{7.8}$$

Finally computers may calculate (7.7) and (7.8) in terms of finite sums, with an approximation as good as desired.

7.7 CT images

Figure 7.10 shows an anterior view of the entire aorta from a CT chest–abdomen–pelvis angiography. Obtained after injecting an iodinated contrast agent, it shows a suprarenal aortic aneurysm as well as an aneurysmal dilation of the mid-descending thoracic aorta. Figure 7.11 is a magnified view of the aneurysm. The 3D reconstructions of Figure 7.10 and 7.11 derive from a data set of 370 source images. In Figure 7.12 three source images of the chest, abdomen, and pelvis.

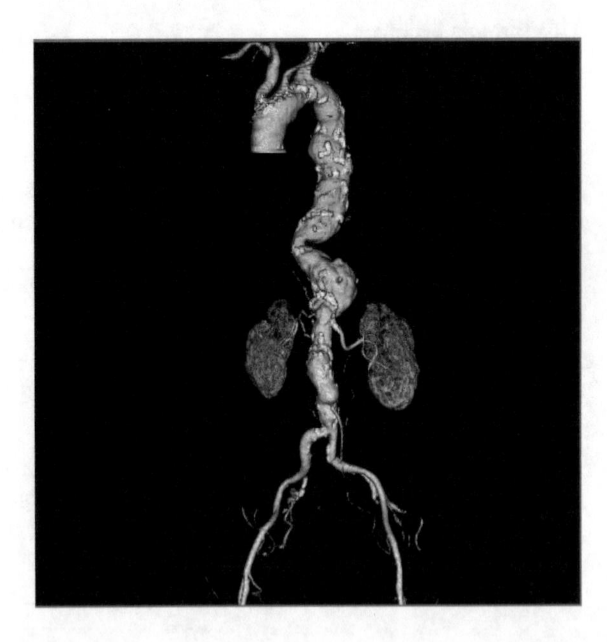

Figure 7.10 An anterior view of the entire aorta. (*Courtesy of L. Logan, Department of Radiology, Stanford University*)

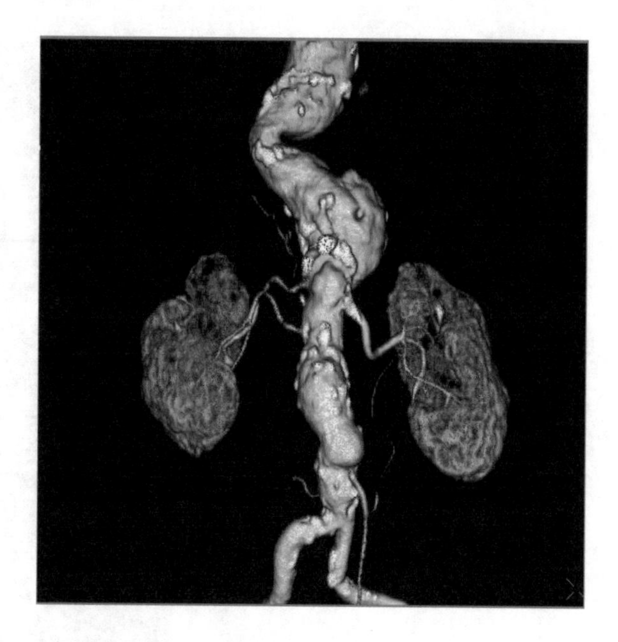

Figure 7.11 A magnified anterior view of the aneurysmal dilation of the suprarenal abdominal aorta in Figure 7.10. (*Courtesy of L. Logan, Department of Radiology, Stanford University*)

7.8 Local tomography and microtomography

It often happens that only part of an object needs to be imaged. For instance the heart might be of interest, not the entire chest. In this case it would be advantageous if only *local data* were needed, that is integrals over lines that intersect the region of interest. In the case of the heart the x-ray tube would still have to rotate around the patient to acquire data from all directions intersecting the heart, but it would not need to translate to obtain data from the entire thorax section. This would imply a significant reduction in x-ray absorption and time.

Unfortunately it has been shown that the reconstruction problem does not have a unique solution in case of local data. While attempts to reconstruct the density f itself had to be abandoned, an original idea was introduced independently by Vainberg et al. [1981] and by Smith and Keinert [1985], that forms the basis of *local tomography*. Their method reconstructs another function, strongly related to f though, having the same singular set. This is a very significant set as far as imaging, being the set of all points of abrupt changes, like those encountered in going from tissue to veins or arteries. The contour between such regions is actually enhanced by local tomography (Fig. 7.13, 7.14). Indeed local reconstruction is expected to increase spatial resolution and to avoid reconstruction artifacts like streaks and shading. Direct information about the size of the jumps of f is not given but in many cases the algorithm can be modified to provide a reconstruction close to the electronic density (Faridani [2003]).

Using x-rays generated at the National Synchrotron Light Source of the Brookhaven National Laboratories (USA), Erik Ritman of the Mayo School of Medicine in Rochester and collaborators obtained, in local reconstruction, images of a

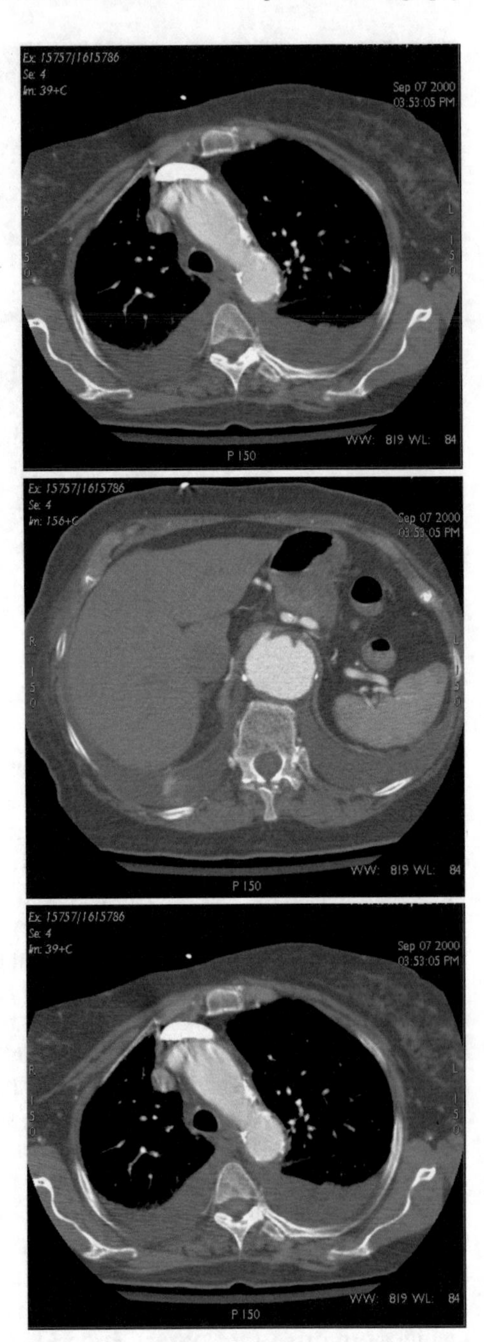

Figure 7.12 A source image of the chest, abdomen, and pelvis for Figure 7.11. (*Courtesy of L. Logan, Department of Radiology, Stanford University*)

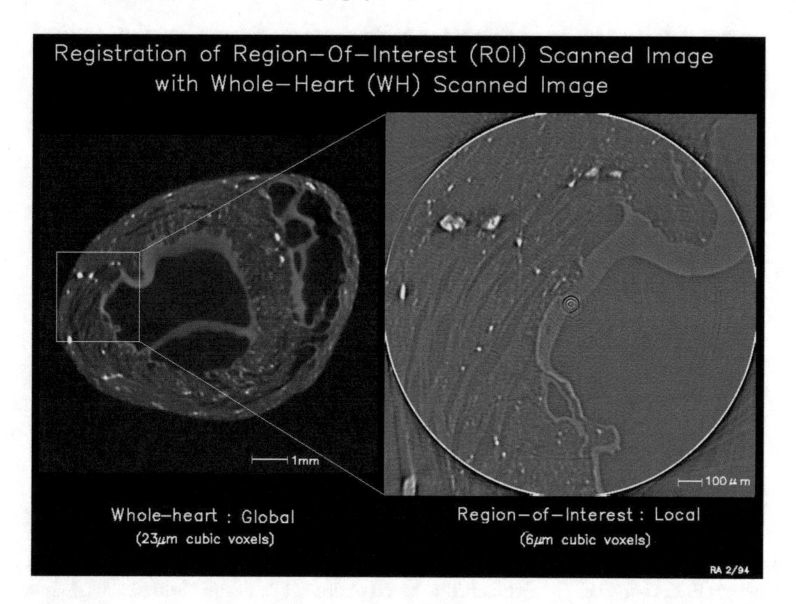

Figure 7.13 The entire cross section of a rat's heart reconstructed with global reconstruction. In Fig. 7.14 the detailed region of interest is reconstructed with local reconstruction. (Ritman E. L. et al. [1997],"Synchrotron-based micro-CT of in situ biological basic functional units and their integration," *Proceedings of SPIE, Developments in x-ray Tomography* 3149, pp. 13-24.) (*Courtesy of E. L. Ritman, Department of Physiology and Biophysics, Mayo Graduate School of Medicine*)

rat's heart with seven micron resolution (Fig. 7.14), a very significant improvement with respect to the about one mm resolution of global tomography as well as MRI. In Figure 7.14 note edge enhancement features of the local reconstruction and the sharper outlines of the opacified microvessels. Also note the radial shading within the global reconstruction, which is not found in the local reconstruction.

Figure 7.15 is a 3D reconstruction of a rat's heart with 23 micron resolution. The heart itself was about one cm in transverse diameter. Local tomography has also been used in nondestructive testing, as in Sivers et al. [1993],Vainberg et al. [1985].

The future of CT will surely bring further improvements concerning the reduction of artifacts, image quality, and space resolution. Assuming a wider point of view, an increasing use of CT scans for industrial purposes can be foreseen, for detailed inspections of complex or critical parts in rocket motors, turbines, and nuclear reactors.

Also the introduction of cardiovascular CT among clinical examinations can be expected. The rapid heart beat — one every second on the average — requires scanning times of the order of the millisecond to avoid artifacts. Since the beginning of the 1980s, prototypes have existed with scanning times of 30–50 ms, further reduced to ten ms in more recent ultrafast scanners. At the same time improvements in space resolution would be greatly beneficial particularly for image reconstruction of arteries. The inner diameter of an artery is on the average 2–4 mm, but smaller

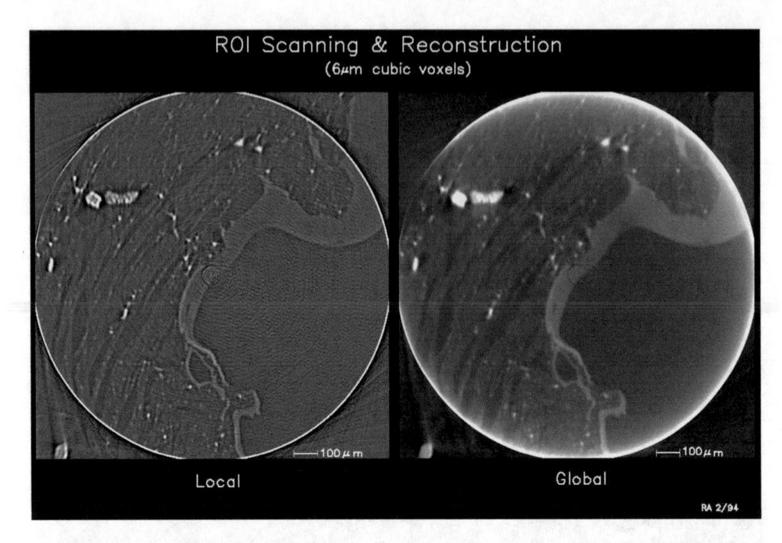

Figure 7.14 Details of the myocardial wall in a rat heart: local and global reconstruction side by side. (Ritman E. L. et al. [1997], "Synchrotron-based micro-CT of in situ biological basic functional units and their integration," *Proceedings of SPIE, Developments in x-ray Tomography* 3149, pp. 13–24.) (*Courtesy of E. L. Ritman, Department of Physiology and Biophysics, Mayo Graduate School of Medicine*)

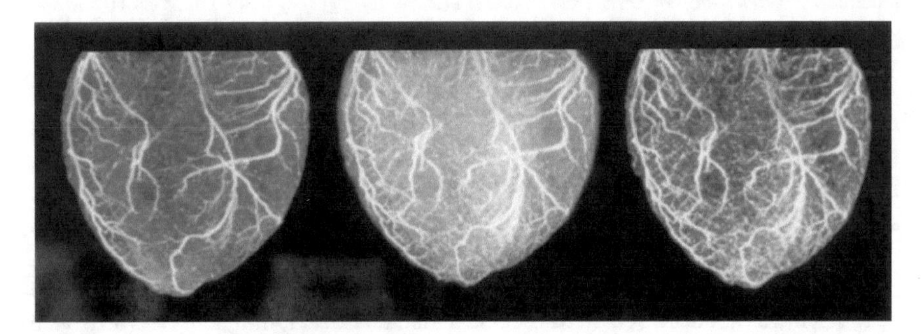

Figure 7.15 3D reconstruction of a rat's heart with 23 micron resolution. The heart had contrast dye in the arteries. (Reprinted with permission from Faridani A., Finch D.V., Ritman E. L., Smith K. T. [1997], "Local tomography II," *SIAM J. Appl. Math.* 57, pp. 1095–1127.) (*Courtesy of E. L. Ritman, Department of Physiology and Biophysics, Mayo Graduate School of Medicine*)

ones might be of interest. A higher resolution would reduce the need for invasive procedures as well.

The speed of progress allows great hope: While in 1963, it took Cormack two days to obtain 256 data on which to base a reconstruction, nowadays CT tests provide reconstructions almost instantaneously.

Chapter 8
Nuclear Magnetic Resonance: Imaging and Spectroscopy

8.1 "We are all radio stations"

After Maxwell's radical work of synthesis, the discovery of the electron in 1897 by Joseph J. Thomson brought questions concerning electrical properties to the subatomic level. The concept of quanta of energy, advanced by Planck in 1900, triggered the development of *quantum mechanics*, also called *wave mechanics* for it describes electrons by wave- functions. Investigations of magnetic properties went along. These investigations will mark the history of nuclear magnetic resonance (NMR).

The starting point might be placed back in 1896 when the Dutchman Pieter Zeeman (1865–1943) performed a spectroscopic experiment that would earn him the Nobel prize for physics in 1902. Visible light spectroscopy had already taught scientists that each element has its own spectroscopic pattern (Section 5.2) which, as a signature, allows its identification. Zeeman wondered about the effect that a magnetic field would have on that signature. So he placed a Bunsen burner in a strong magnetic field and recorded a substantial effect as he observed the spectral lines of the yellow light, emitted by sodium, split into multiple narrow lines (Zeeman effect). A year later in the very same paper in which he calculated the rate of energy radiated by an accelerating electron (Section 5.6), Joseph Larmor advanced a first theoretical analysis of the Zeeman effect (Larmor [1897]).

In Frankfurt, in the years 1921 and 1922, Otto Stern (1888–1969) and his collaborator Walther Gerlach (1889–1979) performed an experiment on an atomic beam (*molecular beams* they are usually called) of silver atoms, with the goal of measuring the magnetic moment of individual atoms (*electronic magnetic moment*) and to test the theory of *space quantization*. According to this theory, advanced by Arnold Sommerfeld (1869–1951), only certain quantized orientations are allowed. The two scientists vaporized silver by heating it in a furnace. The electrons, being charged and in motion, make the silver atoms behave like tiny compass needles. (The silver

E. Prestini, *The Evolution of Applied Harmonic Analysis*,
Applied and Numerical Harmonic Analysis,
DOI 10.1007/978-1-4899-7989-6_8

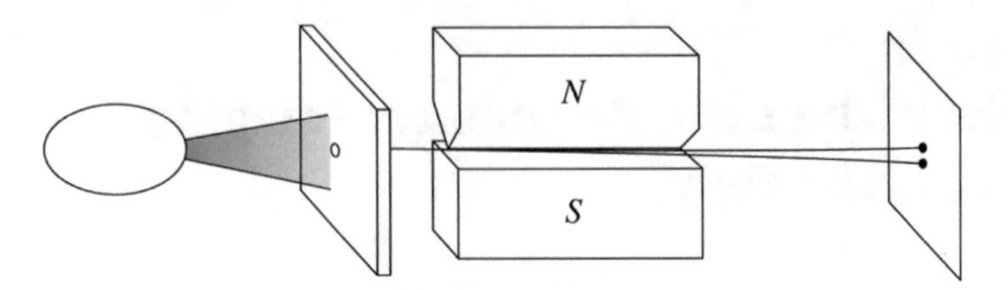

Figure 8.1 The set up of the Stern–Gerlach experiment.

atom has 47 electrons, 46 of which can be visualized as a symmetric electron cloud whose magnetic moment is zero. Hence the electronic magnetic moment **m** is due to the 47th electron only.)

The silver atoms were allowed to escape, through a series of small holes, to be directed horizontally between the pole tips of a magnet (Fig. 8.1). Thus, going through the magnet the atoms experienced a magnetic field **B** directed vertically along the z-axis. One of the tips, say the north tip, was very sharply edged to make the magnetic field strongly nonuniform. Therefore the *field gradient*, a vector that points in the direction of maximum variation of the magnetic field, was directed upward. According to the theory, each atom was expected to be deflected by an amount proportional to the component of its magnetic moment along **B**, to be denoted by m_z. Precisely the formula has the atoms pulled by a force directed along the field gradient and proportional to both m_z and the strength of the field gradient. (If **B** were constant, the field gradient would be the null vector and no deflection would take place.) In the classical model all directions are equally possible for the magnetic moments: Those atoms with magnetic moment parallel to **B** ("up") would be attracted upward, those with antiparallel orientation ("down") would be pushed downward, those at right angle with the field would proceed unaffected, and so on. In other words the overall effect was expected to be the widening of the narrow beam of heated atoms: In the central region of a cold glass detector the experimental finding should have been a condensed and deposited smear of silver in the shape of a continuous distribution such as a vertical line segment. From the length of the segment the strength of the magnetic moment could be calculated.

It came as a shock to see no continuous distribution on the glass (Gerlach [1922]). Only two distinct traces showed and nothing in between, as if the initially randomly oriented atomic moments were given by the superimposed magnetic field only two choices, parallel or antiparallel (*quantized orientations*), leading to two separated beamlets.

The Stern–Gerlach experiment did not allow precise measurements, the deflection being very small. It was for the Hungarian born Isidore Isaac Rabi (1898–1988) to obtain a much greater precision by his method of "magnetic resonance" that would in essence provide the technique for all subsequent experiments in the field.

Magnetism is the main theme in Rabi's work, starting with his dissertation at Columbia University titled "On the principal magnetic susceptibilities of crystals" (Rabi [1927]). Interest in the magnetic properties of certain crystals, with uncommon electrical properties (Tutton salts), was spurred by his attendance at a seminar by the Nobel laureate Lawrence Bragg. The setup of the experiment was suggested to Rabi by a passage in Maxwell's *Treatise on Electricity and Magnetism*, where Maxwell discusses the force of a nonuniform magnetic field on a magnetic body immersed in a fluid with a magnetic salt dissolved in it. If the fluid has a greater susceptibility, the immersed object will move in a certain direction; if fluid has a smaller susceptibility, the object will move in the opposite direction; and if the susceptibilities are the same, there will be no motion.

So Rabi measured the magnetic susceptibility of a crystal by suspending it in a glass tube containing a saturated solution and by placing it between the poles of an electromagnet. He changed the susceptibility of the solution by adding magnetic salt until no motion was observed. (Beyond that, by adding more salt, the crystal would have started turning in the opposite direction.) From the susceptibility of the solution that of the crystal could be measured.

The years 1925–1927 were key in the development of quantum mechanics and Rabi was eager to get over to Europe, right where it was being done. In 1927, when a fellowship was granted to him, he resigned his teaching position at Columbia, having been denied leave of absence, and left to spend 2 years in Europe. He visited all the major centers where the new quantum theory was developed and in particular during 1927–28 in Hamburg he worked with Stern on molecular beams.

In 1929 he was back at Columbia, from where he spread the new physics in the U.S. and helped put together one of the finest physics departments in the country. Here he set about measuring the nuclear magnetic moment, a more difficult task than measuring the electronic magnetic moment, for the first is about a thousand times smaller than the second.

The existence of a small nuclear magnetic moment had been advanced in 1924 by Wolfgang Pauli (1900–1958) to explain the hyperfine structure observed in atomic spectra. The interaction of nuclear magnetic dipoles with unpaired electrons was thought to produce the hyperfine multiplet (Fig. 8.11).

The principle of resonance was first the subject of a theoretical investigation by P. Güttinger [1931], by E. Majorana [1932], in greater generality by I. Rabi [1937], and the same year by J. Schwinger [1937]. The Dutch physicist Cornelius J. Gorter [1936] was first to point out that the phenomenon could be used to detect nuclear magnetism, however, his experiments in solid matter failed. It was for Rabi in 1939 to perform the first successful experiment, which earned him the Nobel prize for physics in 1944. He had years of experience with nuclear spins and, with the help of graduate student Sidney Millman and postdoctoral fellow Jerrold Zacharias, he had built a sophisticated molecular beam apparatus. The idea of using a radiofrequency oscillator in the setup of the experiment came from Gorter himself, during a visit at Columbia in the fall of 1937.

Rabi's experiment, performed with lithium, carries the same mark of precision as his former experiment with crystals and can be used for measuring electronic moments as well. The apparatus was in the spirit of Stern–Gerlach, but more refined, employing three magnetic fields. First came an inhomogeneous field **A**, to split the beam; last an inhomogeneous field **B** that deflected in the opposite direction as **A** and was meant for refocusing; in the middle was imposed a homogeneous field **H** parallel to **A**, and orthogonal to that a radiofrequency component was added. The beam was detected by a cold wire that heated up when impinged upon. The refocusing condition was fulfilled only if no reorientation occurred in going through **H**. The experiment started by slowly increasing the radio frequency and showed no change up to a point when a significant drop in the signal was detected. The frequency had reached a critical value (*Larmor frequency*) at which many nuclei reoriented, having given the right amount of energy to flip their magnetic moments from "up" to "down." (The same would occur by slowly changing **H** and holding the radiofrequency fixed.) This led to their defocusing—in passing through the field **B**—and to the bottoming out of the signal. Magnetic resonance had been achieved. By a simple calculation, the nuclear magnetic moment of lithium could then be obtained (Rabi et al. [1939]).

Although Rabi had realized conditions of magnetic resonance just for atomic nuclei streaming through an air-evacuated chamber, he knew that the phenomenon was more general. On December 29, 1939 at a meeting of the American Association for the Advancement of Science held in Columbus, Ohio, he said that all atoms, whether part of the heart tissue of man or a piece of steel, may absorb or emit radio waves. The article in the *New York Post* was titled "We're All Radio Stations, Columbia Scientists Report." What Rabi did not know yet was that one could discover whether a living tissue is diseased or not through NMR. That would take some thirty more years.

Meanwhile in Stanford the Swiss born Felix Bloch (1905–1983) was on the same track. In Leipzig (Germany) he had been Heisenberg's first doctoral student and from Heisenberg, only four years older, Bloch "caught the spirit of research." In his dissertation "The quantum mechanics of electrons in crystal lattices," (Bloch [1928]) he explained why conduction electrons in a metal can indeed be dealt with as an ideal gas of free electrons. This common view, which is equivalent to the assumption that the electrons behave as a plane wave, had attracted his attention. He began by considering a simplified model and "by straight Fourier analysis I found to my delight that the wave differed from the plane wave of the free electrons only by a periodic modulation" (Mattson-Simon [1996]). This work spearheaded the creation of the modern quantum theory of solids and laid the foundations for the development of semiconductors.

In 1932 the neutron was the last discovery, made by James Chadwick, to come out of the Cavendish Laboratory in Cambridge (U.K.), following that of the electron by Thomson in 1897 and of the proton by Rutherford in 1920. It helped to solve an accounting problem— the nuclei were known to be about twice as heavy as could be accounted for by the protons alone—but how electrons, protons, and neutrons fitted inside the atom was a matter of study.

As Heisenberg was addressing this problem, Bloch began wondering about the neutron's magnetic properties. The magnetic moment of the electron is a direct consequence of its charge; in the case of the neutron it had to have an entirely different origin. Bloch was fascinated by the problem and originally believed this neutral particle to have a magnetic moment equal and opposite to that of the proton. When in early 1933 Otto Stern came up with a molecular beam experiment that gave a measure of the magnetic moment of the proton to be about 2.5 magnetons and that of the deuteron nucleus (one proton and one neutron) to be between 0.5 and one nuclear magneton, Bloch knew that the neutron had a negative magnetic moment but not exactly equal to that of the proton.

As Hitler came to power, Bloch left Germany and in the spring of 1934 moved to Stanford. Once there he decided to follow a casual suggestion made by another physicist interested in the neutron, Enrico Fermi, whom he had just visited in Rome. "You should sometimes do some experiments. It is really a lot of fun," Fermi had said and the magnetic moment of the neutron provided the right occasion.

In the summer of 1937, during Fermi's visit to Stanford, Bloch came up with the idea of using a technique closely related to the one Rabi would implement later that September, at Gorter's suggestion, to use a radiofrequency oscillator. Bloch did not publish the idea but mentioned it to Fermi, so Rabi was first in tying his name to nuclear magnetic resonance. The strong neutron beam, needed for direct measurements, was obtained at the 37-inch Berkeley cyclotron and on May 1939 in collaboration with Louis W. Alvarez—a young faculty member at Berkeley—a value of 1.935 ± 0.002 nuclear magnetons was measured by an experimental setup much in the spirit of Rabi's molecular beam apparatus.

Bloch's plans to pursue his subject even further to achieve an accuracy of at least three decimal digits were soon disrupted. The U.S. entered the war during December 1941. After a brief stay at Los Alamos where he joined the group working on the atomic bomb, in November 1943 Bloch moved to Harvard's Radio Research Laboratory in Cambridge. The work there was centered around microwave radars, a new field opened up by the British. Under the direction of the Stanford engineer Frederick E. Terman, Bloch had the opportunity, all-important for his subsequent work, to get acquainted with radio techniques. Rabi too, who had joined MIT's Radiation Laboratory in 1940, was in Cambridge. The two scientists lived close by and had many occasions to discuss scientific issues.

In early 1945 the end of the war was in sight. Still in Cambridge, Bloch had the idea for detecting *nuclear induction* in condensed matter by far simpler electromagnetic methods. In the theoretical paper Bloch [1946], he dealt not only with the so-called Bloch equations but also with two types of relaxation times, "longitudinal" T_1 and "transverse" T_2, that would soon prove to be important for chemical analysis and medical diagnosis as well.

Back to Stanford in the summer 1945, with the help of the graduate student Martin E. Packard and of the expert in radioengineering William W. Hansen, Bloch started to build his experimental apparatus. Calculations had convinced him that a large voltage signal would result; still he was concerned that the signal could be drowned out by the approximately million times stronger random thermal vibrations.

A sample of 100 mg of water sealed in glass was used and a strong static field of about 2,000 gauss was applied, oriented in the z direction, let us say. The nuclear magnetic moment of the oxygen being zero, the effect was limited to the hydrogen nuclei which aligned themselves parallel or antiparallel with the field, with a small surplus in the parallel direction. This is what makes the magnetic method work, but how long the process would take was a matter of concern. The sample was surrounded

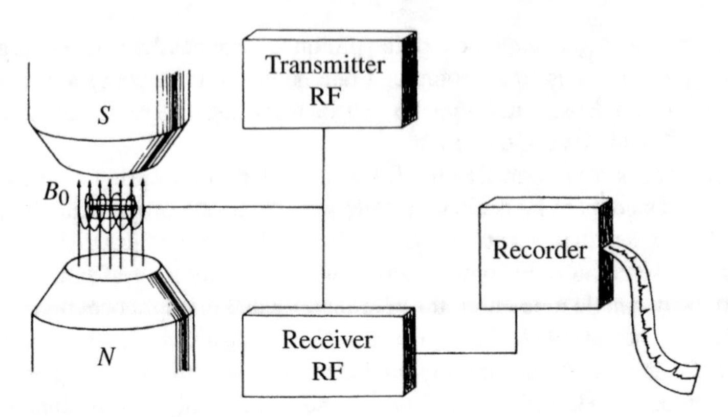

Figure 8.2 Scheme of a nuclear magnetic resonance spectrometer. The sample is placed in a magnetic field. As the radiofrequency transmitter is tuned to the resonance frequency, a signal arises in the receiver. (Adapted from Ruffato C. et al., *RMN in medicina*, Piccin, [1986].)

by a transmitting coil producing flux in the x direction and by a receiver coil in the y direction (Fig. 8.2). As a radiofrequency field in the x direction was superimposed, the parameters were adjusted to meet the anticipated point of resonance. Then the nuclei flopped, inducing a change of the magnetic flux that manifested as a small voltage, at the same frequency, across the terminals of the receiver coil. Bloch's successful experiment dates to early January 1946. About that time he heard that two weeks earlier Edward Mills Purcell (1912–1997) had done a similar experiment with hydrogen rich paraffin. For these achievements they shared the Nobel prize for physics in 1952.

During the war Purcell had also been working on microwave radar at the MIT Radiation Lab and he too drew inspiration from that period for his immediate postwar interests. The task at the Lab was to shorten the wavelength in order to reduce the size of the antenna and to improve the target definition. This went together with a host of technical problems. A curious one arose in reducing the wavelength from the 10 cm of the British to 3 and then to 1 cm: At 1.25 cm sometimes the waves got almost no reflection back. People were very frustrated until the reason was found. The 1.25 centimeter wavelength, chosen for convenience, was actually very inconvenient due to an absorption band of water vapor centered at 1.3 cm. When too much moisture was in the air, all the radar energy was absorbed by the water molecules and little reached the target, and even less came back to the receiver.

This piece of "bad luck" together with knowledge of the molecular beam experiment, such as learned from Rabi's people, suggested to Purcell the idea of an experiment of nuclear magnetic resonance in condensed matter. Two colleagues of the Radiation Lab, Henry Torrey and the electronic expert Robert Pound, joined his project and together they set about building the experimental apparatus. The idea was to place paraffin, 850 cm^3 as they did, in a strong magnetic field so as to have the nuclear magnetic moments align parallel or antiparallel to the field, reaching the so-called *thermal equilibrium*. How long that would take was unknown: weeks, hours, or seconds? A few hours was estimated—so they presoaked paraffin in a strong magnetic field for ten hours, meant a wakeful night for Purcell on the experiment premises—but it turned out to be of the order of 10^{-4}s. The plan was to supply the system with energy at the right frequency that, once absorbed by the lower energy protons (parallel moments), would induce them to become temporarily higher energy protons (antiparallel moments). The temporary imbalance would be detected by the experimental setup and produce a signal, unless thermal equilibrium had not been reached after all or else the relaxation effect was too feeble. In the process they were informed by Rabi of a similar unsuccessful experiment attempted by Gorter, years earlier. They went ahead nevertheless and, after quite a few unsuccessful tries, they could briefly observe the expected signal due to *nuclear magnetic resonance absorption*, as they called it (Purcell, Torrey, Pound [1946]).

While the principal feature of these two experiments in solid matter was in Bloch's words "the observation of transitions caused by resonance of an applied radiofrequency field with the Larmor precession of the moments around a constant magnetic field," Purcell's method "very closely connected to that of Gorter" looked for a "relatively small reaction upon the driving circuit" while in Bloch's experiment "in the first place, the radiofrequency field is deliberately chosen large enough so as to cause at resonance a considerable change of orientation of nuclear moments. In the second place, this change is not observed by its relatively small reaction upon the driving circuit, but by directly observing the induced electromotive force in a coil, due to the precession of the nuclear moments around the constant field and in a direction perpendicular both to this field and the applied RF field. This appearance of a magnetic induction at right angles to the RF field is an effect which is of specifically nuclear origin and is the main characteristic of our experiment" (Bloch [1946]).

These first two experiments allowed an accurate measure of the magnetic moment of the hydrogen nucleus, the proton. Soon after, the technique was employed to obtain information on chemical compounds and about a few decades later had an extraordinary impact in the field of medical radiology.

8.2 Nuclear spin and magnetic moment

The phenomenon of nuclear magnetic resonance is accounted for by the highly technical theory of quantum mechanics, according to which atomic nuclei have a quantized angular momentum, or *spin angular momentum*, given by a vector **L**. Its magnitude L has the same unit as the Planck's constant $h = 6.63 \times 10^{-34}$ kg m^2/s, according to the formula

$$L = \sqrt{I(I+1)}\frac{h}{2\pi}$$

where the *spin quantum number* I—an invariant property of each nucleus—may assume only integer values or half integer, typically 0, 1/2, 1, 3/2, 5/2.

The simplest nucleus, that of the hydrogen atom, is made up of a single proton whose mass (1.66×10^{-24}g) and charge (1.60×10^{-19} coulomb) are taken as the unit of atomic mass and charge respectively. The spin quantum number is 1/2.

A nucleus composed of p protons and n neutrons has total mass $p + n$ atomic units and total charge p, but the vector addition of the spin's angular momentum **L** of the protons involved, each of magnitude 1/2, cannot be predicted in general. Observations show that an odd number of protons or neutrons or both (at least one proton or neutron unpaired) is needed for **L** to be nonzero and for observing magnetic resonance. There are about a hundred such species, relative to which the phenomenon was indeed soon observed.

Also at least one unpaired electron is needed, in atomic species, to observe electronic resonance (*electron spin resonance*, also called *electron paramagnetic resonance*). This is the case of a restricted class of substances called *paramagnetic*. By applying an external magnetic field **B** such materials are magnetized. The resulting magnetization **M**, the net magnetic moment per unit volume, is parallel to **B** and the ratio M/B is called *magnetic susceptibility*. For bulk matter it may be different. For instance silver atoms have an unpaired electron in the last orbital and yet, in bulk matter, the unpaired electron couples together with a neighboring one whose spin direction is opposite, so as to make silver spin-free. Electron spins are about a thousand times stronger and foreshadow nuclear spins, hence only spin-free substances are suitable for experiments of nuclear magnetic resonance.

As every circular electric current produces a magnetic field, so every spinning nucleus ($I \neq 0$) does. The associated *magnetic moment*, or *magnetic dipole*, is represented by a vector

$$\mathbf{m} = \gamma \mathbf{L}$$

parallel to L (Fig. 8.3). The constant γ—termed *gyromagnetic ratio*—is characteristic for each type of nucleus (Tab. 8.1) and implies a better sensitivity to detection when large. Of all nuclei, those of the hydrogen atoms ^{1}H have the largest γ among stable isotopes (Tab. 8.1) and are of special interest since compounds containing hydrogen form a major part of biological systems. In particular the human body is about 75% water. In the water molecule all the electron spins are exactly balanced so that the

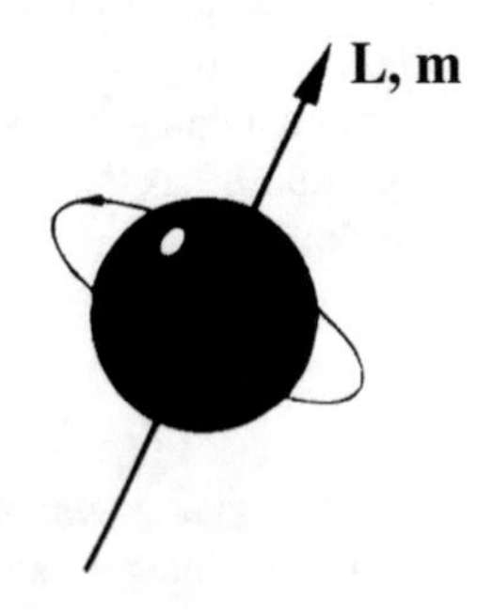

Figure 8.3 An atomic nucleus has a spin angular momentum **L** and also, being charged, a magnetic moment **m** parallel to **L**.

Table 8.1 Properties of important atomic nuclei. (The resonance frequency is measured in MHz; T stands for tesla.) (Hausser K. H., Kalbitzer H. R., *NMR in Medicine and Biology*, Springer, [1991].)

Isotope	Nuclear Spin I	Magnetogyric ration γ [$T^{-1}s^{-1}$]	Resonance frequency at 14.092 T	Natural abundance [%]
^{1}H	1/2	2.6752×10^8	600.0	99.985
^{2}H	1	4.1065×10^7	92.1	0.015
^{3}H	1/2	2.8535×10^8	640.0	—
^{12}C	0	—	—	—
^{13}C	1/2	6.7266×10^7	150.9	1.11
^{14}N	1	-1.9325×10^7	43.3	99.63
^{15}N	1/2	-2.7108×10^7	60.8	0.37
^{16}O	0	—	—	99.76
^{17}O	5/2	-3.6267×10^7	81.4	0.04
^{18}O	0	—	—	0.20
^{19}F	1/2	2.5167×10^8	564.5	100.00
^{23}Na	3/2	7.0762×10^7	158.7	100.00
^{31}P	1/2	1.0829×10^8	242.9	100.00

net molecular moment is zero. On the other hand the water molecule has a very tiny nuclear magnetic moment due to the hydrogen nuclei, the oxygen (^{16}O) having zero nuclear spin.

A convenient unit of measure for nuclear magnetic moments is the *nuclear magneton*, defined to be 1/1,836 times the *Bohr magneton*, itself the magnetic moment of the electron of the hydrogen atom or equivalently 9.274×10^{-24} ampere per square meter (Am^{-2}).

8.3 The swing and the phenomenon of resonance

For an easier understanding, the general phenomenon of resonance is first presented in the gravitational field. A rather common instance is that of a girl pushing a friend on a swing. Obviously, she does not have the strength to lift a body weighing several dozen pounds but knows how to push the swing higher and higher by applying small pushes at the right moment, when the swing is at the lowest point of oscillation. In other words the pushes have to be synchronized with the natural frequency of oscillation of the swing.

Another common instance of resonance occurs with body noise—sometimes heard when driving a car at certain speeds—due to resonance between the natural oscillation frequency of some parts of the body and the frequency of motion of pistons or wheels.

8.4 Driving a resonance: the Larmor frequency

Without an external magnetic field the small nuclear magnetic dipoles of a water sample—we shall restrict ourselves to water from now on unless otherwise stated—are randomly oriented, but when the sample is placed in a static magnetic field they orient themselves in a direction parallel or antiparallel to the field (space quantization), as all nuclei with spin 1/2 do. Two opposite energy levels, negative and positive referred to as lower and higher, correspond to these two orientations.

At the absolute zero (–273 °C) all dipoles are parallel, this being the most stable configuration. Rise of temperature might provide some nuclei with the right amount of energy required to flip to the antiparallel orientation. At infinite temperature the dipoles would split equally between parallel and antiparallel orientation. At room temperature the populations of the two states are almost equal, with the lower state slightly more populated than the upper. For every 10^8 protons with antiparallel spin there are statistically 10^8+1 with parallel spin. This suffices to produce a magnetization $\mathbf{M}_0$ detectable at the macroscopic level and directed as the static magnetic field $\mathbf{B}_0$. Since the populations of the two states are in a fixed proportion—which depends on the temperature T measured in degrees Kelvin and on the Boltzmann constant $k = 1.38 \times 10^{-6}$erg/°C–the formula of the magnitude M_0 is, to first approximation (the high temperature approximation), given by

$$M_0 = N_0 \frac{m^2 B_0}{3kT}$$

where N_0 is the total number of nuclei per unitary volume in the sample (proton density). This is a large number. It suffices to think that in a cubic centimeter of water there are about 10^{22} hydrogen nuclei.

Consider a reference frame $(x,\ y,\ z)$ such that the z-axis is directed along the external magnetic field $\mathbf{B}_0$ and magnetization $\mathbf{M}_0$. If $\mathbf{M}_0$ is dislocated from the vertical to a position $\mathbf{M}$ (Fig. 8.4), then it starts to precess around the z-axis with an angular

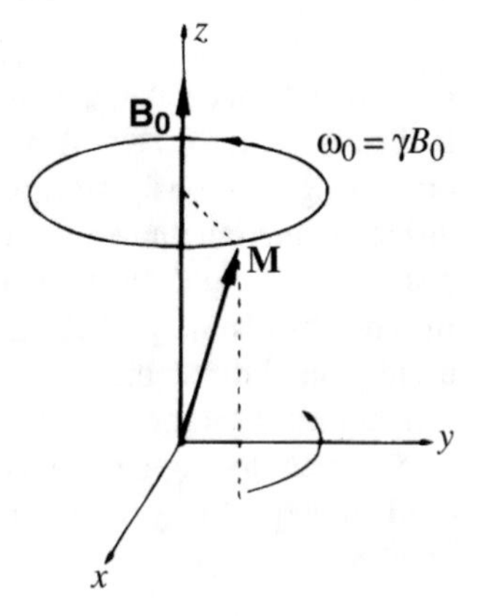

Figure 8.4 Precession of the magnetization **M**.

frequency ω_0, called the *Larmor frequency*, which depends only on the nuclear species and the magnitude of the field $\mathbf{B_0}$ according to the formula

$$\omega_0 = \gamma B_0. \tag{8.1}$$

It is apparent that to a stronger magnetic field there corresponds a greater precession frequency. A spinning top in a gravitational field could be mentioned as a similar phenomenon in classical physics. The spinning top does not fall; rather it undergoes a motion of *precession*: Its axis rotates around the vertical describing a cone. The frequency of precession is proportional to the magnitude of the gravitational field: The same top on the moon would precess with a frequency smaller by a factor of 6 since this is the ratio between the gravitational field on the Moon and that on Earth. With a magnetic field $\mathbf{B_0}$ of magnitude 10,000 gauss (1 tesla), the Larmor frequency of the hydrogen nucleus is 42.57 MHz (42.57 million cycles per second) for the nucleus of phosphorus 31 is 17.24 MHz, and that of sodium 23 is 11.26 MHz, all in the radiofrequency band of the electromagnetic spectrum (Fig. 9.1). With the same $\mathbf{B_0}$ the Larmor frequency of the electron is 28,000 MHz, much lower than the frequencies of x-rays and of the visible light too, indeed in the microwave region.

The dislocation of $\mathbf{M}_0$, mentioned above, can be effectively obtained by applying in the horizontal plane an additional magnetic field $\mathbf{B}_1$ rotating at the Larmor frequency. (Otherwise the two rotating fields would be out of phase and $\mathbf{M}_0$ would wobble only slightly.) An alternating horizontal field, say in the x-direction, oscillating at the Larmor frequency, produces an effect similar to that of a rotating field. It works as well, for the torque on the magnetic moment reverses at each half cycle. In this condition, called *resonance*, a strong interaction takes place: The direction of $\mathbf{M}_0$ suddenly changes. This effect induced by the oscillating field is sometimes referred to as “driving a resonance.”

Hence to observe a chosen nuclear species in isolation from the other nuclear species, it suffices to tune on the corresponding Larmor frequency. In the human body hydrogen is present in water as well as in biological macromolecules. In particular water has a distribution often altered in pathological situations, thus making it possible to distinguish between norm and pathology.

Another species important for NMR is phosphorus 31 which is more involved in questions related to metabolism.

8.5 Relaxation process

In the beam method, each particle can be considered free. In the case of solids, liquids, and gases (confined, as opposed to expanding beams) an interaction takes place between the nuclei and their surroundings which is essential for dissipating the absorbed energy. The temperature of the nuclear spin system otherwise would rise and the surplus number in the lower state would consequently decrease, making $\mathbf{M}_0$ smaller and smaller. When no surplus exists no net absorption can take place (*saturated system*), upward and downward transitions being equally likely. At room temperature at equilibrium, the populations of the two states are very nearly equal. However the lower state is slightly more populated than the upper so that, the two transitions having the same probability, globally upward transitions predominate and a net, yet very small, absorption of energy occurs from the radiation beam. As the populations become equal, upward and downward transitions become equal and no absorption takes place. Only by losing energy may the original equilibrium populations be reestablished.

As the radiofrequency (RF) field $\mathbf{B}_1$ rotating in the $(x,\ y)$ plane at the Larmor frequency ω_0 is applied, the magnetization $\mathbf{M}_0$ is dislocated from equilibrium by an angle that increases with the time duration of the RF pulse and spirals down until it lies in the $(x,\ y)$-plane. It proceeds until it points along the negative z-axis and keeps turning around.

If $\mathbf{B}_1$ is switched off as soon as the magnetization lies in the $(x,\ y)$-plane, then one says that a 90° pulse has been applied. If the time duration of the pulse is doubled, or the strength is doubled, then the magnetization points along the negative z-axis when $\mathbf{B}_1$ is switched off and one says that a 180° pulse has been applied. In general it is possible to turn $\mathbf{M}_0$ in any direction by a suitable duration of the pulse. In a typical NMR experiment a 90° pulse is applied by turning on the RF field. As $\mathbf{B}_1$ is switched off the magnetization returns to its thermodynamical equilibrium by two relaxation processes, the *spin-lattice* and the *spin-spin relaxation process*. The first one concerns the z-component of the magnetization, or *longitudinal magnetization*, which returns to equilibrium by a process characterized by the so-called *spin-lattice relaxation time* T_1. The other concerns the component in the $(x,\ y)-$ plane of the magnetization, or *transverse magnetization*, which goes to zero at a rate of decay determined by the so-called *spin-spin relaxation time* T_2 (Fig. 8.6).

The two processes are independent, for T_1 and T_2 are unrelated. It only holds that $T_2 \leq T_1$ in all cases. In liquids the two relaxation times are about the same and of the order of the second. In solids T_1 could be hours and even days, while T_2 could be of the order of the microsecond.

The time-dependent magnetization induces a small voltage in a detection coil surrounding the sample, which in turn generates an electromotive force oscillating at the Larmor frequency. Amplified and visualized by an oscilloscope it generates the so-called *free induction decay* signal or *NMR signal* (Fig. 8.5) which ends as the

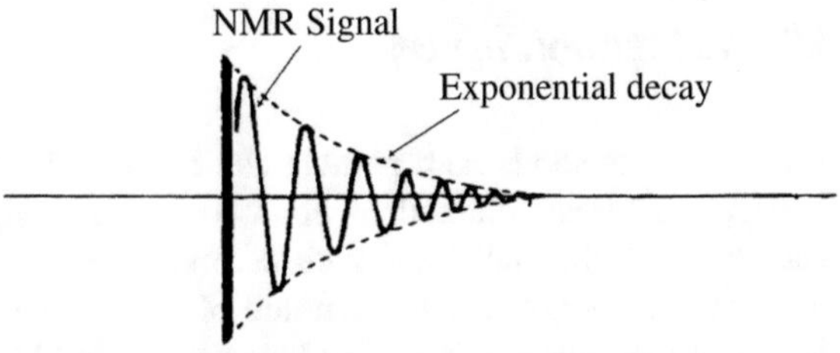

Figure 8.5 The NMR signal for one nuclear species.

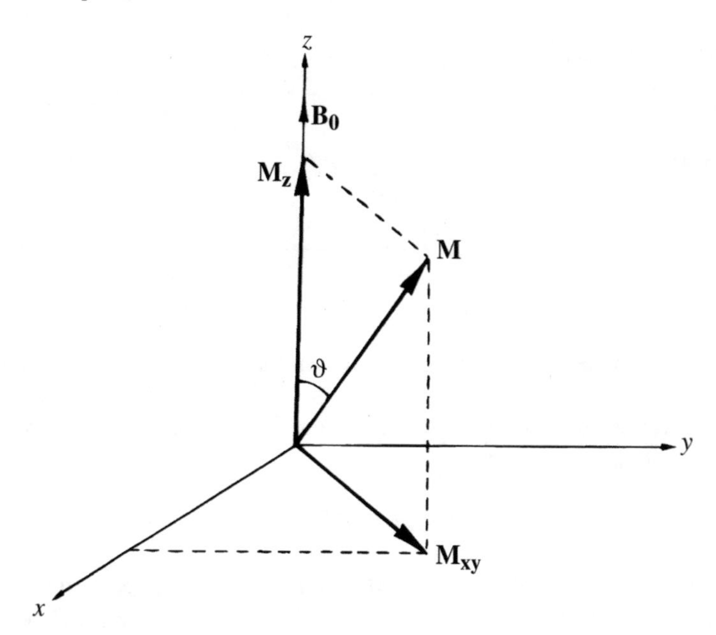

Figure 8.6 The longitudinal $\mathbf{M}_z$ and transverse component $\mathbf{M}_{\mathbf{xy}}$ of the magnetization **M**.

equilibrium position $\mathbf{M}_0$ is reached. The amplitude of the signal is proportional to M_0 and usually decays exponentially. The real part of its Fourier transform is the Lorentzian curve in Figure 3.4, suitably dilated and translated at ω_0.

8.6 Relaxation times

The relaxation process is described mathematically by the *Bloch equations* which form a system of first-order differential equations, one for each component M_x, M_y, M_z of the magnetization **M** (Bloch [1946]). For a system of identical nuclei each experiencing the same magnetic field (i.e., neglecting the shielding effect of the electrons, or chemical shift, as described in Section 8.7), after a 90° pulse, the magnetization longitudinal component M_z, which is zero at time $t = 0$, increases with time and tends to M_0 as indicated by the formula

$$M_z = M_0(1 - e^{-t/T_1}) \tag{8.2}$$

The constant T_1 appearing in (8.2) is the *longitudinal* or *spin-lattice relaxation time* (Fig. 8.7). It is the time required for the z-component of **M** to return to 63% of its original magnitude ($1 - 1/e \cong 0.63$) following an excitation pulse. The spin-lattice relaxation process dissipates energy from the spin system to the *lattice* (surround-

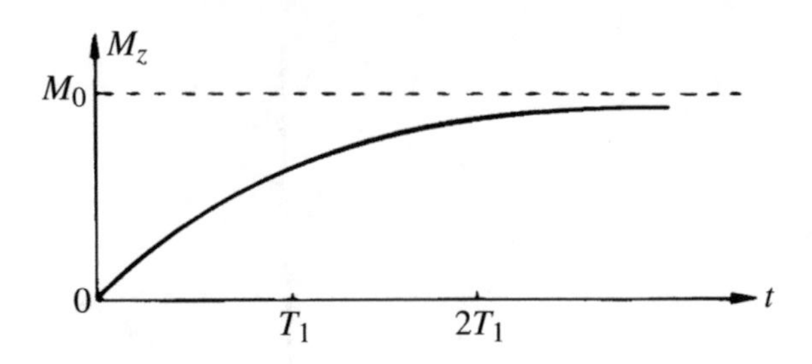

Figure 8.7 The amplitude M_z of the longitudinal component of the magnetization tends to M_0 as time t increases.

ings), by lattice motion (e.g., atomic vibration in a solid lattice, molecular tumbling in liquids and gases).

The transverse component $\mathbf{M}_{xy}$ rotates in the (x, y)-plane as time t goes by. In complex notation it is given by

$$M_0 e^{i\omega_0 t} e^{-t/T_2}. \tag{8.3}$$

The amplitude M_{xy}, clearly equal to M_0 for $t = 0$, goes to zero with time. Its decay is exponential, being given by $M_0 e^{-t/T_2}$. The constant T_2 is the *transverse* or *spin-spin relaxation time*. It is the time required for the amplitude of the transverse component of $\mathbf{M}$ to decay to 37% of its initial value ($1/e \cong 0.37$). As the system returns to thermodynamical equilibrium, the individual spins become randomized or dephased. This is the reason for M_{xy} to go to zero (Fig. 8.8).

In reality small imperfections, which are always present even in the best magnets, make the transverse component and so the free induction signal decay faster than in (8.3). In the literature the constant that controls the decay that follows a 90° pulse, is denoted by T_2^*. So a single 90° pulse does not allow measurement of the true T_2, which can be determined though by suitable pulse sequences. The relaxation times are of the order of the second in liquids and liquid-like materials. In the case of pure simple liquids, T_1 and T_2 are approximately a few seconds. Instead in solid materials T_1 could be hours while T_2 is usually very small, even only a few microseconds. Indeed the spins become dephased very quickly making the signal decay so fast that it is barely detectable. (In particular the magnetization vector $\mathbf{M}$ does not keep the same amplitude, not even approximately, during the relaxation process.) By contrast, in liquids the atoms are in continuous motion and the internuclear magnetic field

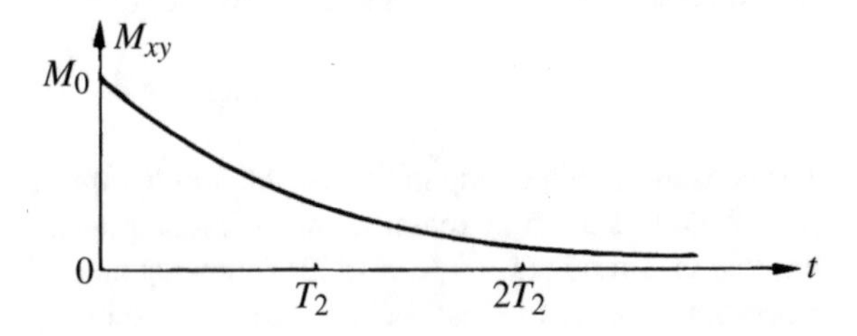

Figure 8.8 The exponential decay of the amplitude M_{xy} of the transverse component of the magnetization.

Table 8.2 T_1 relaxation times (in seconds) at 100 MHz in normal and malignant human tissues. Number of cases analyzed are in parentheses; errors are standard error of the mean. (Damadian R. et al. [1974], "Human tumors detected by nuclear magnetic resonance," *Proc. Nat. Acad. Sc. U.S.*A. 71, pp. 1471–1473.)

Tissue	T_1 tumor	T_1 normal
Breast	$1.080 \pm 0.08(13)$	$0.367 \pm 0.079(5)$
Skin		
Muscle	$1.047 \pm 0.018(4)$	$0.616 \pm 0.019(9)$
Malignant	$1.413 \pm 0.082(7)$	$1.023 \pm 0.029(17)$
Benign	$1.307 \pm 0.1535(2)$	
Esophagus	$1.04(1)$	$0.804 \pm 0.108(5)$
Stomach	$1.238 \pm 0.109(3)$	$0.765 \pm 0.075(8)$
Intestinal tract	$1.122 \pm 0.04(15)$	$0.641 \pm 0.080(8)^b$
		$0.641 \pm 0.043(12)^c$
Liver	$0.832 \pm 0.012(2)$	$0.570 \pm 0.029(14)$
Spleen	$1.113 \pm 0.006(2)$	$0.0701 \pm 0.045(17)$
Lung	$1.110 \pm 0.057(12)$	$0.788 \pm 0.063(5)$
Lymphatic	$1.004 \pm 0.056(14)$	$0.720 \pm 0.076(6)$
Bone	$1.027 \pm 0.152(6)$	$0.554 \pm 0.027(10)$
Bladder	$1.241 \pm 0.165(3)$	$0.891 \pm 0.061(4)$

responsible for the spin-spin relaxation tends to average out, so the free induction signal decays more slowly.

Matter becomes more complicated in the case of water in living tissues. Nevertheless T_1 and T_2 can be accurately measured by suitable techniques, generally consisting of sequences of RF pulses over short periods of time. Numerous studies have produced tables highlighting the differences of T_1 and T_2 for water depending on the different tissues, both healthy and cancerous (Tab. 8.2). These formed the basis of the development of MRI.

Variations from tissue to tissue are due to the different water content but also to a different content of fat cells which are rich in hydrogen and that might be present in high concentration. The corresponding resonant frequencies are only slightly different in case a weak magnetic field $\mathbf{B_0}$ is employed but can be separated in a significant way by applying a strong $\mathbf{B_0}$.

The differences between a normal and a cancerous state, such as discovered by Raymond Vahan Damadian [1971] (Section 8.10), can be explained by the fact that pathological states, in general, give rise to local inflammatory reactions that go together with a higher water content and longer relaxation times.

The NMR signal and the relaxation times T_1 and T_2 are three parameters made available by NMR imaging (Section 8.11). In case the image turns out to be homogeneous with respect to one of them it is possible to obtain contrast by switching to another one.

8.7 NMR spectroscopy: chemical shift and coupling constant

The Larmor frequency being characteristic of the nuclear species, the NMR techniques may be used to detect the presence of particular chemical nuclei in a compound. Also the degree of absorption being proportional to the number of resonating nuclei, it is possible to estimate them quantitatively.

A far greater potential of NMR techniques lies elsewhere. This was soon realized and went together with the availability of highly homogeneous magnetic fields, so that tiny effects are not covered up by the inhomogeneities of the static field. The NMR spectrum of a compound can indeed serve as a "fingerprint" to identify the compound due to a special effect called the *chemical shift*. According to Bloch, Niels Bohr once explained the chemical shift this way: "What these people do is very clever. They put little spies into the molecules and send radio signals to them, and they have to radio back what they are seeing." Indeed in Bloch's own words, "It was not anymore the protons as such. But from the way they reacted you wanted to know in what kind of environment they are, just like spies that you send out" (Mattson, Simon [1996]).

Purcell in his Nobel lecture phrased it this way, "The magnetic field of an atomic nucleus differs slightly from the field externally applied because of the shielding effect of the electron cloud around the nucleus. In different molecules the atom's electron configuration will differ slightly, reflecting differences in the chemical bond." These differences "are interesting to the physical chemist because they reveal something about the electrons that partake in the chemical bond."

This effect was studied theoretically for single atoms by Willis Lamb at Columbia already in 1940. In the years 1949–1950 Norman Ramsey at Harvard further studied the "chemical effects"—as he called them at the time—in different molecules (Ramsey [1950], [1951]).

About the same time these shifts began to be detected experimentally by a number of researchers. Walter Knight [1949] observed them in metals and Arnold, Dharmatti, and Packard [1951] in liquids. In this last paper is found the best known example of the phenomenon: the NMR spectrum of ethyl alcohol CH_3CH_2OH (Fig. 8.9). The hydrogen nuclei of the molecule of ethyl alcohol—protons—if observed in isolation, would of course resonate at the same frequency. Instead because of the chemical

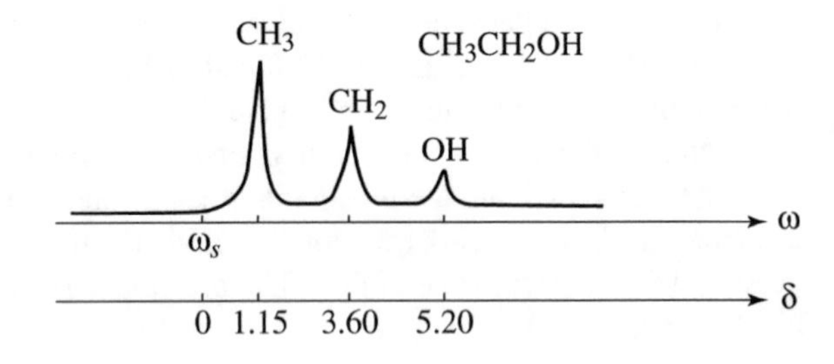

Figure 8.9 Spectrum of ethyl alcohol CH_3CH_2OH as a function of frequency and of chemical shift.

bonds in the molecule they resonate at three slightly different frequencies, all slightly less than the expected one. The areas underlying the peaks are proportional to the numbers 3, 2, 1. The magnetic moments of ^{16}O and ^{12}C being zero, the effect can only be due to the electronic cloud around the hydrogen nuclei of the groups CH_3, CH_2, and OH. (In liquids molecular tumbling continually changes the orientations of the molecules and the effects due to neighboring spins average to zero.) In nature all nuclei are usually found associated with electrons in atoms and compounds. When an external magnetic field is imposed, the surrounding electrons tend to circulate in such a direction as to produce a field opposing the one applied. Thus the nucleus can be said to be shielded from the applied field to an extent that is constant for an isolated atom but varies with the electron density about an atom in a molecule. For instance oxygen being a much better electron acceptor than carbon, the electron density about the hydrogen atom in C–H bonds ought to be higher than that in O–H bonds and weaker is the field experienced by the better shielded hydrogen nucleus in the C–H bonds. Indeed the CH hydrogen nuclei precess with a smaller Larmor frequency than that of OH.

Historically the NMR spectrum was produced and recorded by holding fixed the frequency of $\mathbf{B}_1$ and by slowly increasing the field $\mathbf{B}_0$ to observe the absorption peaks at resonance. Equivalently one may increase the frequency of $\mathbf{B}_1$ at a fixed $\mathbf{B}_0$, as in the modern procedure. In the first case the chemical shift δ is measured in milligauss, in the second case in Hz. Nevertheless such a measure is not appropriate because δ is directly proportional to the applied field $\mathbf{B}_0$ and changes as $\mathbf{B}_0$ is changed. Similarly for the frequency of $\mathbf{B}_1$. Instead the *chemical shift* is conveniently measured with respect to a standard substance by the formula

$$\delta = \frac{\omega - \omega_S}{\omega_S}$$

where ω_S is the Larmor frequency of the standard, which therefore has δ set equal to zero.

The frequencies ω and ω_S are generally very close to each other in size. The pure number δ is then expressed as a multiple of 10^{-6}, that is as parts per million (p.p.m.), the denominator ω_S being of the order of 10^{-6}. For instance in case of ethyl alcohol the chemical shifts of CH_3, CH_2, and OH are respectively 1.15 p.p.m (meaning $\delta = 1.15 \times 10^{-6}$), 3.60 p.p.m., and 5.20 p.p.m., S being the standard reference substance for liquids (Fig. 8.9).

Thus qualitative information is gathered by the fact that the chemical shift for the same nuclei (^1H nuclei) is different depending upon different chemical surroundings. Quantitative information comes from the area underlying each absorption peak, which is proportional to the number of nuclei with that resonant frequency. The chemical shift can be up to 10 p.p.m. in the case of the hydrogen nuclei, but it can reach 300 with phosphorus 31 (^{31}P) because of its higher electronic density.

In spite of being usually rather small, the chemical shift is a most important parameter. Allowing to distinguish between nuclei of the same species in the same molecule, as the spectrum in Figure 8.9 shows, it opened up a new field of research called *high resolution NMR spectroscopy in liquids*, which provides a powerful tool, for the alternative to crystallography investigation (Chapter 6) of the structure of organic molecules like proteins, polysaccharides, lipids, and nucleic acids and study of the dynamics of chemical processes. Many NMR experiments in chemistry, biology, and medicine belong to this area.

For spatial structure determination, which indicates which chemical groups are near neighbors in the molecule, an even smaller effect has to be brought into the picture. In the computation of the magnetic field, experienced by a dipole, the effect of the neighboring dipoles has to be taken into account. This effect, called *dipole–dipole coupling*, has been studied. In the case of liquids these effects average to zero over time because molecular tumbling leads to a fast reorientation of the molecules with respect to the direction of $\mathbf{B_0}$. In solids usually dipole–dipole coupling causes a broadening of the absorption lines.

The "local" field associated with this effect has a small intensity and usually is not directed in the same way as $\mathbf{B_0}$.

At a fixed dipole the local field, due to a neighboring dipole, depends on the distance r between the two and strongly decreases with that distance, as $1/r^3$. Also it depends on the angle θ between $\mathbf{B_0}$ and the vector connecting the two dipoles, as can be seen for instance by drawing the lines of force in the two extreme cases $\theta = 0$ and $\theta = \pi/2$. In particular there exists an angle $\theta \approx 55°$, called the *magic angle*, at which the local field is zero. This is an angle the techniques of NMR in solids take advantage of.

In the case of liquids, however, splitting is often observed, not due to the direct coupling described above but rather to a coupling—mediated by the electron distribution—by which nuclei in the same molecule may influence each other. The phenomenon is called *spin-spin coupling* and, being independent of the size and direction of the external magnetic field $\mathbf{B_0}$, is expressed in terms of a constant J measured in Hz. The simple case in Figure 8.10, where two hydrogen nuclei are joined by a pair of bonding electrons, will be used as an example. If e, one of the electrons, is close to the proton E at a given moment, then the most probable configuration has the magnetic dipole of the electron opposed to that of the proton. By the Pauli exclusion principle the spin of the other electron d must be opposed to e. In the end the dipole of the nucleus D has to be opposed to the one at E so the two are indirectly coupled, that is locked in the paired configuration of Figure 8.10. Paired and opposed configurations are almost equally probable, but not exactly. Hence the resonant line of E is split into doublets centered at the chemical shift by the action of D. Similarly the resonant line of D by the action of E. The equal spacing of the doublets, the so-called *coupling constant* J, is usually measured in Hz. It can reach a few hundred Hz. Its value decreases rapidly as the number of bonds between the nuclei increases (Banwell [1966]).

Coupling between groups like CH_3 and CH_2 is more complicated. It can be shown theoretically that the two nuclei of the CH_2 group split the resonant line of the CH_3 group into three states (a triplet) and the three nuclei of the CH_3 group split the line of CH_2 into four states (a quartet), as verified experimentally. The coupling constant is 7 Hz. When a "fine" structure like the triplet and the quartet (Fig. 8.11) with the

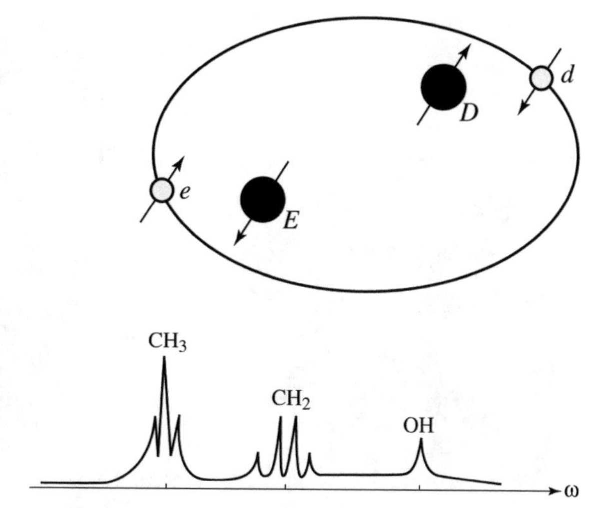

Figure 8.10 The coupling of nuclear spins in the same molecule via bonding electrons.

Figure 8.11 The structure of the NMR spectrum of the hydrogen nuclei for ethyl alcohol CH_3CH_2OH due to spin-spin splitting.

right chemical shift appears in an NMR spectrum, one knows that it comes from the CH_3 and CH_2 groups and moreover that they are near neighbors. No splitting is observed in the resonant line of the OH group. Indeed the OH group is rapidly exchanged and the replacing hydrogen nuclei do not necessarily have the same spin direction, so the net coupling averages to zero.

Knowledge of the above effects concurs with the determination of the complete three-dimensional structure of biological macromolecules. This is an involved process that takes into account still another effect, the "nuclear Overhauser effect," so called after its discoverer in 1953 (Hausser, Kalbitzer [1991]).

8.8 Prion protein

Prion protein (PrP)—made up of 207 aminoacids—is associated with an unusual class of neurodegenerative disease, which include scrapie in sheep, bovine spongiform encephalopathy (BSE) in cattle, and Creutzfeldt–Jacob disease (CJD) in humans.

The normal PrP, whose function is still unknown, is a secreted cell surface protein mainly present in brain cells. So far PrP has been found in a whole range of mammals, marsupials, birds, and a turtle. The three-dimensional structure of the normal form of PrP (Fig. 8.12) consists of three α-helices (Helix A, B, C) and a small flat conformation, called a β-sheet (Strand A and B), made of four aminoacids (two in each strand). The protein is anchored to the cell membrane by a sugar structure (GPI) attached to its C-terminus. In addition the protein has two large sugar structures labeled NGlyc181 and NGlyc197.

The three-dimensional structure of humans for PrP has been resolved by NMR experiments made by Kurt Wüthrich and his group in Zurich working at ETH and

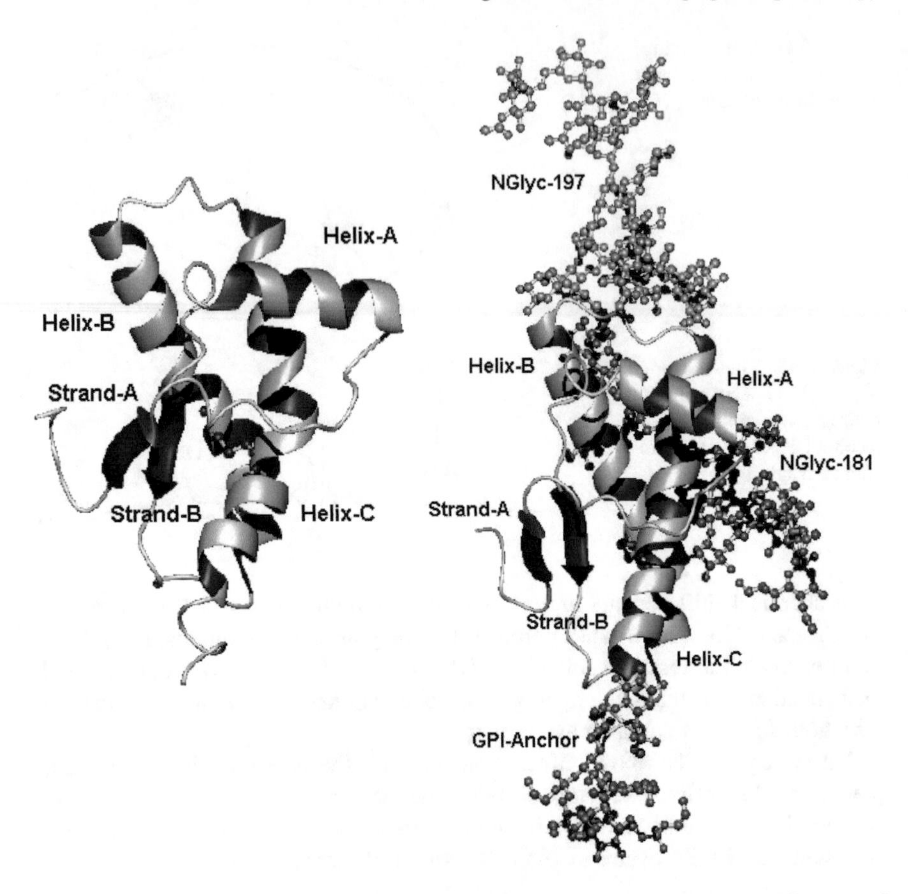

Figure 8.12 Structure of the normal prion protein without and with sugar structures. (*Courtesy of J. Zuegg, School of Medical Research, Australian National University*)

by the Nobel prize winner Stanley Prusiner and his group in San Francisco working at UCSF (Zahn et al. [2000]). It has also been resolved in the case of mouse (Riek et al. [1998]), hamster (Liu et al. [1999]), and bovine (Garcia et al. [2000]). All those PrP show a highly similar structure with only little differences.

Starting from NMR data and using molecular dynamics simulations (Gunsteren, Berendsen [1990]) the sugar structures (GPI and NGlyc) have been added (Fig. 8.12) and their influence on the protein structure investigated (Zuegg, Gready [1999], [2000]). In particular the simulation could show that the sugar structure anchoring the protein to the cell membrane maintains the protein at a distance between 9–12 Å from the membrane surface.

According to the current hypothesis (Prusiner [1982], [1998]) the disease is caused by an abnormal form of the PrP that differs from the normal form only in its three-dimensional structure. The structure of the pathogenic form, which can be precipitated in a clumped form in the brain, has not been resolved yet, being highly insoluble (the resolution of NMR experiments in solids is insufficient to resolve the structure of a protein). Nevertheless experiments showed an increased content of β-sheet structure.

X-ray crystallographic techniques (Chapter 6) have been of little help so far. Crystals of the abnormal malign form are not high quality so that the diffraction data collected from x-ray experiments are only low resolution (Inouye, Kirschner [1997]). Crystallization of the normal prion protein has not been obtained so far.

8.9 Collecting NMR spectra: continuous wave versus pulses

The historical method for collecting spectra is also the most intuitive and the simplest. When the sample is placed into a homogeneous magnetic field $\mathbf{B_0}$, the radio frequency field $\mathbf{B_1}$ is continuously (CW) changed and the absorption lines recorded. This happens only when the scan passes through a resonant line. (Equivalently the RF field $\mathbf{B_1}$ is held fixed and $\mathbf{B_0}$ is slowly changed.) It is a time-consuming procedure—for instance it could take about half an hour to collect the spectrum of ethyl alcohol (CH_3CH_2OH)—requiring an extremely stable magnetic field over time and heading toward saturation. To make things worse, in the presence of a poor signal-to-noise ratio (*sensitivity*) most of the time is spent collecting off-resonance noise. As Weston Anderson wrote in an article in *Radiology Today* of 1992: "Early work on practical applications was so disappointing that researcher Martin Packard quipped 'You could smell the cork on the bottle and almost make the same analysis as those early systems.'"

The idea of using a broadband excitation was advanced and a patent was filed already in 1956 by Russel Varian of Varian Associates in Palo Alto (California). The idea was there but the procedure still took a long time, so long in fact as to appear impractical. Ten years later Richard Ernst and Weston Anderson [1966], of Varian Associates, used a pulse (it may be noted that pulsed NMR experiments were already described in Bloch [1946] with the different purpose of measuring the relaxation times), since a short pulse has a wide frequency band, according to the indeterminacy principle (2.17). If a radiofrequency oscillator, operating at the Larmor frequency ω_0, is switched on and quickly switched off—typically the pulse duration is 10^{-5}s—the sample is exposed to a wide range of frequencies centered at ω_0.

One nuclear species, in ideal conditions, produces a signal decreasing exponentially (Fig. 8.5) and oscillating at the Larmor frequency. If many nuclear species are present, the signal is the sum of many modulated exponentials. The problem of disentangling them is solved by calculating the Fourier transform whose real part is the sum of many Lorentzian curves each centered at the corresponding Larmor frequency. At the time this was a long calculation that put the feasibility of the method into question. Indeed in the same article mentioned above, Anserson wrote "what good is a 10-second data collection if you couldn't see the spectra for several days?" The advent of cheap computers and the algorithm of the fast Fourier transform (Section 3.9) of 1965 made the difference. The resulting signal, the so-called free induction decay (FID), is sampled and the recorded values quantized. According to the sampling theorem, the sampling time is chosen to be $1/2\omega_{max}$ where ω_{max} is the highest frequency to be observed.

A greater sensitivity is obtained by applying the time-averaging method, considered in Ernst [1965]. It consists of generating a sequence of n identical pulses that give rise to n identical signals. The idea then is to average those signals, for the true signal can be expected to add coherently, while the noise tends to average out because of incoherence (Section 3.6). Indeed it can be shown theoretically that the signal-to-noise ratio increases as the square root of the number n of successive measurements or "scans." All of the above accounts for the acceptance and almost exclusive application of the pulsed NMR nowadays.

Ernst returned to his native Switzerland in 1968, to the Federal Institute of Technology (ETH) in Zurich from which he had graduated in chemistry in 1956, some 28 years after Bloch. There in the summer of 1974 his group made the first experiment with a two-pulse sequence and produced a two-dimensional spectrum by Fourier transform (5.1). This technique allows overlapping effects to be separated, as explained in Ernst et al. [1987]: "In many circumstances resonances that overlap in conventional one-dimensional spectra can be unravelled in this manner." To explain why with an intuitive example, one can compare the profile of a chain of mountains and the much more precise two-dimensional view that can be acquired from an airplane flying above them.

At about the same time Ernst realized that the same principle could be applied to the NMR imaging (MRI) method proposed by Paul Lauterbur of New York State University at Stony Brook.

For his work on pulse Fourier transform methods, in the context of NMR high resolution spectroscopy and MRI, Ernst was awarded the Nobel prize in chemistry in 1991.

8.10 The road to NMR imaging in the work of Damadian

The idea of obtaining a signal from hydrogen nuclei in biological tissues had occurred early. "Soon after the first successful NMR experiments at Stanford, Bloch obtained a strong proton NMR signal when he inserted his finger into the RF coil of his spectrometer," just as Purcell and Ramsey at Harvard in 1948 had inserted their heads, with the only recorded sensation being the "EMFs generated in the metal fillings of their teeth as their heads moved in and out of the magnet" (Andrews [1988]). Nevertheless the idea, that an image as complicated as the interior of the human body was in principle obtainable escaped the early researchers.

This idea came from a medical doctor Raymond V. Damadian. In September 1969 he felt confident enough to write in a grant application at the Health Research Council of the City of New York, "To the best of my knowledge, it is generally true that all malignant cells have been marked by elevated cell potassium values and depressed Ca^{++} levels. *I* am very much interested in the potential of NMR spectroscopy for early nondestructive detection of internal malignancies." He continued, "I will make every effort myself and through collaborators, to establish that all tumors can be recognized by their potassium relaxation times or H_2O-proton spectra and proceed with the development of instrumentation and probes that can be used to scan the body externally for early signs of malignancies."

Damadian, who had been uncertain before entering college between a career as a professional violinist and a scientist, chose the latter and enrolled with a fellowship at the University of Wisconsin to major in mathematics with a minor in chemistry. After receiving his M.D. degree from the Albert Einstein College of Medicine of Yeshiva University in the Bronx, he decided to be an internist and moreover to do research. Even though he did not know if "anything would come out of it," he knew he wanted to try his hand "at an effort to relieve patient suffering on a broader scale, through research, than would be possible as a practitioner" (Mattson, Simon [1996]).

His early research was centered around the balance between sodium and potassium in the body which is critical for the maintenance of tissue electricity, hence for life itself. In individual cells there is much more potassium than sodium, while for extracellular material the opposite is true. This raises the question of the discriminating ability of the cell. It was assumed to be the work of an enzyme (a protein). To search for it Damadian set up a plan in two stages. First he was going to develop mutants of the *E. coli* bacteria lacking the ability to transport potassium across the cell membrane; second he would compare one by one all proteins of the mutants and the original strain to uncover the defective one. He succeeded with the first part and concentrated on the second during two years as a drafter at the Aerospace Medical Division of Brooks Air Force Base in Texas. He arrived in 1965—the time of escalation of the Vietnam war with the bombing of North Vietnam below the 20th parallel as ordered by President Lyndon Johnson—and brought with him the mutants. He was allowed to care for them provided he also conducted experiments on hydrazine, a liquid propellant of interest to the Air Force. His efforts with the mutants though did not have success, for he did not come across any defective protein.

His research had him focused on cell metabolism which, as he was well aware, is altered by cancer. How to detect that by direct chemical analysis was an obvious question in his mind. Damadian, who had the habit of spending time in libraries browsing through books, had his attention alerted when he chanced to read, in a chemistry text, of experiments conducted by Nicolaas Bloembergen a couple of decades earlier, showing a prolongation of the relaxation times T_1 and T_2 that were associated with a decrease of viscosity in glycerin. The matter was not completely new to him, since he had audited a course in quantum physics taught by Purcell during the academic year 1963–1964 when he was working at Harvard Medical School. Right away he thought of comparing signals of cancerous tissues and normal tissues, but his experiments with the only available CW machine to measure the relaxation times indirectly failed, unhappily for him.

Then in April 1969 at a meeting of the Federation of American Societies for Experimental Biology he met Freeman Cope, "who had made successful NMR measurements of sodium in brain" and "wanted to measure potassium in biological tissue." Cope was concerned however that the weaker magnetic moment of potassium would make it difficult to detect in tissue. Damadian said he could provide some bacteria from the Dead Sea "that contained 20 times the normal complement of potassium" and so was asked to join the project. Indeed in late summer 1969 when the two scientists operated the pulsed NMR spectrometer on the bacteria at NMR Specialties in New Kensington (Pennsylvania), they obtained a signal immediately. "It was the first time potassium had been measured in a biological tissue" (Mattson, Simson [1996]).

Eleven days later Damadian wrote the proposal to the Health Research Council of the City of New York already mentioned and in another ten months, in June 1970, he had collected the data on the relaxation times of the hydrogen nuclei in normal and cancerous rat tissues. The second ones were markedly prolonged. Damadian's paper "Tumor detection by nuclear magnetic resonance," published in 1971 by *Science*, provided the basis for image contrast between normal and pathological tissues. The same paper carries data showing that even normal tissues differ markedly among themselves in relaxation times and moreover states that "in principle, nuclear magnetic resonance techniques combine many of the desirable features of an external probe for the detection of internal cancer."

To accomplish that, he needed of a much larger spectrometer, large enough to accommodate a human body. Also he needed some method for space localization, unheard of in NMR spectroscopy where test tubes filled with chemicals spun at high speed to achieve homogeneity. He had an idea and in March 1972 filed a patent, issued in February 1974, the first U.S. patent of a new "shoe" called "Magnetic Resonance Imaging or Spectroscopy." Meanwhile he went ahead and by 1977 he had built "Indomitable," the first whole body MR scanner, and produced the first human scan, which took 4 hours and 45 minutes. It consisted of a cross-sectional view of a human thorax and was made by 106 picture elements, the values of which were manually entered in a hand-drawn matrix.

Maybe then it could start healing the helpless feeling that must have gripped him since an early age when his grandmother, living with his family, died of cancer after a long suffering. He had understood this terrifying process and made "every effort" to detect it early enough that it could possibly be stopped.

8.11 The principle of image formation

Damadian's method of space localization was shortly superseded by a superior method proposed by Paul Lauterbur, a chemist of the State University of New York at Stony Brook. After witnessing Damadian's experiment repeated on rat tissue afflicted by a different tumor, Lauterbur had the idea of achieving space localization by superimposing a nonuniform magnetic field on the static magnetic field, to make different parts of the sample resonate at different frequencies. The idea of using a linear gradient for space determination in one dimension was not new (Carr [1993]). It was the first time though that it was proposed to obtain a 3D image as complicated as the interior of the human body. Lauterbur [1973] demonstrated the validity of his idea by producing the first two-dimensional image: a map of two capillary tubes 1-millimeter in size, containing water. His method made use of the one-dimensional Fourier transform applied repeatedly to obtain four backprojections on which the reconstruction was based in the spirit of x-ray tomography.

A more efficient procedure (Kumar, Welti, Ernst [1975]), linked to a different choice of the gradient and whose pivotal idea is attributed to Ernst, is based on the three-dimensional Fourier transform. The gradient system makes use of a set of three independent coils that generate spatially varying and time varying magnetic fields: Gradients G_x, G_y, and G_z in the x, y, and z direction respectively are applied for successive, consecutive times t_x, t_y, and t_z.

Then there followed a contribution due to Peter Mansfield [1977] of the University of Nottingham and embodied in his "echo-planar" method. After applying a gradient for a time interval of suitable duration, the sign of the gradient is inverted for a time interval of the same duration. This produces an echo in the signal that gives the name to the entire method. Inherently three-dimensional like the preceding one, it reduced the measuring time.

Many combinations of gradient and pulse sequences have been proposed since then. The choice depends, among other considerations, on the tissue under study. This might be the head, which can be held still during the entire scan procedure, but it might also be the constantly beating heart, which presents a more complicated task.

The principle of image formation will be illustrated in a form that is the simplest both conceptually and for collecting data. For simplicity the hydrogen nuclei will be assumed to be in ideal conditions, that is, isolated and in free space. In particular, any effect coming from the presence of the host tissue is neglected (a reasonable assumption in most cases except for fatty tissues). In this case the NMR signal contains only the Larmor frequency of the hydrogen nuclei and decays exponentially (Fig. 8.5). To within a constant factor it is given by the formula

$$s(t) = M_0 e^{-t/T_2} e^{i\omega_0 t}, \quad t > 0. \tag{8.4}$$

The real part of its Fourier transform is the Lorentzian curve in Figure 3.4, suitably dilated and translated at ω_0. The desirable situation calls for the protons in the proximity of different points of the sample to resonate at different Larmor frequencies. In this case indeed, if an RF pulse is chosen to contain those frequencies, then in the NMR signal all and only those Larmor frequencies appear. This is evidenced by the Fourier transform. For instance assume that at two points A and B of the sample the proton density is the same but the Larmor frequencies ω_A and ω_B are different. Furthermore assume that ω_A and ω_B are present in the RF pulse. The corresponding NMR signal, even in this simple case, cannot be easily decoded (Fig. 8.13), were it not for the fact that it is already known that $s(t) = s_A(t) + s_B(t)$. Instead the information offered by the Fourier transform (Fig. 8.13) is quite clear. If the proton density in A is different than in B, this too is evidenced by the Fourier transform (Fig. 8.14). Hence to distinguish even the smallest regions one from the other it suffices to use a suitable number of different pulses.

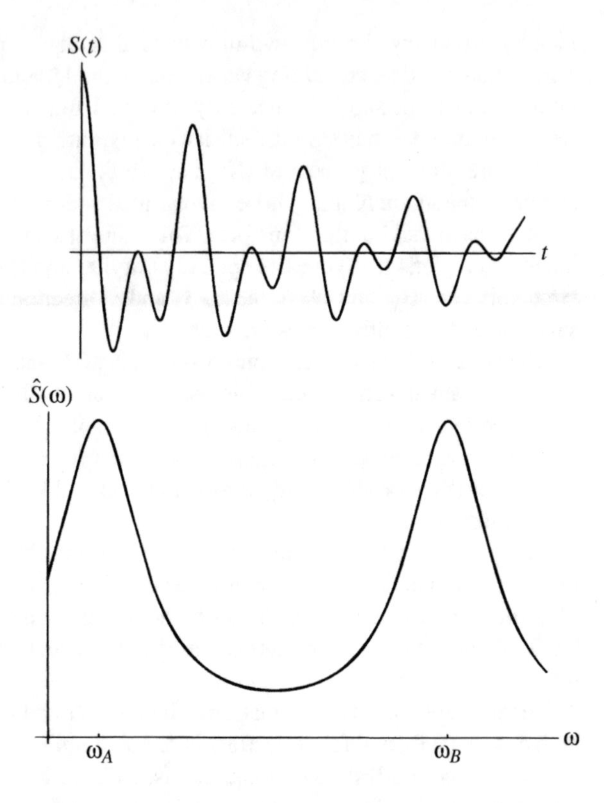

Figure 8.13 The signal above does not have an immediate interpretation, whereas the information offered by the Fourier transform below does.

To have the hydrogen nuclei in the various parts of the body resonate at different frequencies, a magnetic field $\mathbf{B}$ is superimposed on the static magnetic field $\mathbf{B_0}$. Directed in the same way as $\mathbf{B_0}$ it depends linearly upon the spatial coordinates $x, \; y, \; z$ and has a small magnitude compared to B_0, since that suffices. For simplicity the *field gradient* is assumed independent of time. If $G_1, \; G_2, \; G_3$ denote its components along the $x, \; y$, and z axes, then the total magnetic field at a point $(x, \; y, \; z)$ has magnitude

$$B_0 + G_1 x + G_2 y + G_3 z$$

and so the Larmor frequency at the point $(x, \; y, \; z)$ is given by

$$\omega(x, \; y, \; z) = \gamma B_0 + \gamma(G_1 x + G_2 y + G_3 z).$$

Following a 90° pulse, for a fixed field gradient $\mathbf{G}$, the NMR signal, the sum of all elements of the volume V under examination, depends only upon time and contains all Larmor frequencies weighted according to the local magnetization $m_0(x, \; y, \; z)$ as follows

Figure 8.14 The NMR signal having the Fourier transform above allows reconstruction of the proton density below.

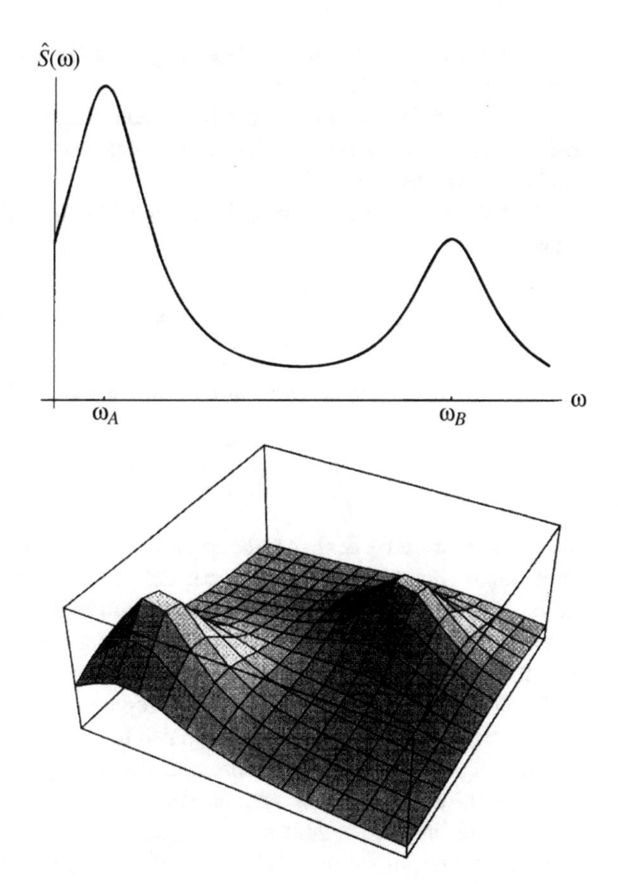

$$s(t) = M_0 \int_V m_0(x,\ y,\ z) e^{i\omega(x,y,z)t} dx\, dy\, dz.$$

Here for simplicity the exponential decay has been omitted or, equivalently, relaxation effects neglected. This in first approximation is correct since T_2 is a long time compared with the time duration of the field gradient, which are of the order of 10 ms, as well as that of the pulses, which is of the order of microseconds.

By choosing the reference frequency to be γB_0 ("on-resonance" condition) the Larmor frequencies become $\gamma(G_1 x + G_2 y + G_3 z)$. With $\mathbf{K} = (\gamma G_1 t,\ \gamma G_2 t,\ \gamma G_3 t)$ and $\mathbf{x} = (x,\ y,\ z)$ the above formula becomes

$$S(\mathbf{K}) = \int_V m_0(\mathbf{x}) e^{\mathbf{iK \cdot x}} d\mathbf{x}. \tag{8.5}$$

Equation (8.5) is the fundamental relationship of NMR imaging. It states that functions $S(\mathbf{K})$ and $m_0(\mathbf{x})$ are related by the inverse Fourier transform in three-dimensional space. The $\mathbf{K}$-domain is explored by changing the direction of $\mathbf{G}$ and sampling time t. This allows the local magnetization m_0 to be determined with high accuracy by the FFT algorithm.

Note that (8.5) is similar to the fundamental equation of x-ray crystallography (6.2). In the present case though only radiofrequencies are involved and the phase can be determined.

For instance if $\mathbf{G}$ is set in the z-direction, that is $\mathbf{G} = (0, 0, G)$, then the recorded signal is

$$s(k) = \int_{-\infty}^{\infty} \int_{-\infty}^{\infty} \int_{-\infty}^{\infty} m_0(x, y, z) e^{ikz} dx\, dy\, dz$$

with $\mathbf{K} = (0, 0, k)$. Hence the Fourier transform is

$$\hat{s}(z) = \int_{-\infty}^{\infty} \int_{-\infty}^{\infty} m_0(x, y, z) dx\, dy \tag{8.6}$$

and the spectral data $\hat{s}(z)$ may be regarded as the "projection" of the local magnetization on the z-axis, that is the axis defined by the gradient's direction. For this reason (8.6) is referred to as the "projection profile," following a terminology of x-ray tomography (Fig. 7.5). However it has to be remarked that, unlike x-ray tomography data, the NMR data are acquired in the Fourier transform domain in the first place.

The dependency of the signal upon NMR parameters like the two relaxation times T_1, T_2 and the proton density, which is tied to local magnetization, can be changed by choosing suitable pulse sequences. Consequently the parameter's weight in the image can be adjusted for optimal contrast.

Several techniques, notably sequences of impulses, have been proposed to improve space resolution, signal-to-noise ratio, absence of artifacts, and recording time (Hinshaw [1983], Ru ato et al. [1986], Hausser, Kalbitzer [1991], Callaghan [1993])

8.12 MRI-guided interventions

Images of the brain are an important instance of MRI (Figs. 8.15, 8.16). On the one hand the radiofrequency radiation goes through the skull, and more generally the bone, undamped for all practical purpose. On the other hand T_1 and T_2 weighted images provide optimum contrast resolution for soft tissues in general (Fig. 8.15). It is an important fact that MRI does not require any mechanical movement of the subject or of the apparatus. For instance to obtain heart images with good resolution the periodic heart beat is a problem that requires synchronization with the EKG. In the abdomen again peristaltic motions are a problem that may be overcome by giving to the patient antiperistaltic substances. Moreover if there is a need to improve contrast, ferromagnetic or paramagnetic substances (contrast agents) that modify the relaxation times may be used.

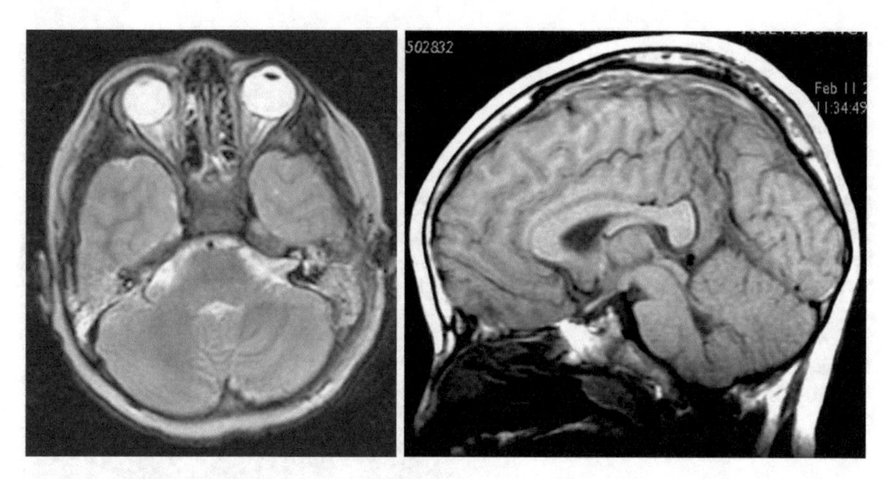

Figure 8.15 Axial T_2 and sagittal T_1 weighted images of the brain of a seven years old boy. (*Courtesy of M. Safilos, Department of Radiology, Stanford University*)

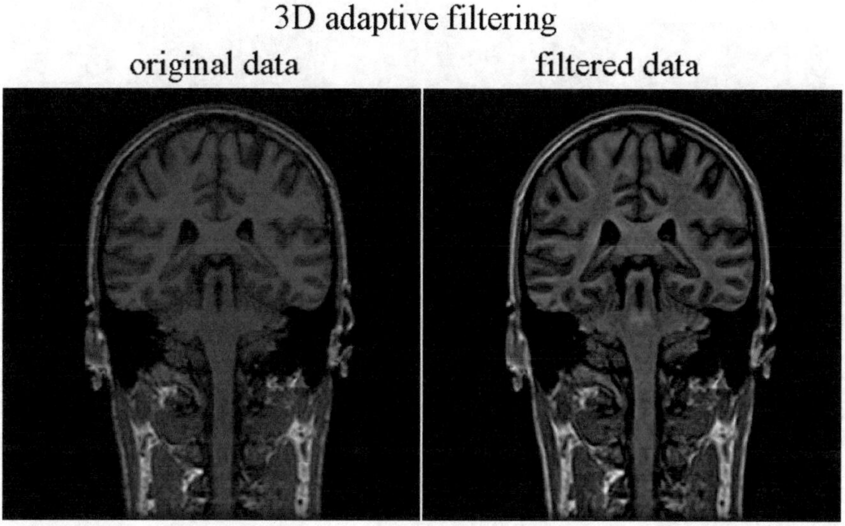

3D adaptive filtering

original data filtered data

Figure 8.16 Unfiltered and filtered coronal images of the head. (*Courtesy of Surgical Planning Laboratory, Harvard Medical School*)

MRI appeared among the techniques for clinical examinations in the 1980s. By 1988 Ferenc Jolesz had the idea of an open configuration MRI system to provide access to patient's various anatomic regions so as to allow MRI guided interventions (Fig. 8.17). The technology needed was not fully developed at the time, but in only 2 years the first MRT (Magnetic Resonance Therapy) system operating at 0.5 tesla was built by General Electric in Schenectady, New York. The clinical trials began in 1994 at the Brigham and Women's Hospital of the Harvard Medical School

Figure 8.17 The open magnet configuration allows access to the patient while the MRI test is performed. 3D models help surgeons to determine the safest routes to their target in presurgical planning. Real time scans help to check how far surgery has advanced while surgery is carried out. (*Courtesy of Surgical Planning Laboratory, Harvard Medical School*)

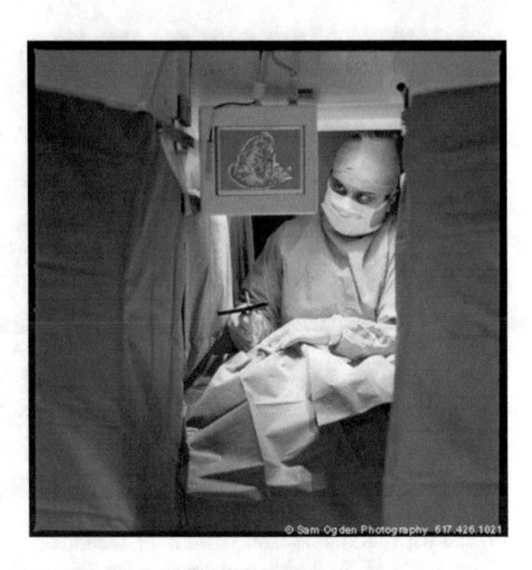

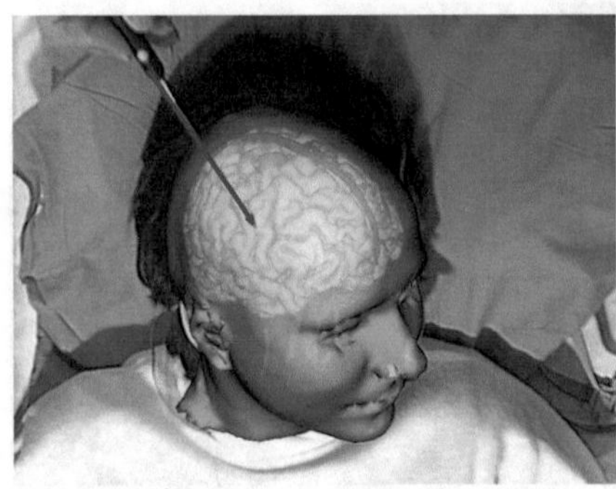

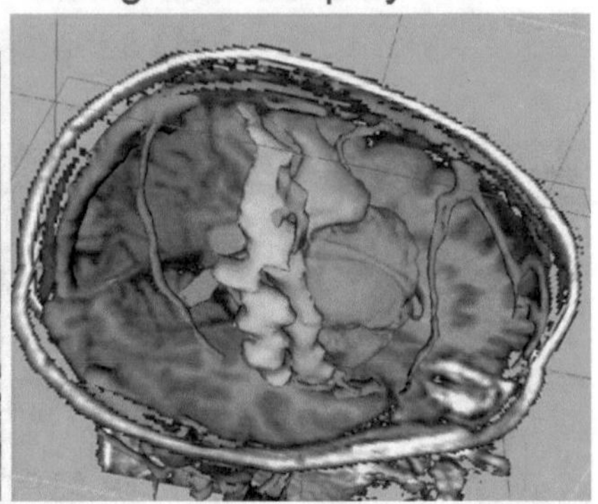

Figure 8.18 With the patient's head held still (left) the NMR image of the brain affected by a tumor, indicated by arrows, is obtained (right). (*Courtesy of Surgical Planning Laboratory, Harvard Medical School*)

(Jolesz [1995], Grimson et al. [1999]). The image in Figure 8.18 was obtained by a combination of structural, functional MRI and MR angiography. Selected structures can be highlighted by different colors. Also deserving mention is the first surgical intervention on the brain using laser-induced interstitial thermotherapy (Vogl et al. [1995]) without trepanation of the skull, which took place in October 1996 at the University of Zurich.

8.13 Functional MRI

MRI not only provides anatomic images possessing good soft tissue contrast but also images depicting brain function. As such it is an important tool for neurosurgeons, for the localization of the sensorimotor cortex for presurgical planning, as well as for neuroscientists. Starting from the first demonstration by Belliveau and colleagues [1990], MRI has been used to detect and assess cerebral pathophysiology and to understand the neural basis of cognitive functions such as vision, motor skill, language, and memory (Glover [1997], Mosely, Glover [1995]). Changes in activation or metabolism are not directly observed. Rather regionally increased blood flow is the key target to demonstrate sites of neural activity. Indeed it has been known for more than a century that local cerebral blood flow increases in regions of the brain that undergo different stimulation because of the increased neuronal metabolism: local oxygen levels in capillaries and draining veins begin to decrease while carbon dioxide levels increase. This triggers an increase of the blood supply to the affected region that, within 1–2 seconds, results in an oversupply of oxygenated blood.

Local increase of blood flow and microvascular oxygenation produce small alterations in the intensity of T_1 and T_2 weighted images of the order of a few percent.

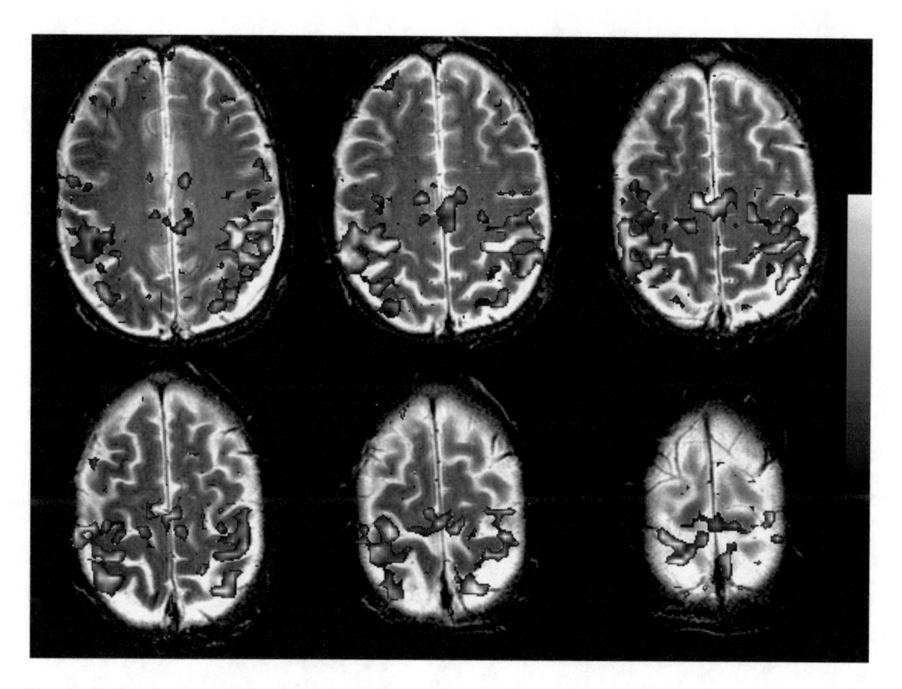

Figure 8.19 Functional images of sections at increasing height, starting from the middle of the head, depicting neuronal activation in the primary motor cortex and supplementary motor area consequent to a bilateral finger tapping task (20 on/off sequence for 5 cycles). (*Courtesy of G. Glover, Department of Radiology, Stanford University*)

Proton density is less sensitive to the phenomenon of blood increase and therefore is not used.

"In a typical brain activation experiment, the neuronal responses during two states of stimulation ("A" and "B") are compared by gathering images during both states and estimating the difference between them. In a test, to map sensorimotor cortex for example, the subject may perform finger apposition for approximately 25 seconds, rest for 25 seconds, and then repeat the sequence while images are acquired continuously for a total imaging time of several minutes (Fig. 8.19). In the difference process, metabolic activity that is unchanged between the two states is eliminated and only those brain regions that show differences from the stimulation are highlighted." (Moseley, Glover [1995]). Finally statistical tests are used to generate the activation map because the very weak changes makes difficult to discriminate between real differences and spurious ones. Acquisition of many images is therefore necessary. This typically requires just a few minutes.

In closing this chapter one cannot avoid remarking that all this stemmed from the desire to obtain precise measurements of nuclear magnetic moments.

Chapter 9
Radioastronomy and Modern Cosmology

9.1 Astronomical observations: "six thousand years for a witness"

The electromagnetic radiation that comes from space covers all wavelengths from gamma rays to radio waves (Fig. 9.1). Much is stopped by water vapor and by gases like oxygen (O_2), ozone (O_3), and nitrogen (N_2) in the atmosphere, Earth's blanket and umbrella, which extends more and more rarefied to at least 1000 km.

The atmospheric filter shows two "windows" through which electromagnetic radiation can reach Earth and be observed from it, the *optical window* and the *radio window*. By the same token, only through these windows can radiation emanating from the Earth's surface escape.

The optical window covers wavelengths from 3,000 Å to 300,000 Å. It includes the visible radiation, the closest ultraviolet, and part of the infrared with bands in which the atmosphere is opaque (Fig. 9.2). The optical observations make use of telescopes, photographic plates, other light detectors, and the human eye which is sensitive only to radiation of wavelengths from 4,000 Å to 7,000 Å. This is light, from violet to red. Observations in the visible range have existed since the dawn of recorded civilization and before. No wonder since the view of the starry night sky fills the heart with awe and curiosity even today.

Observations led to cosmological ideas. Very influential ones were advanced by the ancient Greeks and are to be found in the *Almagest*, the encyclopedic work divided in thirteen books written by the celebrated astronomer Claudius Ptolemy (2nd century A.D.). He lived in Alexandria, Egypt, and enlarged Hipparchus' star catalogue of about 850 entries to about a thousand. The Earth's centered view of the solar system—Ptolemy's main original contribution—predicted the motion of the planets with good accuracy with respect to the observations and was accepted for more than 1,300 years.

E. Prestini, *The Evolution of Applied Harmonic Analysis*,
Applied and Numerical Harmonic Analysis,
DOI 10.1007/978-1-4899-7989-6_9

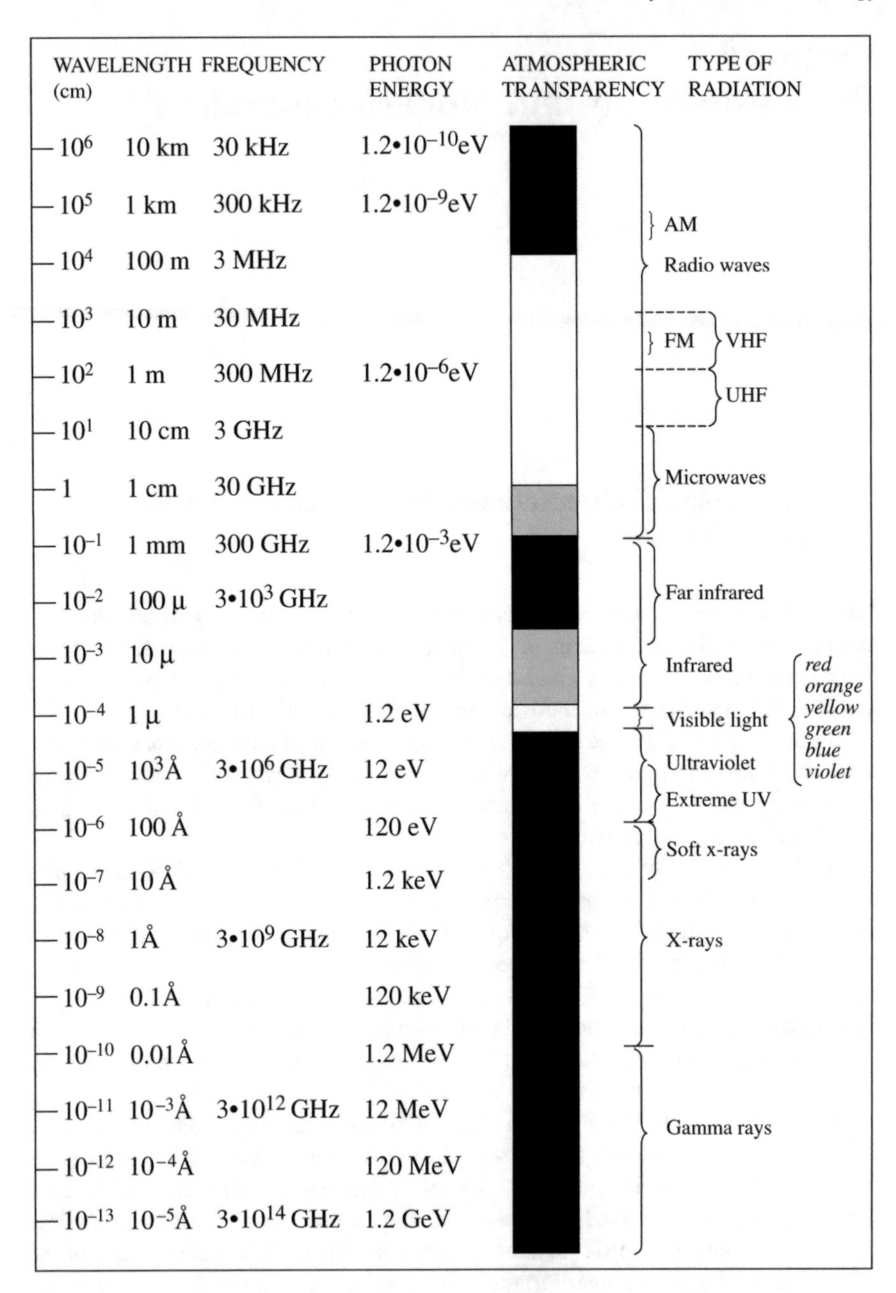

Figure 9.1 The electromagnetic spectrum on a logarithmic scale and the atmospheric transparency.

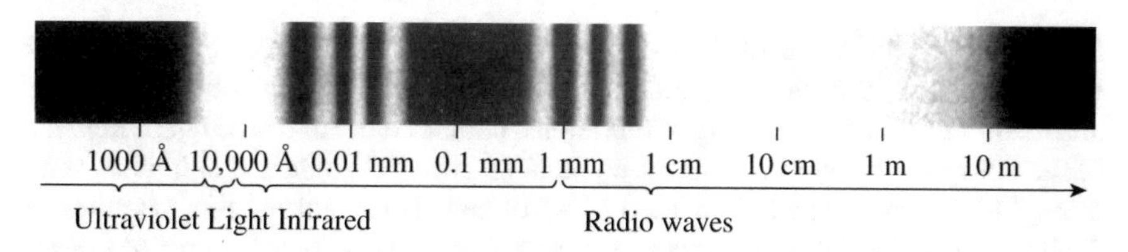

Figure 9.2 The optical window, the radio window and the atmospheric transparency.

Then came the revolutionary work of Nicolaus Copernicus (1473–1543). Born in Poland he studied astronomy in Krakow, then in Bologna, Italy—where he made his first recorded observations—and later medicine in Padua. In the Greek literature he had found, advanced by Aristarchus of Samos as far back as the 3rd century B.C., the idea of the Sun as center of the Universe. The "absurdity" of a moving Earth became convincing after many years of mathematical calculations. The system he devised, assuming circular motion and constant speed, still retained complexity. He exposed his heliocentric theory in *De Revolutionibus orbium coelestium* ("On the Revolutions of the Celestial Spheres"). It was published in the year of Copernicus' death due to his hesitation dictated by prudence.

In the framework of this theory Johannes Kepler (1571–1630), imperial mathematician at the court of Rudolph II in Bohemia, published in 1609 *Astronomia nova* or *Commentarius de stella Martis* ("New Astronomy" or "Commentary on the Planet Mars"). He had been assigned to the theory of Mars by his predecessor in the post, the Danish astronomer Tycho Brahe (1546–1601). Having tried to fit the naked eye observations – painstakingly made and recorded over many years by Tycho Brahe and his school—in every combination of circles he could think of and not succeeded, Kepler abandoned the circular scheme and solved the enigma of the irregularities of Mars' orbit. He recorded "How intense was my pleasure from this discovery can never be expressed in words. I no longer regretted the time wasted. Day and night I was consumed by calculations." Going through massive calculations he had come to the conclusion that the orbit was an ellipse and not a circle. Extended to all planets this is Kepler's first law. In the same book the second law or "area law" is hinted at. By 1619 with the publication of *Harmonicē Mundi* ("The Science of the Harmony of the World") he had stated all three mathematical laws, named after him, that planetary motion obeys. An exhilarated Kepler could then write about the book, "It can wait a century for a reader, as God himself has waited six thousand years for a witness."

At the appearance in 1632 of the persuasive Galileo Galilei's *Dialogo sopra i due massimi sistemi del mondo, tolemaico e copernicano* ("Dialogue Concerning the Two Chief World Systems, Ptolemaic and Copernican"), written in Italian for maximum diffusion, the matter exploded in controversies. Judged on the grounds of faith, the "Dialogue" found its place in the *Index* of prohibited books—together with those of Copernicus and Kepler—its author tried for heresy.

In 1609 Galileo (1564–1642), having heard news of a device that made distant objects appear closer by means of lenses (*refracting telescope*)—ordinary binoculars are made with a pair of them—quickly constructed such a device himself and improved it until it reached a magnification of about 30. Galileo pointed it to the sky and made a series of astonishing observations, published in 1610 in *Sidereus nuncius* ("The Starry Messenger"). He had seen mountains on the Moon and individual stars in the Milky Way. He had discovered a host of new fixed stars and four satellites of Jupiter, ... and the Galilean telescope was soon in demand all over Europe. Moreover the *Tabulae Rudolphinae*, published in 1627 by Kepler, gave the planetary position with unprecedented precision, bringing the error from the order of a degree down to a minute of arc.

By 1663 the *reflecting telescope*, which uses a concave mirror to obtain the image, was proposed. Newton built the first one in 1668. With such an instrument William Herschel (1738–1822) made the first systematic survey of the sky, laying the foundations of modern stellar astronomy.

Astronomical observations were made exclusively through the optical window until the 1930s when the radio window was discovered. It spans wavelengths from the millimeter to about 20 meters, depending upon the variable conditions of the ionosphere. This is the upper part of the atmosphere, above 50 km, where the number of ions—mainly due to the action of solar radiation on the air's neutral atoms—is large enough to affect radio waves. The radio window is much larger than the optical one: thirteen octaves against six.

Radioastronomy allows the probing of regions of space where there is literally nothing to see. Radio signals penetrate the clouds of dust, which obscure the view of optical instruments, as well as clouds and bad weather on Earth. Moreover, this science of long wavelengths reveals objects that are not an appreciable source of light and emit only or mostly radio waves. Pulsars and radio galaxies are examples. It must be said though that weak radiation can go undetected due to interference by radio signals generated on Earth, like TV and radio broadcasting. Much care goes into protecting the radiotelescopes from this, starting with the choice of the place of installation.

Between light and radio waves is found the infrared radiation. This radiation too allows dark regions to be probed, regions not hot enough to emit light. Having a wavelength longer than the typical dimensions of dust particles in space, much of it goes through the cosmic dust clouds. Some important observations can be made from Earth-based instruments located on high mountains—such as Mauna Kea on the island of Hawaii—the atmospheric absorption filter being much reduced in parts of the band, as for example at wavelengths close to one micron.

Atmospheric attenuation is complete for the short ultraviolet and x-rays. From Earth no observations can be made in this range, not even from the highest mountains. Gamma rays, though extremely scarce, become detectable at very high energies.

It was only in the 1960s, with the launch of the first spacecraft, that the whole electromagnetic spectrum was opened to observations. Noteworthy in those early times, particularly for the X-ray band, was the *High Energy Astronomical Observatory #2*, also named the *Einstein Observatory* (Fig. 9.3). This satellite started sending data in 1979, the centennial of the birth date of the famous scientist.

Figure 9.3 The Einstein Observatory (HEAO-2) before launching from Cape Kennedy by an Atlas Centaur rocket in November 1978 (NASA). (*Source*: *Istituto di Astrofisica Spaziale, CNR*)

X-rays, as well as gamma rays, cannot be focused by lenses or mirrors in the conventional way. Due to their penetrating power they normally go straight through. Consequently the X-ray telescopes differ radically in design from optical telescopes, relying on highly polished metal surfaces mounted so as to make grazing incidence at angles less than 4°. Through a series of such reflections the radiation is guided to a focus to be detected, converted into radio signals, and transmitted to radio telescopes on the ground. "Detecting x-rays from objects which are so hot that almost all of their radiation is in x-rays, not light, this satellite has done for x-ray astronomy what Galileo's small telescope once did for optical astronomy" (Field, Chaisson [1985]).

9.2 How radioastronomy was born

The first half of the twentieth century has been a period of enormous and stimulating progress for science, especially physics. The frontier of scientific knowledge was set in motion in the realm of the extremely small as the working of atoms began to be understood, but the frontier was advancing also on the side of the very large with the discovery of the structure of our galaxy, of the existence of many more galaxies, and of the expansion of the Universe.

Besides, in the middle of the century a fantastic discovery was made in astronomy, namely the existence in the electromagnetic spectrum of the radio window. To be true, matters had started a little earlier, in 1931, but not by the work of astronomers. It was the work of engineers attempting to identify, with a view to suppressing, the annoying disturbances that affected their telecommunications.

Physicists and astrophysicists in particular had a good reason for that. Since the time radio waves had been discovered in a laboratory by the German physicist Heinrich Hertz, several attempts had been made to detect possible solar emissions. Thomas Edison had tried and also English, German, and French scientists had carried out experiments but found nothing and drew the conclusion that the Earth's atmosphere was working as a shield.

Closer to the truth was Charles Nordmann, a French student of the Meudon Observatory. In 1901 he made an attempt, again without success, in the area of Chamonix on the slopes of Mont Blanc at a height of 3100 m, using an antenna 200 m long. Knowing that the eleven year cyclical activity of the solar spots was at its minimum, he suggested repeating the experiment in a period of maximum activity. Nothing followed.

Planck's radiation law of 1900 allowed calculation of the energy of the radiation emanating from a body, once its temperature was known. In the case of radio waves from the Sun, the emission was calculated to be so weak as to make the measurement unthinkable. The calculation was based on the belief that the Sun's temperature could not be greater than 6000 K.

Radio emission from the Sun, generally scarce, is due to the *corona*, its outer atmosphere. Made of sparse gases extending for millions of miles from the apparent optical surface of the Sun, the corona is only visible during total solar eclipses. The solar radio emission reaches values significantly higher from sunspot regions and during the spectacular but brief solar "flares." The temperature of the regions of the corona involved has been calculated to be about 20 million K. Some of the emission is nonthermal, being due to ejected relativistic electrons.

It was 1931 when a young American made the first important discovery about radio emission of heavenly bodies. Karl G. Jansky (1905–1950) was working as a radio engineer at the Bell Telephone Laboratories in New Jersey, a major world center for industrial research. He was assigned the job of finding the origin of noise in the recently inaugurated transatlantic radiotelephone. To that purpose he built a rotating radio antenna array (Fig. 9.4), designed to operate at a wavelength of 14.6 m (20 MHz). In 1932 he ascertained that thunderstorms were the main cause of the

Figure 9.4 The rotating radio antenna array used by Karl Jansky in the discovery of radio radiation from the Milky Way. (NRAO/AUI)

disturbances, but he also encountered a hiss-like noise, weak and steady, that changed direction progressively during the day, making a complete tour of the horizon in 24 hours and coming in strongest about four minutes earlier every successive day. "Just as the stars," noticed the astronomer Melvin Skellett, a bridge companion. Indeed the Earth's sidereal rotation period is four minutes shorter than a solar day. Jansky, who had already ruled out the Sun, came to the conclusion that the source was located outside the solar system and in 1933 published several articles on the subject (Jansky [1933]). Given the precision of his measurements it could be in the direction of the constellation Hercules or Sagittarius, toward the center of the Milky Way. He could not determine though if the origin was inside our galaxy or not. (Subsequent studies showed the source to be the central regions of the Milky Way.)

A press release from Bell Labs in the spring of 1933 made Jansky popular right away. The cosmic hiss was broadcast and millions of people listened to it. The *New York Times* devoted a front-page column to the discovery and the *New Yorker* magazine noted that it was the first time a noise was searched for, that far away: "It has been demonstrated that a receiving set of great delicacy in New Jersey will get a new kind of static from the Milky Way. This is believed to be the longest distance anybody ever went to look for trouble."

Since the hiss did not have any influence on radio communications, Bell Laboratories lost interest and Jansky too as a consequence. His last article on the subject dates from 1935. As for astronomers, under their silent domes, from their observatories not yet entered by electronics, what interest could they have for radio hisses and cracklings due to who knows what, coming from who knows where? It can be said that, with rare exceptions, the discovery that makes Jansky the father of radioastronomy did not have any follow-up in the contemporary scientific community.

During the following ten years there was only one active radioastronomer in the world, Grote Reber. He worked as a radio engineer during the day. At night he tended to his passion, radioastronomy, building in 1937 on his own and at his own expense a radiotelescope with a bowl-shaped antenna. Fifteen meters high and ten meters in diameter it was the biggest of its kind in the world and was superior to Jansky's for angular resolution and for receiving on a wider frequency band. He placed it in his backyard, at 212 West Seminary Avenue, Wheaton, Illinois, to the neighborhood's curiosity.

Reber, who got into the habit of making his observations between midnight and dawn, to avoid unwanted interference, remained active in the field for more than 30 years. He confirmed Jansky's conclusions and published in 1944, in the *Astrophysical Journal*, the first contour map of the radio brightness of the Milky Way. He also discovered intense regions of radio emission, later identified as the sources Cassiopeia A and Cygnus A and moreover detected radio waves from the Sun.

Reber believed in a thermal origin for the phenomenon even though this contrasted with Planck's law, the radiation being much more intense at low rather than high frequencies.

9.3 The most "mortifying" episode of English naval history and radar

Jansky and Reber's discoveries did not fall on fertile ground. The radioastronomers to come would start all over again and would owe to them little or nothing. They came from the development of *radar*, one of the most important technical innovations of World War II.

England had to face a terrifying escalation of air attacks by bombers, V1 buzz bombs, and V2 missiles. Its well-structured community of physicists was assigned the task of finding a way to give the alert, in case of attack, in time to deploy the Spitfires for the air defense of the British Islands. The episode that had the British double their efforts dates from February 12, 1942 when the German cruisers Scharnhorst and Gneisenau, covered by adequate noise, sailed from Brest to a safe harbor in the North Sea, crossing the Channel without being detected. Referring to an episode of June 4, 1667 the *Times* of London wrote, "Nothing more mortifying to the pride of our sea power has happened in home waters since the seventeenth century" when a fleet of Dutch warships, after bombarding Sheerness on the island of Sheppey, proceeded up the Thames as far as Gravesend, 22 miles from London.

A system of radars, the *chain home*, was operational by 1939. (Work on it had begun in 1935 under the direction of Robert Watson–Watt.) It signaled the attacks of the enemy and at times something else. On February 27 and 28, 1942 a dozen radars signaled an intense noise, impossible to eliminate. It was correctly attributed to the exceptional activity of solar spots, but the intensity of the signal—it had to be remarked—was a 100 times higher than that predicted by Planck's law for a body at 6000 K. In 1944, it was the turn of the latest radars. They gave rise to an impressive number of false alarms due, as was discovered, to flying objects at an altitude of about 100 km: meteors in the high atmosphere. Finally in an attempt to intercept the V2's, much more sensitive new receivers were installed. They did not distinguish themselves from the preceding ones. This time the cause was found in the cosmic hiss that Jansky had discovered.

The development of the two-antenna interferometer, which would become an instrument of fundamental importance in radioastronomy, is also due to an Englishman, Martin Ryle (1918–1984). (The sea-interferometer, which antedates the 2-element arrangement, used only one antenna placed on a cliff overlooking the sea. For there is a virtual antenna, namely the image of the real antenna in the sea, due to the reflecting sea surface.) For this invention and astronomic research he was awarded in 1974 the Nobel prize for physics together with Antony Hewish who had discovered in 1967 the first pulsar, by analyzing the data of a powerful interferometer that Ryle had just built.

During the war Ryle worked for the Telecommunications Research Establishment, a research center on radars. (At that time the word radar did not exist in England. It was a later American palindromic word, first noted in November 1942 in the *New York Times*. Its inventor is apparently not known.) Subsequently he moved to the Cavendish Laboratories in Cambridge where an excellent group of radio physicists was already active. Around 1950, using an instrument with precision from 1° to 3°, he localized about 50 radio sources in the sky (Ryle, Hewish [1955]). None was related to a star of good luminosity maybe because, as it would be discovered later, the great part were due to noise. This led to the conjecture that a new kind of stellar objects existed. They were named radio stars.

Then using a German antenna recovered at the end of the war, 7.5 m in diameter, Ryle could determine positions in the sky with a precision, unparalleled at the time, of one arc minute. The optical telescope of the Palomar Observatory (California) could finally see the two most powerful radio sources: Cassiopeia A located inside our galaxy and Cygnus A, an elliptical galaxy in the constellation Cygnus, about a billion light years away. Both emitted far more in the radio band than in the optical one. Virgo A, an elliptical galaxy and weak source of light, five million light years away and third most powerful radio source in the sky, was discovered in Sydney, Australia. Since the Milky Way is a spiral galaxy whose main dimension is a hundred thousand light years, the belief that point radio sources were nearby stars was seriously shaken.

These discoveries, which were bound to have a profound effect on the conception of the Universe and on cosmological theories, began to attract astronomers to this new field on a large scale.

Radio waves were used to investigate the solar system as well. The first radio echoes from another planet were obtained in 1958 from Venus, in spite of the dense cloud cover surrounding it. They allowed an accurate measurement of its distance from Earth and improved the value of the *astronomical unit*, the mean distance of the Earth from the Sun. Radio studies also showed valleys and mountains on Venus's surface and revealed the sandy nature of our Moon's surface before any landing took place.

9.4 Radio waves and radiotelescopes

It is important to underline that radio waves have nothing to do with sound and cannot be perceived by the ear. They are modulated during radio transmissions for broadcasting purposes, but it is not sound that is transmitted. Indeed sound in air requires a physical vibration of the air molecules. In the case of radio waves or light waves or any other kind of electromagnetic radiation, no bulk movement of matter is detected.

Besides infrared radiation, which is perceived as heat, light is the only kind of radiation perceived by our senses. We are not aware that it propagates in waves because of the extremely high frequency: some thousands of millions of millions of oscillations per second, which is far too many, given the time of response of the eye.

Many astronomical objects emit electromagnetic radiation of various kinds: radio, infrared, visible, ultraviolet, and so on. Radio waves are characterized by wavelengths from a few millimeters up. Neither the human eye nor photographic emulsion are sensitive to them. So they are detected by other means, for instance with the aid of an antenna where radio waves induce a weak current as in every conducting material. Electrical filters in the receivers, connected to the antenna, may be employed to select one frequency at the time so as to determine which frequencies—or equivalently wavelengths—are present in the radiation.

A dipole antenna pointing toward the sky can detect radio waves but cannot determine the direction they come from. For this purpose a telescope is needed. It might consist of an array of dipoles or of a surface, made of conducting material, that reflects radio waves following the same rules of the reflection of light by a mirror. In the second case the radiotelescope consists of a paraboloid, called a "dish," made by solid metal or by a metallic fine mesh. Usually the dish is mounted so it can be steered and pointed in any direction of the sky (Fig. 9.5), just as an optical telescope. Paraboloids have the property of reflecting all rays coming in parallel to their axis to a single point called the "focus." Rays having directions close to the axis are reflected to points in the focal plane near to the focus. A radiotelescope in the form of a paraboloid differs from an optical telescope in that the focal plane device records only one image pixel. So the focal plane must be scanned in two dimensions to build up the radio image sequentially, instead of simultaneously as in optics where a film or other two-dimensional detector is used.

Figure 9.5 The radiotelescope in Effelsberg, Germany, 100 meters in diameter. (*Courtesy of Max Planck Institut für Radioastronomie, Bonn, Germany*)

The current induced in the elements at the focus of the dish is conducted to a receiver to be amplified, filtered, and recorded by instruments usually located in a room at the base of the telescope. Following both the antenna and the bandpass filter there is another detector that is a nonlinear device. In early times such a detector was a point contact between a conductor and a crystal of galena. It was later replaced by vacuum tubes. Now silicon p-n junctions are used, not very different in principle from early crystal detectors. By analogy the detector in an optical system is a silver iodide molecule suspended in a photographic emulsion, another nonlinear device.

9.5 Resolving power

The resolving power of a system—whether radio or optical—measures the ability to *resolve* two close objects, meaning the ability to record separate images of them (compare Figures 9.6 and 9.14).

The phenomena of diffraction and interference cause the image of a celestial point source, made with a paraboloidal antenna, to be a tiny spot surrounded by faint concentric rings. Such rings, clearly visible under good conditions in the case of an optical telescope (Section 5.4), are not as pronounced in case the image is obtained by a radio dish. The central spot is called the *diffraction disk* of the image. No details smaller than that can be resolved.

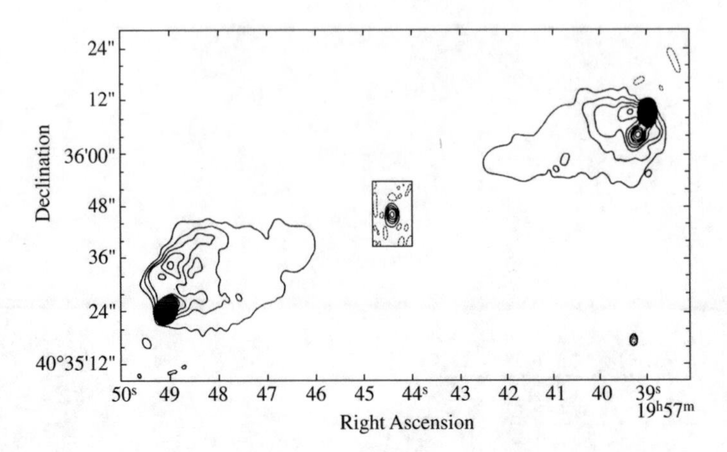

Figure 9.6 Map of the source Cygnus A using the Cambridge Five-kilometer telescope at 5 GHz. This map showed for the first time the radio nucleus associated with the central galaxy and the high brightness at the outer edges of the radio lobes. The resolution is 2.0 × 3.1 seconds of arc. (Hargrave P.J., Ryle M. [1974], "Observations of Cygnus A with the 5-km radio telescope," *Mon. Not. Roy. Astr. Soc*. 166, pp. 305–327. ©Blackwell Science Ltd.)

In the case of an extended source the diffraction patterns of various parts of the image overlap, destroying the finest details. In the case of two stars whose images are closer together than the size of the diffraction disk of either, a single spot perhaps slightly elongated will show in the image. In astronomical practice the *resolution* of a radiotelescope—the same holds for a lens or the reflecting mirror of an optical telescope—is the smallest angle α between two stars for which separate images are produced. It is given in seconds of arc by the formula

$$\alpha = 2.1 \times 10^5 \frac{\lambda}{D}, \tag{9.1}$$

which requires the wavelength λ and the aperture D of the paraboloid or lens to be measured in the same units, for example centimeters.

In the case of the human eye, the resolution—almost 1 minute of arc—is not as good as that of an ideal lens of the same size (λ close to 0.5 microns and D equal to a few millimeters) due to the insufficiently fine structure of the retina.

Radio waves are much longer than light waves. For example radio waves of 20 cm are about 400,000 times longer, so to resolve the same angle a radiotelescope ought to be 400,000 times larger than an optical telescope.

The largest radiotelescope (Fig. 9.7) is located near Arecibo on the island of Puerto Rico. Made by a spherical mesh of fine wire, it stretches over a natural valley in the mountains. With a diameter of 305 m (effectively illuminated 225 m) it is unmovable and takes advantage of the Earth's rotation, which brings into view of the telescope a 40° belt of the sky at one time or another. It operates at wavelengths from 3 cm to 1 m with resolving power from 0.45 to 15 minutes of arc. The largest

Figure 9.7 The radiotelescope in Arecibo, Puerto Rico, 305 m in diameter, is part of the National Astronomy and Ionosphere Center, which is operated by Cornell University under a cooperative agreement with the National Science Foundation.

steerable radiotelescope, about 100 m in diameter, is located in Effelsberg, near Bonn, Germany (Fig. 9.5). Then comes the 91 m radiotelescope in Greenbank, Virginia (USA), followed by the 76 m at Jodrell Bank, England and many smaller ones.

The radiotelescope in Effelsberg operates at wavelengths from 3.5 mm to 74 cm with resolving power from 15 seconds to 25 minutes of arc. It is an engineering masterpiece: the deformation due to movement of the dish pointing in different directions of the sky, as well as that due to wind and heat, is very small, otherwise the precision obtained by its large dimension would be significantly affected. To give an example, it resolves points on the Moon 800 km apart. This may seem crude. Even only the human eye does better, separating points 100 km apart while the best optical telescope can resolve, always on the Moon, points 1.8 km apart due to atmospheric turbulence and inhomogeneities. Unimpaired by that, the Hubble Space Telescope, with a diameter of 2.4 m, can separate points 40 m apart. The limits of radioastronomy appear evident, the more so since the construction of telescopes of even greater dimensions is technically difficult and economically strenuous. Nevertheless, as will now be described, the best resolution that can be achieved in radioastronomy is several orders of magnitude superior to that of the best optical telescope.

9.6 Interferometry

An ingenious technique was found for obtaining radio images with high resolution by pioneer Martyn Ryle. Radio signals can be transmitted along electrical wires. This allows elements kilometers apart to be connected. Starting with the signals from two (or more) elements, the technique of *interferometry* generates a third signal that shows a fine structure described as interference fringes. The resolution is that of a single continuous antenna equal in length to the element spacing, therefore potentially bounded only by the Earth's diameter of 12,800 km or the separation of satellite elements. In Section 9.8 the mathematical model will be described.

The most impressive radiotelescope array is the Very Large Array (VLA) located near Socorro, New Mexico (USA), in the dry St. Augustin plain, the least polluted by parasitic emissions (Fig. 9.8). Operated by the National Radio Astronomy Observatory (NRAO), it is open to foreign researchers. The facility has been a workhorse of contemporary astronomy since its opening in 1978 and greatly contributed to studies of the fine structure of many cosmic objects. It is composed of 27 telescopes, each of aperture 25 m, all of which can be moved along rails laid out in a "Y" configuration, with a total span of about 36 km. It receives wavelengths from 90 to 1.3 cm with resolving power from 3.6 to 0.05 seconds of arc, comparable to the best optical telescopes.

Many arrays do not use separate parabolic dishes. For example the Molonglo Radio Telescope, built in 1963 near Canberra, Australia, consists of two wire

Figure 9.8 The Very Large Array near Socorro, New Mexico (USA), operated by the National Radio Astronomy Observatory. (*Source*: *Istituto di Astrofisica Spaziale, CNR*)

meshes—each a mile long and parabolic in cross section—that meet at right angles at their centers. The incoming radio waves are reflected to a long antenna—one for each arm—set at the focus of the parabolic cross section and running down the entire length of the arm. This kind of configuration is known as a “Mills cross” after B. Y. Mills who proposed it in 1958 and built the first one at Fleurs near Sydney, Australia. Crosses of comparable dimensions were built also in the northern hemisphere, in Italy and the Soviet Union. Since 1969 at Medicina, near Bologna, a radiotelescope has been operating, known as the Northern Cross after the original project that assigned it the shape of a Mills cross with arms 1200 m long. It actually looks like a “T.” The East-West arm is a semiparabolic cylinder 564 m long and 35 m wide while the North-South arm, 640 m in length, is made by an array of 64 parabolic cylinders, each 23.5 m × 8 m in size, set 10 m apart. It operates at the wavelength of 73.5 cm.

With two telescopes very widely spaced on Earth, a greatly superior resolution could be obtained. If the distance were equal to the Earth’s diameter, very fine details could be resolved, up to 0.0001 seconds of arc, comparable to the arc subtended by a tennis ball on the Moon. This is actually feasible since time can be measured with extreme precision by atomic clocks. The technique is called *intercontinental interferometry*. Two radio telescopes very far away can “simultaneously”—that is within a few millionths of a second—receive and record radio waves emanating from the same source. The signals, recorded together with the receiving time, can then be taken to a common facility and processed as if the whole procedure were taking place in real time.

Italy built in the 1980s, mainly for observations of intercontinental interferometry, two twin telescopes 32 m in diameter one located at Medicina and the other at Noto, near Catania in Sicily. They operate at wavelengths from 1.3 to 20 cm.

9.7 The principle of image formation in a radiotelescope

By means of an aperture, such as the paraboloid of a radiotelescope, the two dimensional distribution of the radiation intensity of a celestial body can be made into an image with a resolution imposed by the aperture dimensions.

Due to the phenomenon of diffraction at infinity there is a Fourier transform relation between the electric field that exists on a remote plane, parallel to the aperture plane, determined by the 2D intensity distribution to be reconstructed and the electric field on the aperture plane of the paraboloid. Then in turn there is a Fourier transform relation between the electric field on the aperture surface and the field produced in the focal plane of the paraboloid, after reflection from the curved surface. So, the field on the focal plane constitutes an image of the radiating body. This is shown by the mathematical model that follows which, being independent of the wavelength, also holds for an optical telescope with a reflecting mirror. Hence the principle of image formation rests upon the inversion formula of the Fourier transform in two dimensions (Section 5.3).

For simplicity the mathematical model will be presented in the case of a one-dimensional aperture and will depend upon an angle θ, through the function $\sin\theta$. In reality the aperture is two-dimensional. It may be the object under observation in the sky as well as the dish of the radiotelescope. The corresponding model then depends upon two angles $\theta,\ \phi$ through the functions $\sin\theta$ and $\sin\phi$.

The one-dimensional aperture will be represented by a segment described by the variable x, once the origin is fixed in O and the unit is conveniently chosen to be equal to the wavelength λ of the observed radiation. The electric field at a point x and time t is given by the formula $b(x)e^{i\omega t}$ where ω denotes the angular frequency under reception and $b(x)$, called the *aperture function*, is complex valued having an amplitude and space phase.

At time $t = 0$ the infinitesimal element between x and $x+dx$ induces at a far away point Q—in the direction θ at a distance R from the origin (Fig. 9.9)—an electric field, called the *far field*, proportional to $C_R\ b(x)dx$. The constant C_R, dropped in the sequel, accounts for the fact that the radiation power is distributed over the entire spherical surface of radius R and thus becomes fainter and fainter as R increases. Moreover the electric field at Q shows a phase delay that depends upon the length of the radiation path—denoted by r—by means of the wave cycles in it. Recalling that the wavelength λ has been chosen as the unit of measure, the resulting formula for the far field at Q due to the infinitesimal element dx is then

$$b(x)dx\ e^{-2\pi i r}.$$

Integrating over all the infinitesimal elements the instantaneous electric field in Q is obtained.

A suitable approximation leads to the interpretation of the far field. Since Q is placed at a great distance R, the approximation $r \cong R + x\sin\theta = R + xs$ holds with $s = \sin\theta$ (Fig. 9.9). Then $e^{-2\pi i r} = e^{-2\pi i R}e^{-2\pi i x s}$ and so the far field $E(s)$ at Q is given with good approximation by

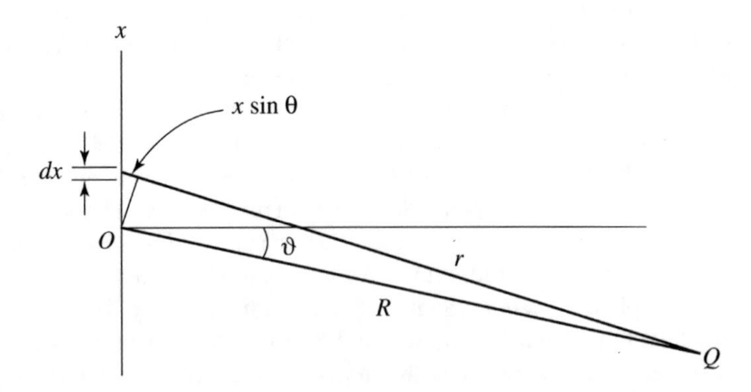

Figure 9.9 Construction to deduce the relation between the local electric field and the electric field measured by instruments at a far away point Q.

$$E(s) = \int_{-\infty}^{\infty} b(x) e^{-2\pi i x s}\, dx \tag{9.2}$$

where the constant $e^{-2\pi i R}$ has been taken out of the integral and dropped, together with C_R. In (9.2) the domain of integration is extended to $(-\infty, +\infty)$ having set $b(x) = 0$ outside the aperture. Now it is easily recognized that the far field $E(s)$ at Q is the Fourier transform $\hat{b}(s)$ (2.15) and that it depends on the direction θ through $s = \sin\theta$. Because of this, the far field $E(s)$ is also called the *angular spectrum.* Its square amplitude $P(s) = |E(s)|^2$ is called the *angular power spectrum.*

The above model has shown that the electric field E impinging on the paraboloid is proportional to $\hat{b}$, the Fourier transform of the electric field $b(x)$ on the far away object under observation. The electric field on the paraboloid is then reflected to its focal plane. The phenomenon "at infinity," described mathematically above, takes place again. So by the same argument the electric field recorded on the focal plane is proportional to $\hat{E}$. This is b by the inversion formula.

Let us test the above mathematical model in the simple case of a point source, represented mathematically by the Dirac delta function. The radio waves, originally spherical and of constant amplitude, in reaching the telescope far away can be regarded as plane waves with good approximation. The electric field on the paraboloid is therefore uniform and it is recognized to be the Fourier transform of the delta function (Section 3.4).

Now if the paraboloid were infinite, by the inversion formula, on the focal plane there would be exactly the point source, mathematically the Dirac delta function. In reality the telescope dishes have finite, yet large, dimensions. The electric field is constant on the dish (always in case of a point source) and zero outside. Therefore in the focal plane the amplitude of the electric field has a central maximum weakly degraded by much weaker surrounding maxima, as in the optical case of Figure 5.3. Hence the picture obtained by the telescope is an image slightly degraded. More precisely it is a diffraction pattern.

In the description of the principle of image formation, certain ideal conditions have been assumed. The electric field reflected by the paraboloid has been taken equal to the impinging field. In reality the two are different due to phenomena of distortion introduced, for example, by lack of uniformity of the very same reflecting surface. To avoid a serious drawback of this nature the telescope in Effelsberg, when operating at short wavelengths, makes use only of the central part of the dish, which is more uniform.

Also it has to be considered that the signal may have a very small intensity. For example, an extragalactic radio source of average power produces on Earth a signal comparable to that of a 100 watt lamp located on the Moon. Consequently the signal is easily plagued by noise in the receiver. For techniques devised to address this problem as well as others, like "blurring," see Bracewell [1995], Gonzales, Wintz [1987], Andrews, Hunt [1977].

9.8 Aperture synthesis

In case two (or more) elements are available, their signals, recorded at the same time and summed together, lead to an improved resolution, due to interference fringes similar in nature to those obtainable in optics with a system of two apertures (Figs. 5.7 and 5.8). The technique that allows simulation of a full dish by two smaller dishes is called *aperture synthesis*. It requires only two telescopes, movable along rails, and longer observation times during which the radiosource is assumed not to vary.

With an interferometer made by two equal apertures set apart by u—measured in multiples of the wavelength λ—let θ be the direction of observation with respect to the line orthogonal to the axis of the instrument. In the case of the two equally illuminated apertures, if the origin is chosen midway so that one is set *at* $-u/2$ and the other at $u/2$, the total electric field is

$$b(x - u/2) + b(x + u/2)$$

where $b(x)$ here denotes the field on one of the apertures placed at the origin (Fig. 9.10). Consequently the sum of the signals on the focal planes is

$$E(s) = \hat{b}(s)(e^{-2\pi isu/2} + e^{2\pi isu/2}) = 2\hat{b}(s)\cos\Psi/2$$

where $\Psi = 2\pi us = 2\pi u\sin\theta$. The corresponding intensity distribution or *angular power spectrum* is

$$P(s) = |E(s)|^2 = 4|\hat{b}(s)|^2(1 + \cos\Psi)/2.$$

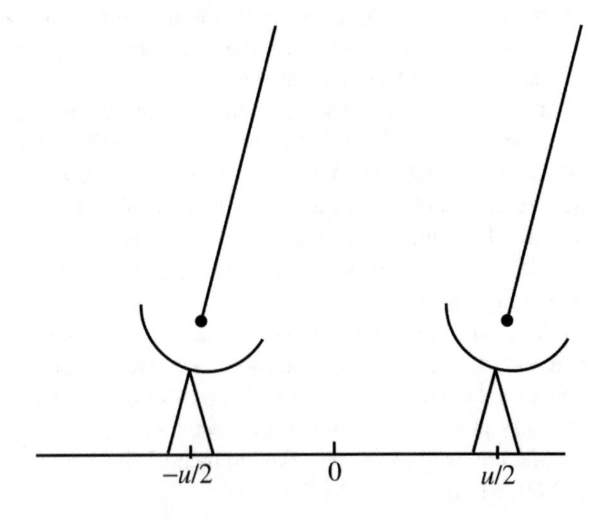

Figure 9.10 A simple interferometer.

There it appears the factor $1 + \cos \psi$, introduced by the geometrical setting. Such a factor, which depends upon the direction of observation θ, is approximately a periodic function in θ. The period, or separation between two maxima, can be calculated to be $1/(u \cos \theta)$ by the approximations

$$
\begin{aligned}
1 = \Delta\Psi/2\pi &= u \sin(\theta + \Delta\theta) - u \sin \theta \\
&= 2u \sin \frac{\Delta\theta}{2} \cos \frac{2\theta + \Delta\theta}{2} \cong u\Delta\theta \cos \theta.
\end{aligned}
$$

Note that $u \cos \theta$ is the projection of the *baseline* u on the direction orthogonal to the direction of observation and equals u if the direction of observation is vertical ($\theta = 0$). As already calculated, in the case of a point source, $\hat{b}(s)$ is constant and the fringes, described mathematically by $1 + \cos \Psi$, clearly show (Fig. 9.11 a)). An extended source, yet small enough to be considered uniform, can be thought of as made up of many point sources. Each will contribute with its own $\cos(u \sin \theta)$, which tend to annihilate each other, since the θ values are different. Hence, the fringes of an extended source are expected to be weaker than those of a point source of equal flux (Fig. 9.11 b)). The mathematical model confirms this and moreover it proves that the modulation (fringes) disappears when the angular extension α of the source equals the period $1/(u \cos \theta)$. Indeed the interferometric pattern reduces to the dotted line in Figure 9.11.

Thus, the angular dimensions of a source can be calculated from the observations. For example, assume the one-dimensional radiosource to be extended in the direction of the baseline. With the two telescopes pointing toward the source so that $\theta = 0$,

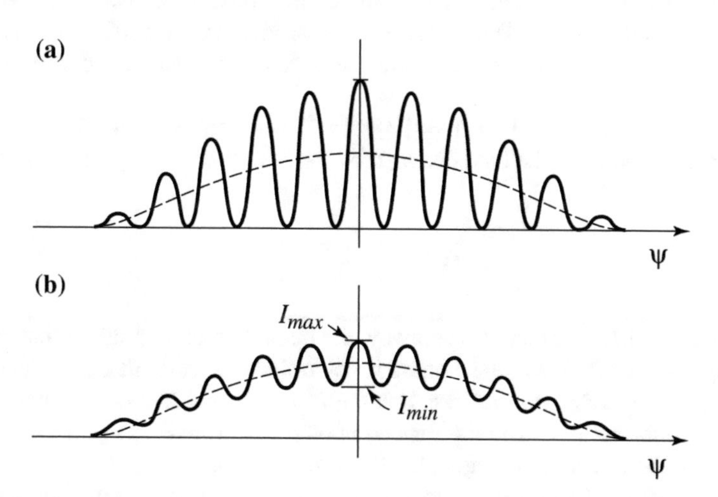

Figure 9.11 Radio interferometric pattern a) in the case of a point source, b) in the case of an extended source.

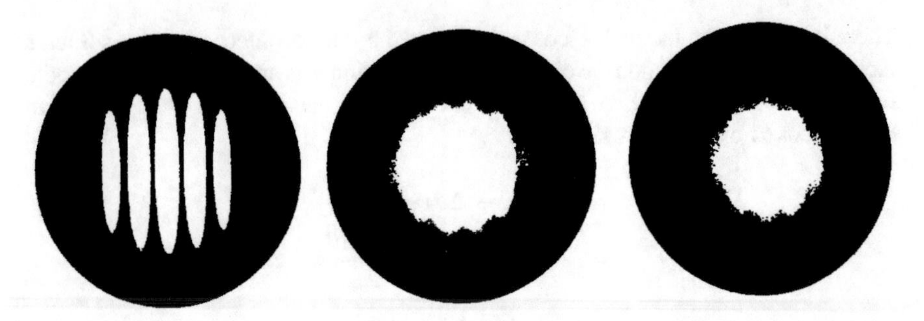

Figure 9.12 Optical interferometric pattern due to an extended source, as the angular dimension increases. (H. Lipson, *Optical Transforms*, Academic Press, 1972)

their distance is augmented until the fringes are no longer visible. If u_0 is the length reached by the baseline, then $\alpha = 1/u_0$ is the angular dimension measured in radians. The analogous phenomenon in optics is shown in Figure 9.12. Actually it ought to be recalled that the first interferometric techniques in astronomy go back to 1890, to the optical work of Michelson. Then in collaboration with F. G. Pease, Michelson was able to measure the diameter of some large nearby stars such as α Orionis, Arcturus, and Betelgeuse, starting from the interferometric data (Michelson, Pease [1921]).

A deeper understanding can be gained by pushing the mathematical model further ahead. Two functions, the brightness $B(\theta)$ and the complex visibility V, need to be defined. The *brightness* measures the power received per square meter of receiving surface, per unit solid angle, and per unit of frequency band (centered at the frequency under observation). It is a function of the direction of observation θ and it is measured in W m^{-2} sterad^{-1}Hz^{-1} Clearly nonnegative, brightness is a particular case of the definition of intensity and is the function that has to be determined with the best possible resolution.

The *complex visibility* $V(u_p) = |V|e^{2\pi i u_p \Delta\theta}$ is a complex-valued function that can be calculated from the data obtained by the interference fringes: the amplitude is

$$|V| = \frac{I_{max} - I_{min}}{I_{max} + I_{min}}$$

where I_{max} and I_{min} denote, respectively, the maximum and minimum intensity of the fringes (Fig. 9.11), while the phase is determined by the shift $\Delta\theta$ of the fringes with respect to a reference position that the fringes would have if the source were a point source. Both $|V|$ and $\Delta\theta$ depend upon $u_p = u\cos\theta$, the projection of the baseline on the direction orthogonal to the direction of observation.

The remarkable fact is that, up to a multiplicative constant, the complex visibility evaluated at the "frequency" u_p is the inverse Fourier transform of the brightness. Therefore by formula (2.15)

$$B(\theta) = \int_{-\infty}^{\infty} V(u_p)e^{-2\pi i u_p \theta} du_p = \int_{-\infty}^{\infty} |V(u_p)| e^{2\pi i u_p \Delta\theta} e^{-2\pi i u_p \theta} du_p.$$

Once a sufficient number of values $|V|$ and $\Delta\theta$ are measured by the interferometer, that is once the complex visibility at a sufficient number of points is available, the brightness follows by an FFT.

The procedure to measure the angular dimension α of a uniform radiosource (defined for $|\theta| \leq \alpha/2$), mentioned above, follows from the formula. The inverse Fourier transform of the brightness, that is V, has its first zero at $u_p = 1/\alpha$ (2.15). Now V is zero precisely when $I_{max} = I_{min}$, which is the case of no fringes mentioned above. The method is similarly extended to two-dimensional sources.

9.9 Synthesis by earth's rotation

The Earth's rotation provides variations of the u_p in the aperture synthesis described above. In the simple case of a one-dimensional source, its movement on the sky provides different values of u_p and correspondingly of $V(u_p)$ for the synthesis without any need to change the baseline. In the concrete case of a two-dimensional source the method requires the antenna spacing to be varied in one direction only, along the East-West line. For a source of high declination, the position angle of the baseline projected onto a plane orthogonal to the direction of the source rotates 180° in 12 hours. By observing one day, holding fixed the distance between the antennas, and by changing the distance from one day to another, a complete set of two-dimensional data for the synthesis is obtained.

This procedure, proposed by Ryle (Ryle, Hewish [1960]) was implemented first at the Cambridge One Mile telescope. The maps, obtained for the strong sources Cassiopeia A and Cygnus A, exhibited an unprecedented degree of structural detail.

The exploitation of the Earth's rotation was not a new idea, having been used for years in solar studies (Christiansen, Warburton [1957]). A principle of image reconstruction similar to the one used in CT scanners (Chapter 7) was proposed in the radioastronomical context for the first time in 1942. A strong and variable radio emission from the Sun in the wavelength of a meter was detected that year. The question related to the source is: was it the entire solar disk or was it only parts of the disk, such as the areas involved in the eruptions?

The angular diameter of the Sun is about 30 arcmin, so that a better resolution—say 3 arcmin, approximately 10^{-3}rad—was needed. At the wavelength of 1 m a radiotelescope of 1 km was necessary, and at a wavelength of 10 cm, one of 100 m. The biggest in existence was 5 m only.

Using an array of telescopes the issue was solved in favor of localized emissions in conjunction with eruptions. As far as resolving power the array is equivalent to the long and narrow antenna that envelopes it. Hence with an array 100 m long and 2 m wide, set along the East–West direction, the resolving power is approximately 3 arcmin in this very same direction and in the orthogonal one is approximately 3°,

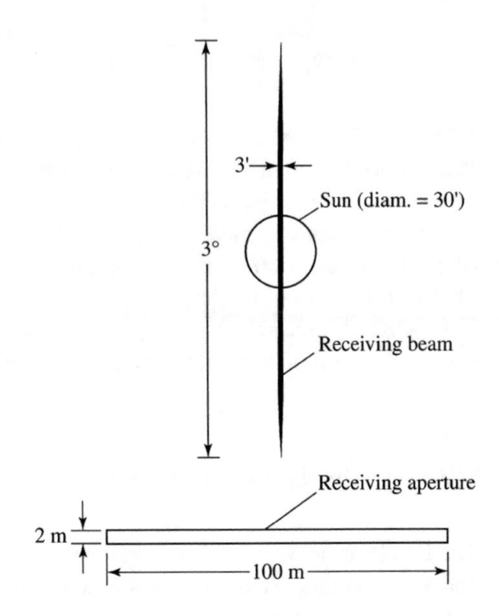

Figure 9.13 A long and narrow antenna laid in the East–West direction provides the integral of the distribution of solar intensity over a thin strip (in black) that moves from West to East and covers the whole solar disk by effect of the Earth's rotation. (R. N. Bracewell [1979], "Image reconstruction in radio astronomy," article in *Image Reconstruction from Projections*, G. T. Herman (Ed.), Springer.)

for a radiation of 10 cm wavelength. Inside such a strip on the celestial dome the array does not separate points (Fig. 9.13). Therefore, it can only provide integrals of the distribution of the solar intensity over such narrow strips that are approximately line integrals. If the array is held fixed on the ground, then the strip crosses the solar disk from West to East due to the Earth's rotation and several line integrals can be obtained. These data are similar to the projections made available by X-ray CT scanners. Only the line of integration is not the direction of the radiation anymore but rather orthogonal to it.

Different "projections," in a sufficient number, provided the necessary data for a reconstruction of the solar disk with the required resolution (Bracewell [1956], [1979]).

9.10 Quasars, pulsars, and the Big Bang

Ordinary stars, such as the Sun, are grouped together in galaxies all formed 10 billion years ago, as it is estimated. Stars are the seat of transformation of matter by a process that takes place continuously so that some die and others are born. Now while radio emissions are among the most spectacular manifestations of solar activity, especially in conjunction with flares, it is a fact that the Sun and other stars in its class are not strong radio sources but rather emit a large fraction of their energy in the visible spectrum.

Nonetheless sources of energy much more powerful than stars are found in the universe. In the visible range they might appear as stars but their energy emission, originating in a small volume—of the size of our solar system for example—is comparable to that of a hundred or even a thousand galaxies.

From the beginning they appeared to be extraordinary heavenly bodies. Only when the radio source 3C 273 was located with high precision by Cyril Hazard (Hazard et al. [1963]) was the optical telescope at Palomar Observatory successfully pointed to one. Yet the light spectrum did not resemble anything known. Finally on February 5, 1963 Maarten Schmidt, working at the same observatory recognized in 3C 273 the classical lines of hydrogen and other known elements, but all very strongly red shifted. From the shift the distance from Earth could be deduced, as well as the speed of motion, by Hubble's law. In one case the distance turned out to be two billion light years. These objects, *quasars* and *radiogalaxies*, were the farthest ever detected. The first were discovered in the 1950s; nowadays they are counted in the thousands. A typical example is Cygnus A (Fig. 9.14) whose distance from Earth is estimated at 700 million light years. The distance placed them far outside our galaxy, and since it took their light a billion years or so to reach Earth, the objects are being observed as they existed when the Universe was much younger.

One thing remained to be explained, their enormous energy output, and the more so in view of the gigantic distance. This took a long time during which several theories were advanced. It is now explained by the existence of a black hole surrounded by captured matter that forms a so-called *accretion disk*. The electromagnetic energy originates from the fall of matter into the black hole. As the matter falls, converting gravitational potential energy into kinetic energy, a lot of kinetic energy goes into hot radiating plasma as the infalling particles collide.

In a number of these objects two jets are observed, in the direction orthogonal to the accretion disk (Fig. 9.14, 9.15, 9.16). It is speculated that the inner regions, the closest to the black hole, rotate at higher speed than the exterior ones. The rotation could be so fast as to offer resistance to the fall of matter (just as the orbiting of the Earth stops it falling into the Sun). The two jets, emitted violently in the direction of minimum resistance, would serve the purpose of alleviating the enormous pressure. The jets accumulate matter outside the galaxy in two symmetrical volumes where the greatest part of the radio emission is generated.

The accretion disk is thought to be opaque in the exterior regions to the infrared, optical, and X-ray radiation originating in the most inner regions. If the direction of observation is close to the plane of the accretion disk, then the radiation is absorbed and attenuated: This is the case of a radiogalaxy. If on the contrary the direction of observation is orthogonal to the above plane or close to the orthogonal, then the radiation of the central region appears very intense: This is the case of quasars. Only one jet emanating from the central source is generally well evidenced in this case, the opposite one being much weaker.

The reason the nucleus of a galaxy hosts a black hole is not well understood. Yet it is believed that when two galaxies encounter one another and merge—not such an exceptional occurrence as may be thought—phenomenal changes in the structure take place that might lead to a great quantity of matter to fall on one of the nuclei, thus giving rise to the described effect.

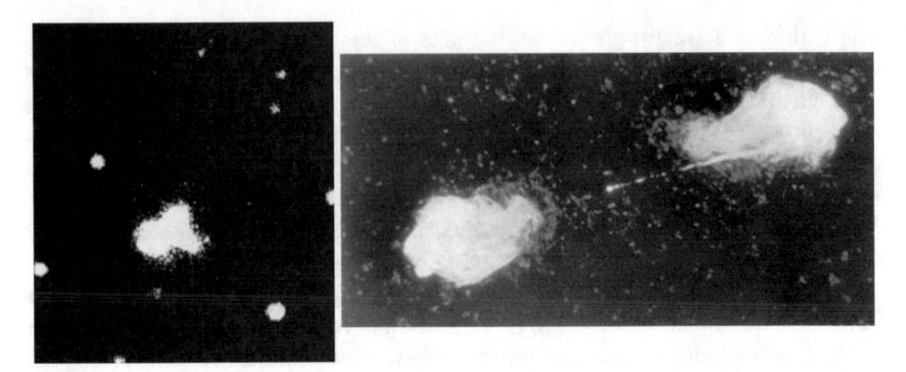

Figure 9.14 On the left the radiogalaxy Cygnus A in the visible spectrum (Palomar Observatory). On the right the same in the radio spectrum, made with the VLA at 6 cm wavelength (4.9 GHz), shows two "clouds" fed by a radio nucleus (the bright spot at the center) that coincides with the center of the optical galaxy. The resolution is 0.4 arcsec (Perley, Dreher, Conwan [1984]). The display of the image involves a nonlinear process to enhance contrast of the fine structure to emphasize the jet from the central galaxy to the top right lobe and the filamentary structure in the main lobes. (*Source: Istituto di Astrofisica Spaziale, CNR*)

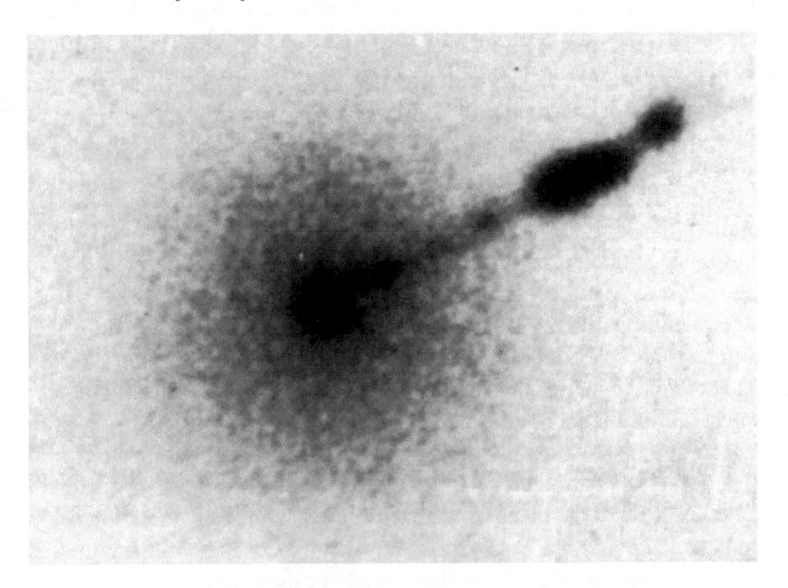

Figure 9.15 The elliptical radiogalaxy M87, 40 million light years away, shows a radio jet extending for about 6000 light years. With this galaxy the correspondence between the image of the jet in the radio, optical, and X-ray bands is extremely precise. (*Source: Istituto di Astrofisica Spaziale, CNR*)

Synchrotron radiation (Chapter 5) is the main mechanism of radio emission in the universe. The observer does not receive a series of pulses but a continuum due to the mixing of pulses generated by different electron trajectories.

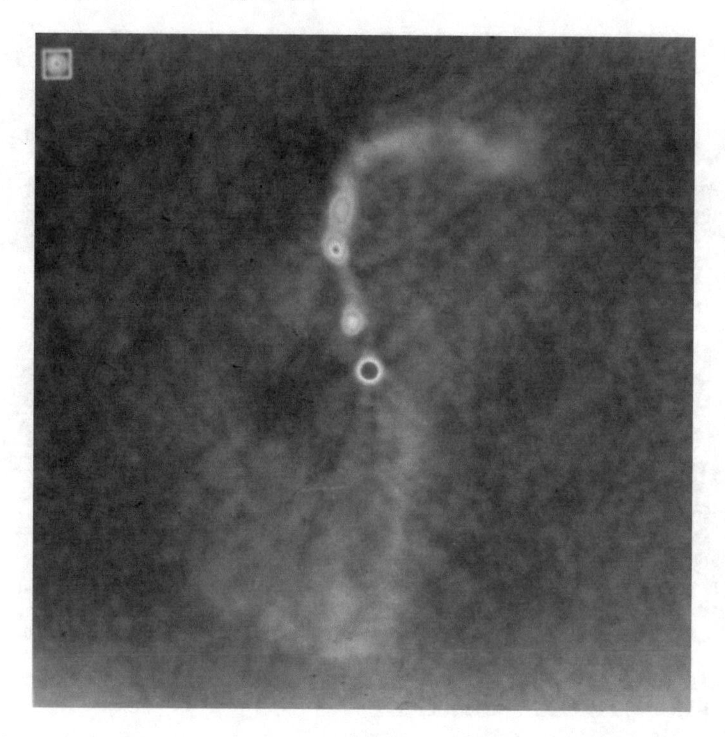

Figure 9.16 Radio image of the quasar 2300-189 obtained at 20 cm wavelength (VLA). The point source of radio waves at the center of the picture coincides with the object visible in optics. One observes, looking above, a fine and irregular jet feeding a region of extended and weak emission. The central object emits two symmetrical jets; the one directed toward us is much more visible. (*Source: Istituto di Astrofisica Spaziale, CNR*)

Neutron stars are other extraordinary heavenly bodies. A year after the discovery of the neutron, made in 1932 by James Chadwick, few physicists advanced the idea that protons and electrons could fuse together and give rise to neutrons in the most dense of the stars. The associated conditions were so strange though—for example a thimble of neutron star matter would weigh a hundred million tons on Earth—that many scientists ignored it all.

Thirty five years later, in November 1967, Jocelyn Bell, a graduate student at the University of Cambridge, analyzing the data from the new telescope she had seen under construction, noticed a regular sequence of rapid, sharp, and intense radio pulses emanating from a precise spot in the constellation of Vulpecula. The regularity of the pulses was amazing, one every 1.3373011 seconds. She reported the phenomenon to her advisor Anthony Hewish. Could they be signals from an intelligent civilization of extraterrestrials? Half jokingly, the source was dubbed LGM for "little green men" by the astronomers of the group. Working in great secrecy, they discovered four such objects in widely separated directions in the sky and published in a short time two sensational articles carrying detailed properties.

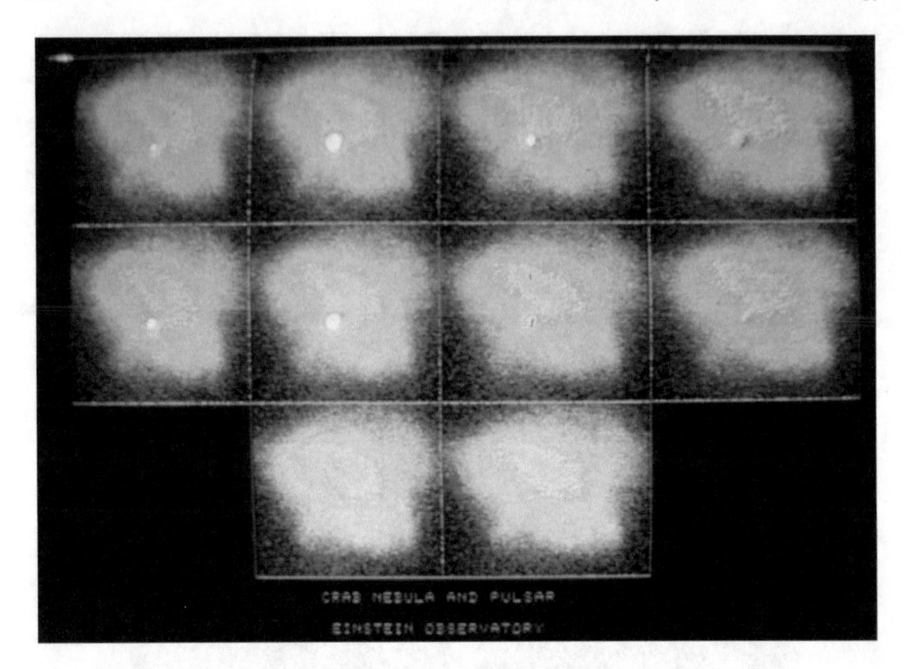

Figure 9.17 Images of the Crab nebula with a pulsar on and off, taken by the Einstein Observatory. (*Source: Istituto di Astrofisica Spaziale, CNR*)

The authors advanced the idea that these stars had a mass comparable to our Sun's mass, were extremely small (e.g., 10 km in radii), and in rapid rotation. Conceivably, neutron stars. If a relatively slow star such as the Sun, which makes a full rotation in 27 days, were to contract to a 10 km radius, then the period of rotation would become a small fraction of a second by the law of conservation of momentum (the same law an ice skater takes advantage of to spin faster). Moreover if it emitted radio waves at one or more points on the surface, then the emission would be received on Earth at regular intervals, in the same manner as the beam of a lighthouse is perceived (Fig. 9.17). This is the case of *pulsars*, most of which are neutron stars. They are clocks so precise as to allow studies of phenomena pertaining to general relativity and otherwise inaccessible.

Besides continuous radiation, radiosources may produce *spectral line radiation*, that is radiation of only one specific wavelength. The best-known example of this kind is the radio emission by neutral hydrogen atoms present as clouds in the interstellar medium and subject to inner turbulence and to movements in blocks. At rest with respect to the observer, the hydrogen emits a faint radiation that, according to the laws of quantum mechanics, corresponds to a wavelength of 21 cm or 1420.405 MHz. The phenomenon, predicted in 1944 by the Dutch astronomer Hendrick C. van de Hulst, was observed for the first time in 1951. It marked the beginning of kinematic studies, inside as well as outside the Milky Way, up to then virtually forbidden in optics by absorption in the galactic plane. By measuring the shift toward higher or lower values with respect to the 21 cm wavelength "at rest," the displacement of clouds of hydrogen atoms could be measured according to the radio analogue of the Doppler effect. The Doppler spread within our galaxy reaches several hundred kilohertz.

In 1965, once more by chance, radioastronomy had revealed another important phenomenon: the *cosmic microwave background* radiation (CMB), fossil relic of a primeval age. It was discovered by Arno Penzias and Robert Wilson at Bell Telephone Laboratories in Holmdel, New Jersey. They were working at the 7 cm wavelength with an experimental antenna—in the shape of a horn and originally designed for satellite telecommunications—that had a receiver of exceptional quality. Plagued by noise while taking careful measures of galactic radiation coming from certain directions, they checked the Galaxy, the Sun, the atmosphere, the ground, and even the equipment to find the source and ruled them all out. With a series of exceptional measures they convinced themselves that this supplementary emission, the same in all directions, was coming from the sky, but about the origin they had no idea.

Close by, at Princeton University, the physicist R. H. Dicke was of the opinion that if the universe was expanding, an established fact since Hubble's work, at the beginning when it was very concentrated it had to be very hot. One could then attempt to detect the residual radiation of that distant past. He calculated that it ought to be found in the millimeter or centimeter wavelength and indeed had begun construction of a suitable receiver on the roof of the Princeton biology building. Attempts have been made even prior to that but the measurement were so difficult, the values so uncertain, that interpretation of the data could not be brought into the picture. Under discussion was the theory of the *Big Bang*, formulated before the war. Penzias and Wilson had found the experimental evidence (Dicke overestimated the strength, getting 30K, and also failed to mention that the calculation had previously been done by George Gamow) and were awarded the Nobel prize for physics in 1978.

The "vanished brilliance of the origin of the worlds" that went together with the explosive beginning of the Universe—as postulated by the Belgian cosmologist Abbé George Lemaître (1894–1966)—was really there.

9.11 A "flat" forever-expanding Universe

In December 1998 for ten days the balloon Boomerang (balloon observations of millimetric extragalactic radiation and geomagnetics) flew over Antarctica at an altitude of about 38 km and was able to take pictures of the background microwave radiation at a time when the universe was in its infancy, denser and hotter and some 50,000 times younger than today. For the same purpose the balloon Maxima had flown one night over Texas in August 1998 (Hanany et al. [2000]).

When the temperature dropped below 3,000 K—300,000 years after the Big Bang—hydrogen atoms could form reducing the number of free electrons and making it possible for the photons of light to escape and start their journey. The resulting intense radiation traveled for more than ten billion years across the Universe and reached Earth red shifted. It is by and large of thermal origin and is detected as a faint microwave background radiation with an average temperature of 2.7 K (–270°C).

The measures by Penzias and Wilson showed a remarkable uniformity in all directions in the sky: The cosmic background radiation was isotropic to within ± 1.0 K. Soon after, at the Stanford Radio Astronomy Institute, R. N. Bracewell and student E. K. Conklin made a search for departures from isotropy with an upper limit of 5 mK allowable variation over a 1° field (Conklin, Bracewell [1967a], [1967b]). It was a remarkable result that endured for some years and was influential in cosmological theory.

In 1991 the Cosmic Background Explorer (COBE) satellite revealed temperature variations. With receivers 100 times more sensitive and resolution 35 times higher, Boomerang could detect variations of 10^{-4} K (Fig. 9.19) with resolution of 22.5 arcmin. It allowed the drawing of the first high resolution maps of CMB at 90, 150 (Fig. 9.18), 240, and 400 GHz (wavelength 3, 2, 1.2, and 0.8 mm, respectively) over 3% of the sky (de Bernardis et al. [2000]). While at 400 GHz the map appears uniform because the much stronger emission of our Galaxy is hiding the CMB, at the first three frequencies the maps show temperature variations similarly located (Fig. 9.18). Brighter (hotter) spots are areas of compression, darker (cooler) spots are areas of rarefaction. Hence there appear the seed structures that in the next 10 billion years would evolve into today's stars, galaxies, and clusters of galaxies.

The quantitative analysis of the images gives much information, answering questions going back decades. For instance the radiation's long trip could have taken place along straight lines if the geometry of the cosmos were Euclidean (zero curvature), as well as along curved paths if the geometry were like that of a sphere (positive curvature) or else of a saddle (negative curvature). In the first case the spots in the maps were expected to be about one degree in angular size. They would appear larger for the most part, in the case of positive curvature and smaller in the case of negative curvature. Opening the era of precision cosmology (Hu [2000]), Boomerang's measurement point toward a "flat" Euclidean universe and consequent everlasting expansion.

9.12 Search for extraterrestrial intelligence

In 1959, there appeared in *Nature* a paper by physicists Giuseppe Cocconi and Philip Morrison [1959] advancing the suggestion to search for extraterrestrial intelligence (SETI) by detecting radio signals. Shortly thereafter Bracewell [1960] speculated about the existence of civilizations incredibly advanced, linked in a chain of communications and with long experience in effecting contacts with emerging communities like ours: "Their signals would have the appearance of echoes having delays of seconds or minutes, such as were reported thirty years ago by Størmer and van der Pol and never explained."

The first search for extraterrestrial intelligence, Project Ozma, was indeed carried out in 1960 at the National Radio Astronomy Observatory in Green Bank,

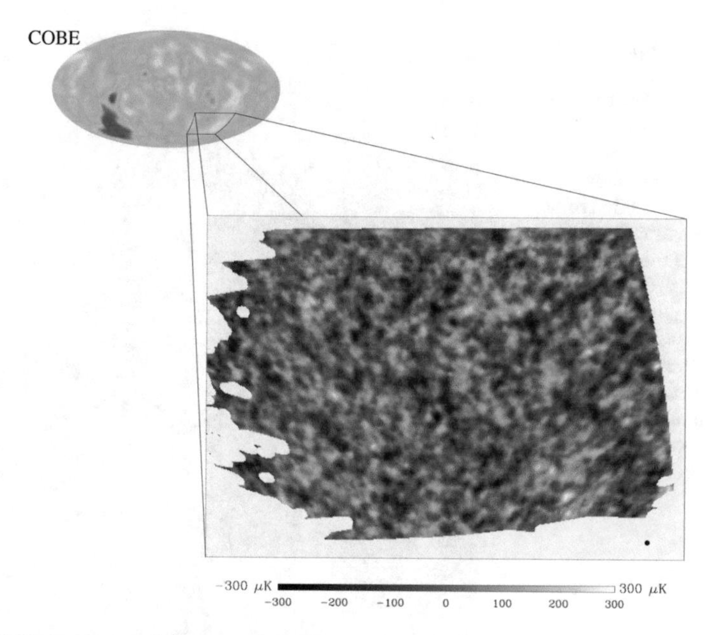

Figure 9.18 Evidence of the structure in the cosmic microwave background was found in 1991 by NASA's COBE satellite which mapped the entire sky (upper left). The Boomerang image (150 MHz) covers 3% of the sky with angular resolution 35 times that of COBE and reveals hundreds of structures that are visible as tiny variations in the temperature of the CMB. (*Courtesy of Boomerang's team*)

West Virginia (USA), by Frank D. Drake. Since then a number of SETI experiments have been performed (Sagan, Drake [1975]). Currently in progress is the SETI Institute program directed by Jill Tarter and the BETA project (Billion Channel Extraterrestrial Assay) lead by Paul Horowitz at the Harvard-Smithsonian Observatory (Fig. 9.20). Started in October 1995 it covers with a radiotelescope, supported by a supercomputer, the band from 1400 to 1720 MHz. It can be imagined as 250 million radios, all tuned to adjacent channels, all receiving simultaneously.

The most likely frequencies for such radio transmissions are thought to be around 1420 MHz—the emission line of the most common element in the universe, hydrogen—and have been thoroughly examined. Even though a number of signals too strong to be noise have been recorded, they were never reobserved so that it can be said that no signals from an extraterrestrial intelligent life has been identified so far. This statement is not as strong as it may sound. As pointed out above, it is relative to a certain range of frequencies and moreover to a certain power of transmitters and up to a certain distance. Besides what is beyond the Milky Way would go undetected independently of power. Any signal originally inside the preferred band would be shifted outside it, upon reception on Earth, by the motion of galaxies.

Thanks to radio astronomy, knowledge of cosmic phenomena in the last fifty years made exceptional progress, bringing to light unsuspected phenomena. This

Figure 9.19 A picture of the inner refrigeration stage that cooled the detectors of Boomerang to 0.3 K, reducing the intrinsic noise of the sensors and thus allowing the detection of the CMB. The stainless steel sphere, 12 cm in diameter, contains the ^{3}He used as a refrigerating fluid. This stage is mounted inside a larger ^{4}He and nitrogen cryostat. The cryogenic system worked automatically for the 10 days of the flight. (*Courtesy of Boomerang's team*)

Figure 9.20 Paul Horowitz at the supercomputer for the BETA project of the Harvard–Smithsonian Observatory. On the lower right a portrait of Fourier. (*Planetary Report*, Vol. XVI, no. 2, 1996.) (*Photograph by Thomas R. McDonough*)

propulsive role is now shared with space astronomy and optical astronomy, which is going through a period of considerable development thanks to the construction of giant telescopes and interferometers. Astronomy has presently at its disposal several tools—from the various observational techniques to numerical simulations—to explore the universe and to try to reconstruct its origin and to foresee its evolution scanned by cosmic times.

Chapter 10
Our Friendly Atmosphere: Methods of Remote Sensing and Climate

10.1 Cannons, sirens, explosions, and atmospheric acoustics is born

It was 1704 when men of church were still active on the forefront of science. A letter was sent to the English clergymen W. Derham (1657–1735), Rector of Upminster in Essex, claiming that "sound is seldom heard at Rome at such distances as in England and northern climates." In [1708] Derham, writing in Latin, published a long paper that clearly speaks of the difficulty posed by sound and the "abstruse phenomena" associated with it. He starts by the disagreement among best authors concerning the velocity of sound, from Isaac Newton to Mersenne, to the Florentine Accademia del Cimento, to finish with the French Observatory. It is clear to Derham that this stems from experiments made over too short distances, which reached at most three miles. He mentions as well inadequacy of instruments and wind as other possible causes. His own experiments with guns at Blackheath, whose flashes he could see from the turret at Upminster church and whose report he could hear almost in all weather, proved "the motion of sound to be the same with what the Accademia del Cimento had determined." He goes on investigating echoes and the motion of sound in different weather and time of the year and provides a "Table of Experiments on Sound" according to the directions of the various winds such as transverse, favoring, strong favoring, snow favoring, rather favoring. The section titled "Motion of Sounds in Italy" is taken by the issue raised in the letter mentioned above. The letter was written by Mr. Richard Townley who affirmed that, when he was in Rome and the great guns of the Castle St. Angelo were fired, the report was much more weak on Trinità dei Monti that at any other place of the same distance. His brother also affirmed that, being once at the Castle Gandolfo which stands on a hill 12 miles from Rome, the report of the same great guns was weak and subdued. About this issue Derham, having contacted his Italian acquaintance S. Averrani working in Florence at the order of the Grand Duke of Tuscany, goes on to mention Averrani's experiments distinctly heard at some 55 miles away, as well as credible information that Averrani

E. Prestini, *The Evolution of Applied Harmonic Analysis*,
Applied and Numerical Harmonic Analysis,
DOI 10.1007/978-1-4899-7989-6_10

had about the sound of the great guns at the sieges of Messina as well as Genoa. Derham concludes "From these observations Averrani is apt to think that there is no difference in this matter between Italy and the northern climates." The beginning of atmospheric acoustics is placed by Brown and Hall [1978] with this very same letter of 1704 to Derham.

In Chapter 4 it has been described the early history of acoustics when vibrating strings attracted interest and string instruments were built. The nature of sound though remained obscure for a long time. Did the sound proceed from the source as a particle or as a wave? The first solid grip on the matter had been put in place by Galileo (1564–1642) with his scientific method, followed by Mersenne—a French disciple of him—and by a number of French mathematicians. Experiments to measure the speed of sound in air had already been carried out in the middle of the sixteen hundreds, though the correct value of 332 m/s in dry air at 0 °C was measured for the first time in 1750 in Paris using a cannon.

Evangelista Torricelli (1608–1647), another of Galileo's disciples, made the fundamental contribution of the mercury barometer. Born of a very poor family, Torricelli was raised at Jesuit and Benedictine Colleges. He stayed with Galileo, confined by the Inquisition to his villa in Arcetri, during the last 3 months of Galileo's life serving as his secretary and assistant. After Galileo's death, he succeeded him as professor of mathematics at the Florentine Academy. Torricelli gave many contributions to physics and mathematics but it is mainly for the invention of the barometer in 1643 that he is remembered (Fig. 10.1). The pump makers of the Grand Duke of Tuscany were experiencing difficulties in raising water above 10 m. Torricelli employed mercury—14 times denser that water—filled with it a glass tube approximately 1 m long and inverted the tube into a basin of mercury. The column fell at about 76 cm, leaving above a "Torricelli vacuum." (A similar experiment with water would require a tube at least 11 m long). After much observation he concluded that the variation of the height of the column of mercury from day to day was caused by changes in atmospheric pressure. At that time the experiment created a sensation, for another reason tough: nobody had ever entertained the thought that vacuum could exist. The *horror vacui* was an established belief from Aristotle. Nevertheless Aristotle was right: vapors of mercury were present in the Torricelli vacuum for, strictly speaking, vacuum cannot be obtained. Blaise Pascal (1623–1662), informed of Torricelli's experiment, repeated it with wine and a glass tube about 14 m long, and further pursued the topic of vacuum and atmospheric pressure. The "torr," a unit of pressure named after Torricelli (760 torr = one atmosphere), was replaced by the "pascal" in the SI system of units starting in 1971 (one pascal = $7.5 \cdot 10^{-3}$ torr).

The sustained development of atmospheric acoustics began in the second half of the nineteenth century when the refraction of sound was investigated by G.G. Stokes [1857] and O. Reynolds [1876]. Sound, just as light, undergoes reflection as well as refraction while passing from a surface separating media of different properties. Refraction obeys Snell's law (10.6). When the two media—say air and water, or else layers of air with different temperature and humidity—are in motion relative to each other due to water currents or winds, models for the refraction of sound can be much more complicated.

Figure 10.1 A postage stamp of the former Soviet Union commemorating Torricelli. (https://en.wikipedia.org/wiki/Evangelista_Torricelli)

A different phenomenon takes place when sound, or light for this matter, strikes an obstacle of finite size. Then some sound may be reflected (echo), some of its energy may be transmitted to the obstacle, and some sound might get around the obstacle undergoing diffraction (Figs. 5.3, 5.5, 5.7, 5.8). In brief the original sound is said to be scattered by the obstacle.

In 1874 the Irish born John Tyndall (1820–1893) attacked the problem of the scattering of sound (Tyndall [1874]) by making experiments in the laboratory and over the sea on the south coast of England, using a siren fitted with a huge horn

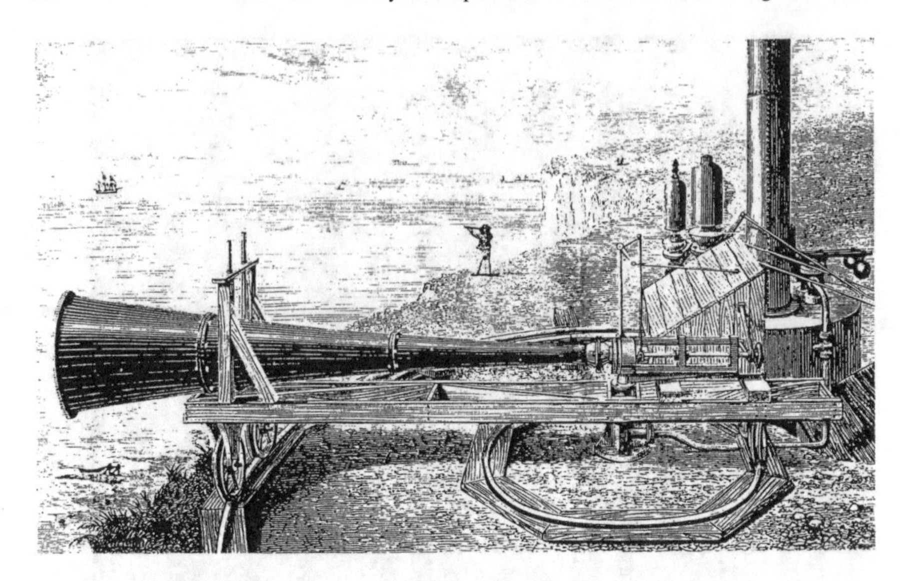

Figure 10.2 Tyndall's apparatus for studying sound propagation in the atmosphere. (Tyndall [1875])

(Fig. 10.2). On occasion he obtained a long persisting echo, taking place in a clear and apparently still atmosphere. He explained it as the effect of temperature fluctuations ("flocculence" in Tyndall's words, i.e., "turbulence" in modern words). His conclusions were not accepted by his peers and this line of research went silent for almost a century.

Tyndall started as a railway engineer in Ireland and later in England. When railway construction slackened he accepted in 1847 a position as mathematician and surveying teacher at a boarding school in Hampshire out of "desire to grow intellectually." In 1848 he moved to Germany, at the time ahead of Britain in experimental chemistry, and in 1850 received his doctorate from the University of Marburg. After his return to England he was elected Fellow of the Royal Society and in 1853 was named Professor of Natural Philosophy at the Royal Institution in London. As such he was a colleague of Michael Faraday, with whom he shared modesty and devotion to science, and succeeded him at Faraday's retirement. Following in Davy and Faraday steps, Tyndall—since the 1860s one of the world's most famous living scientists—spent a significant amount of time disseminating science to the general public at hundreds of public lectures in London and in 1872 even in the U.S. He drew large crowds paying fees to hear him lecture on the nature of light. Tyndall was an experimenter and laboratory apparatus builder. Among his realizations is an apparatus to measure the radiant heat absorption by gases.

10.2 Tyndall, Arrhenius, and some surprising atmospheric gases

In 1859 Tyndall began investigating the radiative properties and absorptive powers of gases such as water vapor, carbon dioxide ("carbonic acid" in Tyndall's words), ozone, and hydrocarbons. An important discovery of him was the vast difference in the ability of "perfectly colorless and invisible gases and vapors" to absorb or else transmit radiant heat. He noted that oxygen, nitrogen, and hydrogen are almost transparent to radiant heat while water vapor, carbon dioxide, and ozone are best absorbers (Fig. 10.3). The stronger absorber among these, and so the most important gas controlling the Earth surface temperature, he claimed to be water vapor. Without it, the Earth would be "held fast in the iron grip of frost."

Perhaps Tyndall's "iron grip of frost" was also in the mind of the Swedish Svante Arrhenius (1859–1927), Nobel Prize for Chemistry in 1903, who—aside from electrochemistry, his main scientific interest—attempted to find an explanation to the onset of ice ages. His calculations and assumptions showed that with no CO_2 in the atmosphere the surface temperature of the Earth would fall by 21 °C. The resulting cooler atmosphere would contain less water vapor, creating an additional temperature decrease of approximately 10 °C.

Later he was first to recognize that "the slight percentage of CO_2 in the atmosphere may, by the advances of industry, be changed to a noticeable degree in the course of few centuries" stimulating plant growth and providing more food for populations, as he speculated. At that time nobody believed that human activity could significantly affect average global temperatures. Presently, it is on record that CO_2 had risen at a rate much faster that Arrhenius had predicted, followed closely by acid rains, deforestation, and acid oceans.

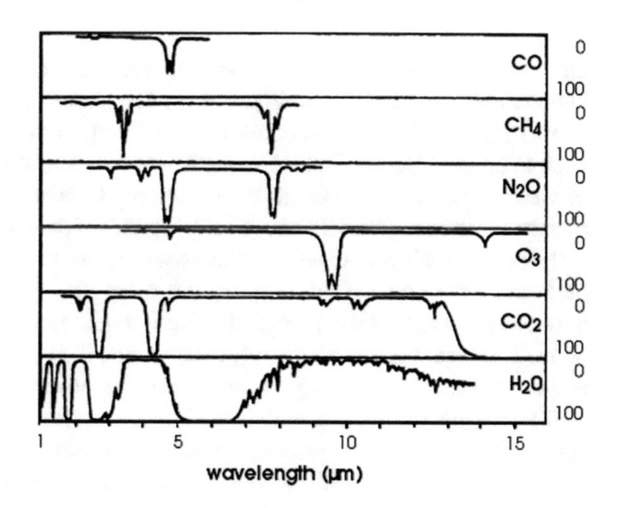

Figure 10.3 Percentage of radiation absorbed by some atmospheric gases: CO (carbon monoxide), CH_4 (methane), N_2O (nitrous oxide), O_3 (ozone), CO_2 (carbon dioxide), H_2O (water vapor). (http://earthobservatory.nasa.gov/Features/Tyndall)

10.3 The discovery of the stratosphere

In 1902 Léon Teisserenc de Bort—having resigned his position of chief meteorologist at the Centre de la Météorologie Nationale in Paris—discovered the stratosphere by measurements made with unmanned instrumented balloons, launched from his private observatory in Trappes, near Versailles. He described the stratosphere as an "isothermal" layer. Hence he was led to conclude that the atmosphere was divided into two layers: up to about 11 km the *troposphere,* where temperature and weather change, and above the *stratosphere* with constant temperature and unchanging weather conditions. Observations of meteor trails showed instead that, in the upper part of the stratosphere and up to 50 km approximatively, temperature increased with height (a phenomenon due to absorption of solar radiation by ozone, as it was later established). Researchers focused on it as well as on refraction, trying to explain some anomalies of sound propagation during explosions.

Due to low atmospheric density research balloons cannot fly much higher than 30 km, so several experiments used explosions leading to the unexpected discovery of sound waves of very low frequency (Gowan [1929]), now called *infrasounds.* Finally an acoustic technique, using rocket grenades, was developed at the White Sands Proving Ground (New Mexico) to obtain temperatures and winds up to 80 km "by measuring the velocities of sound and the deflection of sound waves from seven or nine grenades successively ejected and exploded from an Aerobee rocket (Fig. 10.4) along the nearly vertical upward leg of its trajectory" (Stroud et al. [1956]).

Meanwhile, starting with Taylor [1915], the theory of turbulence had been gradually introduced to study the dynamics of the atmosphere at small scale.

10.4 Our friendly atmosphere

Our atmosphere is a rather thin shell that screens solar wind, harmful X-rays, and gamma rays (Figure 9.1) coming from space. At the same time, it keeps the Earth warm by preventing the energy reaching the Earth from the Sun to be mostly radiated back to space. The atmosphere extends from the surface of the Earth to thousands of kilometers. Up to 100 km its composition is approximately uniform and consists mainly of nitrogen (N_2) and oxygen (O_2), traces of the important water vapor (H_2O), carbon dioxide (CO_2), as well as other gases. Above 100 km the atmosphere becomes progressively thinner. Lighter gases are more abundant, like atomic oxygen (O) between 100 and 600 km, helium (He) between 600 and 1000 km, and hydrogen (H) above 1000 km. Further above only solar wind—the flux of particles, mainly protons and electrons, of large enough velocity to escape from the solar corona—is found.

The energy, originating within the Sun by nuclear fusion, transfers to the solar surface and from there radiates into space. An extremely small amount of such energy is intercepted by the Earth. About one-third is then scattered back to space either by clouds or by the surface itself. The remaining energy powers winds as well as oceanic currents and sustains biospheric processes. While the source of the energy

Figure 10.4 Aerobee Rocket. (White Sands Missile Range Museum)

is ultimately the Sun, the resulting thermal processes in the atmosphere make up a complex chain. Schematically the atmosphere is bathed in two radiation fields. The first, originating in the Sun, has the majority of the spectrum in the visible and ultraviolet (Fig. 9.1). The second one, emanating from the surface of the Earth and its lower atmosphere is mostly infrared. This makes the atmosphere heated from below by the moist radiating surface of the Earth and perturbed locally by the energy of sunlight in daytime. The transfer of energy by infrared radiation—the dominant mode of heat transmission between the different atmospheric layers—is mainly due to trace constituents of the atmosphere like H_2O, CO_2, and O_3 but also N_2O and CH_4. They are able to absorb the infrared wavelength radiated by the surface of the Earth, trap it and return it to the ground. In spite of being present only as traces, these gases have a phenomenal effect. Without them the temperature at the surface of the Earth would be about 40 °C colder. If in excess they are responsible for the so called *greenhouse effect.*

The atmosphere can be subdivided into layers depending upon height and related thermal structure. Up to an average of 13 km from the Earth surface lies the *troposphere.* Here the air is rather unstable for the temperature rapidly decreases with altitude. Colder air tends to sink, displacing warmer air upwards. Like a kettle heated from below the troposphere turns over (from the Greek *tropos* meaning "turn over"). Most of the weather of the planet is confined to the troposphere: clouds are generally not found elsewhere. Above lies the *stratosphere* extending up to 50 km. Here the temperature starts to increase with altitude, reaching back about that at the surface of the Earth. The stratosphere appears to be stretched out in layers (*strata* means "layers" in Latin). Here it is to be found an appreciable concentration of ozone that protects the Earth surface by strongly absorbing ultraviolet radiation. Above lies the *mesosphere* (in Greek *mesos* stands for "middle"), where the temperature resumes decreasing with altitude. At about 85 km the mesosphere ends. Still above, where the temperature rises again with altitude, lies the extremely rarified *thermosphere.*

The upper part of the troposphere is called *free atmosphere.* Its complementary lower layer makes up the *boundary layer* (BL), defined dynamically as that part of the troposphere directly influenced by the presence of the Earth surface. The thickness of the BL is rather variable in time and space, ranging from few tens of meters at the poles to 2–3 km at the equator. The point is that solar radiation, scarcely absorbed by the BL, is almost completely absorbed by the ground. It is the ground that, by warming and cooling, forces in the BL rapid changes—lasting about an hour or so—via transport processes of heat and pollutants. The oceans absorb solar radiation as well, about 90 % of it. Evaporation results, involving approximately 1 m of water per year over all the Earth's oceans. The latent heat stored as water vapor in the atmosphere accounts for 80 % of the "fuel" that drives atmospheric motions. It is mostly in the boundary layer that air flow—classified as mean wind, waves, and turbulence—takes place, as well as transport processes like movements of water and oxygen to and from life forms such as plants.

The *mean wind* is responsible for very rapid horizontal transport, commonly of the order of 2 to 10 m/s. Friction causes the mean wind to slow down near the ground. *Waves*—generated by the mean wind flowing over an obstacle, or by thunderstorms, or by explosions—transport energy. *Turbulence,* the gustiness superimposed on the mean wind, churns and mixes the atmosphere distributing water vapor, smoke, and other substances at all elevations. It takes place mostly within 3 km from the surface of the Earth and reaches its maximum around midday when solar radiation heats up the Earth surface. This, together with disturbances around surface obstacles, makes low level winds extremely irregular. At night the Earth surface cools rapidly. When the air near the ground becomes cooler than the air above it, a stable *temperature inversion* is created and the wind speed and gustiness decrease sharply.

In the free atmosphere instead motions are almost *geostrophic* (the wind is directed parallel to isobars) and turbulence is only occasionally present. A thermal inversion, that limits exchanges, is usually present between the BL and the free atmosphere.

10.5 Turbulence: Navier–Stokes equations and Kolmogorov theory

The phenomenon of turbulence in all its manifestations is not completely understood. In the following the main contributions by Navier and Stokes with their equations, by Reynolds with his number (10.1) and decomposition (10.3), by Richardson with his energy cascade, by Kolmogorov with his theory are presented. The topic is further pursued in Section 10.9.

The first to use the word turbulence (*"turbolenza"*) in a scientific context appears to be Leonardo da Vinci (1452–1519) in the year 1500, in a report written for the Republic of Venice that had sought for his advice (Cod. Atlanticus f. 638dv). Leonardo studied the motion of currents in rivers and made experiments placing obstructions in water. With his discerning eyes he observed the results (Fig. 10.5) and, somewhat anticipating the Reynolds decomposition (10.3), wrote: "Observe the motion of the water, which resembles that of hair, which has two motions, of which one is caused by the weight of the hair, the other by the direction of the curls; thus the water has eddying motions, one part of which is due to the principal current, the other to the incident and reverse motion."[1]

With a hiatus of more that 300 years, quantitative analysis was started. First in 1823 the French Claude L. Navier and subsequently the Irish George G. Stokes formulated the equations, carrying their name, that describe fluid motion. Then in the 1880s the British scientist Osborne Reynolds (1842–1912) introduced a pure number, later named after him (*Re*), providing a criterion to establish whether the flow of a fluid (liquid or gas) in a pipe is laminar or turbulent. In the laminar regime the fluid flows in layers that slide one over another, so the streamlines do not cross. In the turbulent case, fluctuations (eddies) are present and the streamlines are tangled, so turbulent flows have the ability to mix fluids much more effectively than laminar flows. This is important in many instances as in pollutants dispersal in the atmosphere as well as to mix reactants in chemical processes. Reynolds observed that the transition from laminar to turbulent regime in a pipe occurs for *Re* numbers between 2000 and 2500 (unless the tube surface is very smooth for, in such a case, *Re* could be higher). Below this range the flow is laminar, and above turbulent. In between a

[1]"Nota il moto del vello dell'acqua il quale fa a uso de' i capelli, che hanno due moti, dei quali l'uno attende al peso del vello, l'altro al liniamento delle sue volte; così l'acqua ha le sue volte revertiginose, delle quali una parte attende a l'impeto del corso principale, l'altra attende al moto incidente e refresso." (Cod. Windsor).

Figure 10.5 Water striking water (1507–1509), 29x 20.2 cm. (Royal Library, Windsor Castle, U.K.)

transition takes place from one regime to the other. For example, a laminar flow in a pipe would experience perturbations from inlet flow pressure fluctuations. The flow remains laminar if viscous forces damp out these perturbations. If the flow rate is increased beyond a certain rate, viscous forces can no longer suppress perturbations and sporadic bursts of turbulence start to appear. Eventually transition from laminar to fully turbulent flow takes place. In Figure 10.6 Reynolds experiment is sketched: a dye streak is injected into flow through a pipe having smooth transparent walls.

Later the Reynolds number was also used for types of flow involving moving objects immersed in a fluid as, for instance, airplanes and also in atmospheric phenomena like hurricanes. The Reynolds number is defined as

$$\frac{[\text{some average flow velocity}]\text{x}[\text{some characteristic length}]\text{x}[\text{fluid density}]}{[\text{fluid viscosity}]} \tag{10.1}$$

In the case of a pipe the characteristic length is the diameter of the tube and the velocity is the average velocity of the flow. In case of an airplane flying at a cruise velocity of about 800 km/h the Reynolds number is of the order of 10^6, the characteristic length being the length of the wing in the direction of motion (wing chord) which is of order of a few meters. In case of a submarine moving at 50–60 km/h the Reynolds number is higher: it is of the order of 10^9, mainly due to the higher density of water with respect to air. In case of hurricanes it is of the order of 10^{11}. Indeed the typical dimension of hurricanes is of the order of 100 km and the

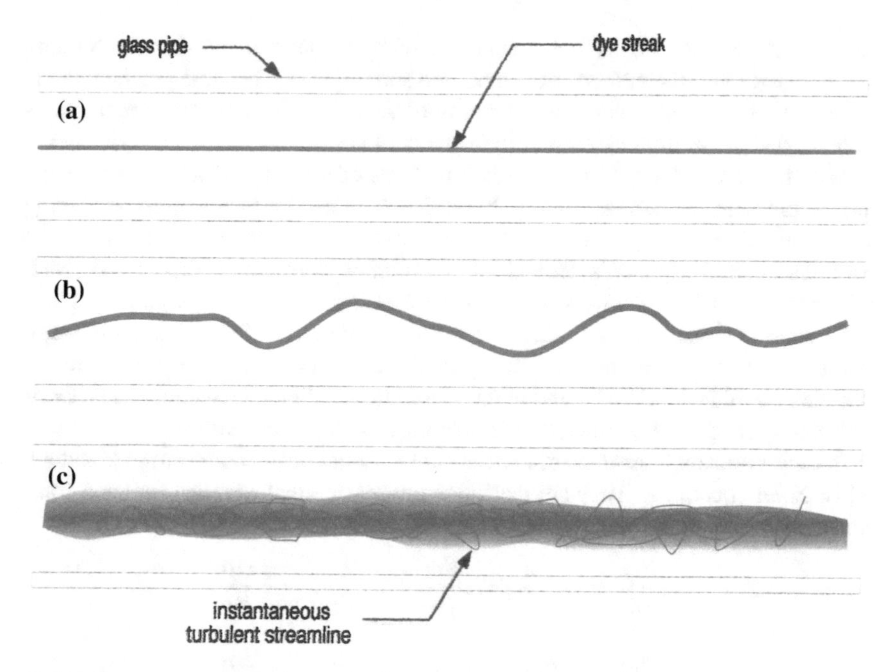

Figure 10.6 (a) laminar flow, $Re < 2000$; (b) transitional flow (still laminar), $2000 < Re < 2300$; (c) turbulent flow, $Re > 2500$, the instantaneous streamlines change direction "erratically." (McDonough [2007])

velocity—tangential in this case—could be in the range of 100–200 km/h. For a fly the Reynolds number is of the order of 100.

Turbulence, strictly a three-dimensional phenomenon, can be visualized as consisting of fluctuations, or vortices, or eddies, of many different sizes next to each other. The atmospheric boundary layer is a collection of vortices. Examples of large eddies are the thermal plumes of warmer air rising from the ground heated from the Sun (Fig. 10.11). The largest among them range from 100 up to 3000 m in diameter, reaching the depth of the boundary layer. Smaller size eddies are apparent when looking at swirling leaves. The smallest atmospheric eddies are of the order of few millimeter in size. The larger the eddy, the larger its kinetic energy and the longer it lasts: large hurricanes can last a couple of weeks. In case of an airplane the smallest eddies are the ones close to the wing surface. Their typical size is of order of one-tenth of a millimeter.

Large vortices are unstable and break up into smaller vortices to which they yield part of their kinetic energy. This process keeps taking place and goes under the name of "energy cascade." It ends when the smallest eddies die off, viscosity being effective in dissipating all their kinetic energy to surrounding molecules. The energy so transferred turns into an increase of the molecular velocity (heat). The energy cascade, characterized by energy transfer to successive smaller scales and by viscous dissipation, is an idea originally advanced by the British scientist Lewis Richardson (1881–1953). He was also the first to apply mathematical techniques to obtain weather predictions (Richardson [1922]) by numerically solving the Navier–Stokes equations. His method, once modified and improved, became practical in the 1960s with the advent of computers since until then about 3 months of computations were needed to predict the weather in the next 24 hours.

With respect to a rigid body, the key features of a fluid are lack of own shape and the pressure gradient as driving force. The *Navier–Stokes equations* describing fluid motion are strictly a statement of conservation of momentum. They are the fluid analogue of Newton's second law of motion from which they are derived. The unknown is the velocity $\mathbf{v}(\mathbf{x}, t)$, where $\mathbf{x} = (x, y, z)$ denotes a point in space and t denotes time. Indeed to describe a fluid motion velocity makes more sense than position, the usual unknown in classical mechanics (e.g., the equations of the vibrating chord in Section 2.1 and of the vibrating membrane in Section 4.11). Once the velocity field is determined then trajectories can be visualized.

In case viscosity is neglected, Navier–Stokes equations become simpler and are called *Euler equations*. They are the following set of equations, one for each component v_1, v_2, v_3 of the velocity $\mathbf{v}(\mathbf{x}, t)$

$$\begin{aligned} \rho(\frac{\partial v_1}{\partial t} + v_1\frac{\partial v_1}{\partial x} + v_2\frac{\partial v_1}{\partial y} + v_3\frac{\partial v_1}{\partial z}) &= -\frac{\partial p}{\partial x} \\ \rho(\frac{\partial v_2}{\partial t} + v_1\frac{\partial v_2}{\partial x} + v_2\frac{\partial v_2}{\partial y} + v_3\frac{\partial v_2}{\partial z}) &= -\frac{\partial p}{\partial y} \\ \rho(\frac{\partial v_3}{\partial t} + v_1\frac{\partial v_3}{\partial x} + v_2\frac{\partial v_3}{\partial y} + v_3\frac{\partial v_3}{\partial z}) &= -\frac{\partial p}{\partial z}, \end{aligned} \tag{10.2}$$

where $p = p(\mathbf{x}, t)$ denotes pressure—an unknown as well—and $\rho(\mathbf{x}, t)$ density. If density is constant then the fluid is said to be *incompressible*. An example is air encountered by an airplane flying at a speed less that 600 km/h, for then air just moves away. To close the system the equation of conservation of mass is added.

If viscosity must be taken into account then a linear term—the so-called stress tensor—is added to the right-hand side of (10.2) which are then called Navier–Stokes equations. Other terms might be added to the right-hand side. For instance, in atmospheric physics another term is added to account for the gravity force, still another one to describe the Coriolis effects that is the influence of the Earth' s rotation. To account for air humidity, added to the Navier–Stokes equations are the equation of conservation of moisture and the equation of conservation of heat. Indeed water vapor in air not only transports sensible heat associated with its temperature, but it has the potential to release or absorb substantial heat during any phase change that might occur (Stull [1988]).

The case of a *compressible* fluid—density $\rho = \rho(\mathbf{x}, t)$ not constant—makes the problem more complicated and less is known about fluid dynamics. Examples are flights in air at supersonic speed, or atmospheric reentry of capsules whose speed may reach 8 km/s from low orbit and 11 km/s from the Moon.

On the right side of (10.2) is the driving force (pressure difference). The left-hand side describes acceleration. In particular the last three terms, which are nonlinear, describe the so-called *convective acceleration* (Fig. 10.7). It is the presence of these nonlinear terms, accounting for the convective transport of momentum, that makes Navier–Stokes as well as Euler equations generally unsolved by analytical methods. Indeed existence, uniqueness and regularity of solutions of the Navier–Stokes equations is a most important, long standing, open problem in mathematics. Therefore, problems in fluid dynamics are usually treated by numerical methods and simulations that lead to approximate solutions. Figure 10.8 shows results of simulating a jet flame with three methods: (a) Reynolds-averaged Navier–Stokes equations; (b) large eddy simulation; (c) direct numerical simulation (Givi [1989], Kuo and Acharya [2012]). This last one, the most accurate, is not feasible for most practical problems being computationally too demanding even for modern computers. Yet the advent of supercomputers, allowing finer and finer grids, made numerical fluid dynamics progress steadily. Still further progress is needed to lead, for instance, to better weather forecasts.

A complementary approach to the problems presented by fluid motion is based on the theory of turbulence. It provides some guidance, stating conditions that must be satisfied statistically by the flow. Reynolds [1895] had already advanced his decomposition (10.3) when Geoffrey Taylor [1915] and others revived interest in the scattering of sound at the root of which is turbulence, the "flocculence" in Tyndall [1874]. Taylor [1935] explicitly stated as assumption that turbulence is a random process and therefore introduced statistical tools, as well as Fourier transforms and power spectra in the turbulence literature. In 1941 Kolmogorov made a lasting impact on this topic.

Andrey N. Kolmogorov (1903–1987) was born just a few years before the Russian revolution. His father Nikolai Kataev was an exiled agriculturalist. His unmarried mother Mariya Kolmogorova died giving birth to him. He was raised by his mother's sister at the estate of his grandfather near Yaroslavl, about 300 km northeast of

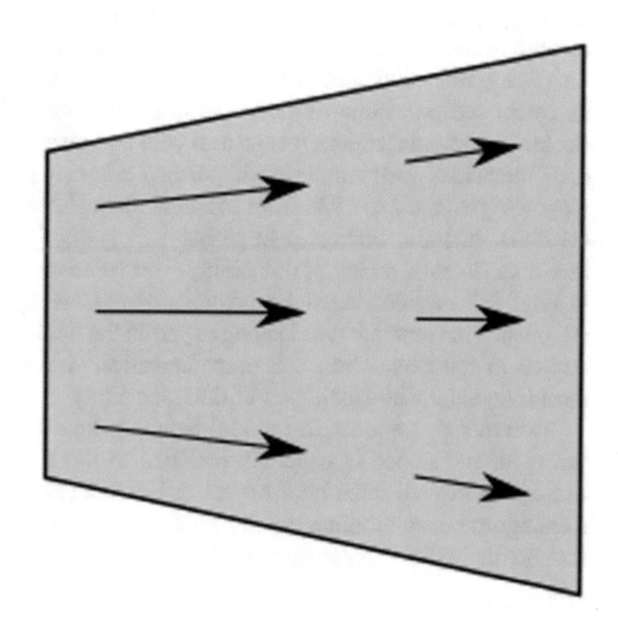

Figure 10.7 An example of convection. Even though the single particles experience a deceleration in time, the fluid's flow may be steady (time independent) as it decelerates with respect to position while moving down the diverging duct. (https://en.wikipedia.org/wiki/Navier-Stokes_equations)

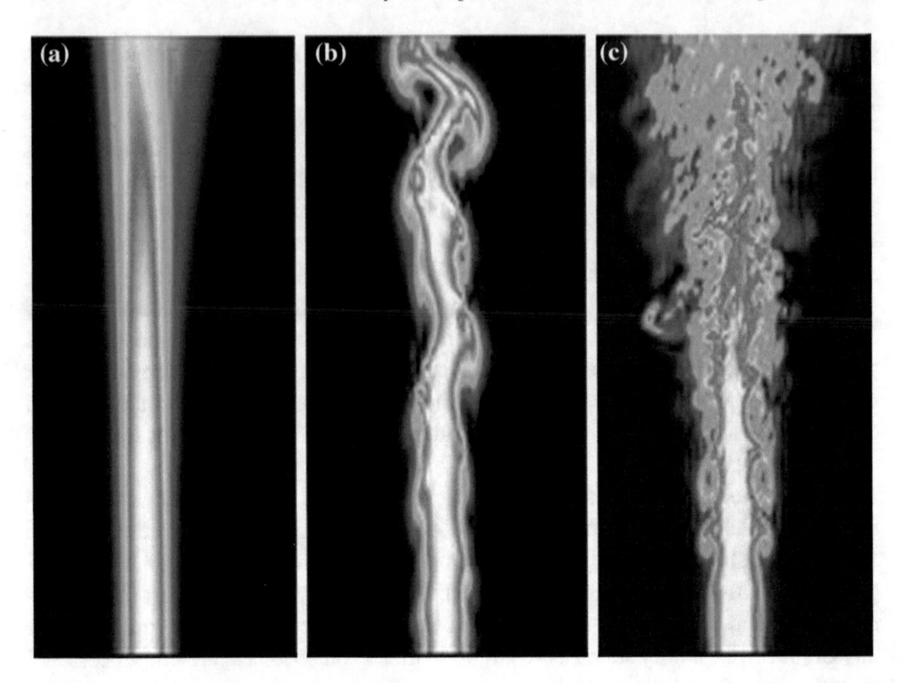

Figure 10.8 Simulations of a flame by three different methods (a) RANS; (b) LES; (c) DNS. (*Courtesy of P. Givi*)

Moscow. Together they moved to Moscow in 1910, where Kolmogorov graduated from high school in 1920. In 1922, at the age of 19, he gave an example of an absolutely integrable function whose Fourier series diverges almost everywhere (Kolmogorov [1926]). This gained him international recognition and raised intense interest among Fourier analysts, eventually leading to a result surprisingly in the positive for closely related classes of functions. This result came more than 40 years later, its celebrated proof due to the Swedish mathematician Lennart Carleson [1966]. (Kolmogorov and Carleson results deal with one-dimensional Fourier series. In several dimensions open problems still persist Fefferman [1971], Prestini [2016].) Kolmogorov graduated in 1925 from Moscow State University where he also obtained his Ph.D. in 1929. Interested in probability theory, in 1933 he published in German his axiomatic theory of probability—based on measure theory—later to appear in English as a book titled *Foundations of the Theory of Probability.* This work established his reputation as leading expert in the field and, together with his contributions to many other branches of mathematics, makes him one of the greatest and most influential mathematicians of the last century.

Even though some studies were previously done in the West (Little [1969]) the theory of turbulence is generally ascribed to the Russian school and especially to Kolmogorov. In three brief papers in Russian (Kolmogorov [1941]), a mark of Kolmogorov deep intuition, he dealt with turbulence for incompressible fluids at a sufficiently high Reynolds number. They were followed by a refinement in [1962].

His theory is built on the energy cascade advanced by Richardson in 1922, and already mentioned. This qualitative description of turbulence marks an important point: viscosity matters only at the smallest scales, whose size is the first quantitative question that comes to mind. By introducing some hypotheses, Kolmogorov answered this question and many more (Pope [2009], McDonough [2007], Lesieur [1987]).

Turbulent eddies can be described as a collection of molecules moving in moderately coherent patterns characterized by several parameters, among which length scale and velocity. As a whole vortices are generally very difficult to simulate and predict, and may be highly asymmetric. Kolmogorov worked under ideal hypotheses, he assumed turbulence to be locally homogeneous (meaning invariant under translations) and locally isotropic. In his theory vortices, not precisely defined, are subdivided into three categories depending on their scale length:

1) The largest scales or integral length scales L. Constrained by the characteristic length of the apparatus (e.g., the diameter of the pipe), the largest scales have higher velocity, last longer and contain most of the energy. This is called the *energy-containing range*. The assumption that $Re = Lv_0/\nu_0$ is large—where ν_0 is the kinematic viscosity (viscosity divided by density)—means that the effects of viscosity are relatively small.

2) The intermediate scales of lengths r in the range $\eta << r << L$. This is called the *inertial range*. Kolmogorov hypothesis of "local isotropy" states that vortices at the small scales $r << L$ are statistically isotropic (invariant under rotations). The possibly asymmetric geometry of the large scales is lost, hence the statistic of the small scales vortices can be regarded as universal.

3) The smallest scales or *Kolmogorov length scales* η. Here vortices are short lived. Their Reynolds number is small enough for dissipation to be effective, their kinetic energy quickly transformed into heat by viscosity. This is called the *dissipative range*.

Kolmogorov determined uniquely—in statistical sense—length, velocity, and timescale of the vortices of the dissipative range, depending on the kinematic viscosity ν and on the turbulent dissipation rate ϵ (average transfer of turbulent kinetic energy per unit mass per unit time). Within a multiplicative constant, i.e., adopting a normalized $Re = 1$, they are given in order by

$$\eta = (\nu^3/\epsilon)^{1/4}$$

$$v_\eta = (\epsilon\nu)^{1/4}$$

$$\tau_\eta = (\nu/\epsilon)^{1/2}$$

Furthermore in the range $\eta << r << L$ vortices are independent of ν and uniquely determined by ϵ (via the formula $(v')^3/r = \epsilon$ where v' denotes the vortex tangential velocity). This means that viscous effects are negligible, thus the name of inertial range.

Kolmogorov *inertial range energy spectrum* determines how the turbulent kinetic energy is distributed. Following the decomposition advanced by Reynolds [1895], the velocity vector is written as

$$\mathbf{v}(\mathbf{x}, t) = \mathbf{w}(\mathbf{x}) + \mathbf{u}(\mathbf{x}, t) \tag{10.3}$$

where $\mathbf{w}$ is the mean velocity and $\mathbf{u} = \mathbf{v} - \mathbf{w}$, associated to turbulence, is the instantaneous difference. An important result of the theory is *Kolmogorov's $\frac{2}{3}$ law.* It deals with $||\mathbf{u}(\mathbf{x} + \mathbf{l}, t) - \mathbf{u}(\mathbf{x}, t)||^2$ that represents the square velocity increment between two points at distance $l = ||\mathbf{l}||$ assumed very small with respect to the integral scale. It states that, in the inertial range, the so-called *second order structure function* defined by

$$D(l) =< ||\mathbf{u}(\mathbf{x} + \mathbf{l}, t) - \mathbf{u}(\mathbf{x}, t)||^2 >$$

where $< \cdot >$ denotes ensemble averaging, satisfies the relation

$$D(l) = C\epsilon^{2/3} l^{2/3} \tag{10.4}$$

with C a universal constant. The above result (10.4), involving the square of a velocity, is clearly related to kinetic energy and the following Kolmogorov $k^{-5/3}$ energy spectrum (10.5) can be derived from it.

A fundamental characterization of a turbulent flow is by kinetic energy distribution (for fluids, such as air, kinetic energy per unit mass is more conveniently used)

$$\frac{1}{2}||\mathbf{u}(\mathbf{x}, t)||^2 = \frac{1}{2}\mathbf{u}(\mathbf{x}, t) \cdot \mathbf{u}(\mathbf{x}, t)$$

For every fixed time t, $\mathbf{u}(\mathbf{x}, t)$may be represented via its inverse Fourier transform

$$\mathbf{u}(\mathbf{x}, t) = \int \hat{\boldsymbol{u}}(\mathbf{k}, t) e^{i\mathbf{k} \cdot \mathbf{x}} d\mathbf{k}$$

By Plancherel theorem the following equality holds $\int ||\mathbf{u}(\mathbf{x}, t)||^2 d\mathbf{x} = \int ||\hat{\mathbf{u}}(\mathbf{k}, t)||^2 d\mathbf{k}$. In polar coordinates, the mean turbulence kinetic energy can be expressed as an integral over the wave number $k = ||\mathbf{k}|| = \frac{2\pi}{r}$, that is in the form

$$\int_0^\infty E(k) dk$$

where the *spectral energy* $E(k)$ denotes the contribution to the mean kinetic energy of all $\mathbf{k}$ lying on the sphere of radius k, corresponding to vortices of length r. Then the law

$$E(k) = C_K \epsilon^{2/3} k^{-5/3} \tag{10.5}$$

describes the so-called *Kolmogorov $k^{-5/3}$ inertial range energy spectrum*, where C_K is called the *Kolmogorov constant.* Proved by Obukov [1941], a Kolmogorov student,

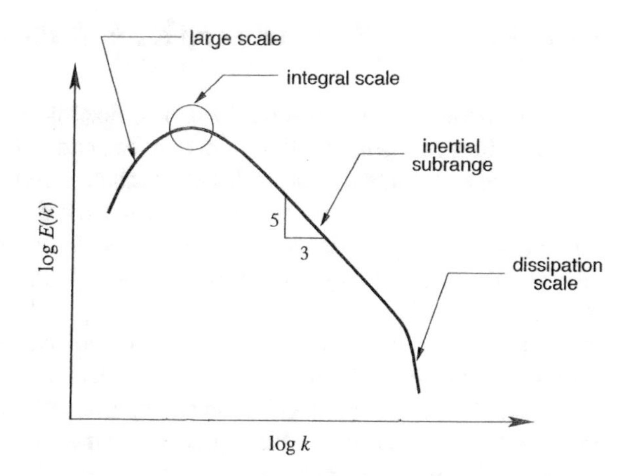

Figure 10.9 Kolmogorov energy spectrum in logarithmic scale. (McDonough [2007])

this result makes $\log E(k)$ to decrease as $-\frac{5}{3}\log k$, a straight line with slope $-5/3$ (Fig. 10.9). This is one of the most famous results of Kolmogorov theory. Much evidence has been found to support it. For instance, turbulent measurements, taken from a submersible vehicle operating in the Knight Inlet on the western coast of Canada, are described in Gargett et al. [1984]: the decay with slope $-5/3$ extends to nearly three decades from $-4 < \log k < -1$. Also from experimental measurements the Kolmogorov constant C_K has been estimated to be in between 1.53 and 1.68.

Another result, the so-called *Kolmogorov* $\frac{4}{5}$ *law,* states that a third order structure function (similar to the second order structure function $D(l)$ but with a third power) is equal to $-\frac{4}{5}\epsilon l$. This result, confirmed by laboratory experiments, may be proved to follow directly from the Navier–Stokes equations (Frisch [1995]).

Then in the 1960s another feature of turbulent flows, still governed by the deterministic Navier–Stokes equations, was brought to attention: extreme sensitivity to perturbations in the initial and/or boundary conditions, requiring unreachable precision to duplicate an experiment. This developed in the "strange attractor" description of turbulence (Lorenz [1963], Ruelle [1989]).

Here is an example of an effect of turbulence that may occur in everyday life. Imagine two friends conversing on a sidewalk when a girl (or boy), wearing a strong perfume, walks by about a meter away. They would smell the perfume right away and would turn quickly to check what is causing it. If the time for the perfume molecules to reach them by molecular diffusion were to be computed it would be found to be about an hour. This long time is due to molecular collisions. The explanation instead is turbulence. Walking by, the girl creates a vortex—of diameter comparable to her size—rotating at about the speed of the girl, say 3.5 km/h. Calculations show that such a vortex reaches the two friends in about a second and that the girl Reynolds number is indeed turbulent, of the order of 10^4.

10.6 Princes of the clouds and lords of the winds

A web of about 800 balloons (radiosondes) floating in the air "covers" the globe. Inflated with hydrogen or helium, the balloons carry small instrument packages to measure the upper air raw data needed for computer based weather prediction models. These *in situ* measurements, taken once or twice a day and recorded in a standard format, are exchanged between countries through international agreements (www.ua.nws.noaa.gov/factsheet.htm). The basic data of temperature, humidity, pressure, wind velocity and direction at different heights (*profiles*) are also made available in terms of averages by scientists operating with sodars, radars and lidars, dedicated instruments that transmit and receive back sound, radio, and light waves, respectively.

In case of *sodar* (sound detection and ranging) targets are temperature fluctuations. While most of the transmitted wave proceeds in the same direction, some is scattered off in all directions by the target. For constructive interference, only temperature inhomogeneities spaced $\lambda/2$ contribute to the backscattered wave (transmitter and receiver colocated) where λ denotes the transmitted wavelength (Section 10.9). Powerful pulses must be transmitted and very sensitive receivers must be designed to pick up the echo. Sodar employs acoustic waves in the audible range: wavelengths are in between 6–20 cm and the corresponding frequencies in the range of kHz. Hence targets are temperature fluctuations at the decimeter scale at different heights. At this scale the sodar is the right probe since the interaction of sound waves with the lower atmosphere is much stronger than that of most part of the electromagnetic spectrum. In addition, relatively simple and inexpensive electronics can be used. Indeed, in case of radar, the same wavelengths are associated to much higher frequencies—in the GHz range (Fig. 9.1)—and require more sophisticated electronics. Nevertheless the strong attenuation of sound waves limits the sodar range to the height of 1.5 km. The method of operation is based on the refractive acoustic index modifications caused by atmospheric turbulence (*Bragg scattering*).

The *radar* signals travel long distances and penetrate clouds. *Meteorological radars* (also called *weather radars*) operate at wavelengths from 1 millimeter to 1 meter, but wavelengths of the order of the centimeter ([3–10] cm) are the most common. Targets are water or ice droplets in clouds. For smaller particles, named *cloud droplets*, shorter wavelengths (λ <3 cm) are used by the so-called *cloud radars*. The methods of operation are based on *Rayleigh scattering* if the target dimension is smaller than wavelength, otherwise on *Mie scattering*. To resolve turbulent structures in *clear air* (cloud and precipitation free), it is Bragg scattering again.

At the even shorter wavelengths of *lidar* the target is aerosol, the system of liquid or solid particles distributed in a finely divided state in the air (Fig. 10.10). Aerosol particles play an important role in precipitation processes, providing the nuclei upon which condensation and freezing take place. Radii of aerosol particles cover the wide range from 10^{-8} to 10^{-4} meters and require the wavelength of light, or shorter as in case of ultraviolet rays. Smoke accounts for most particles with radii smaller than 10^{-7}m, while examples of particles larger than 10^{-7}m are dust and ocean spray.

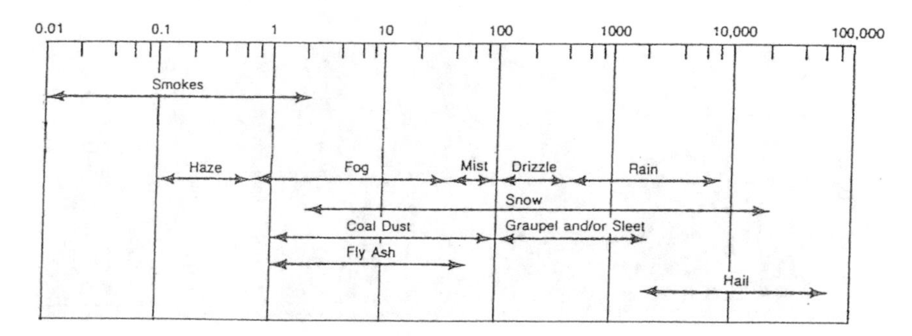

Figure 10.10 Diameter of atmospheric particles in micron (modified). (Reproduced from Pueschel R.F. [1986], "Aerosol measurements in the boundary layer," article in *Probing the Atmospheric Boundary Layer*, (Ed. D.H. Lenchow) 57-86. © American Meteorological Society. Used with permission.)

Furthermore, by *Raman scattering* molecular constituents of the atmosphere can be targeted and detected.

From the delay between the transmitted *pulsed* wave—of sodar, radar, or lidar—and the returned signal, the distance of target (*range*) can be determined. Via the Doppler frequency shift, the radial velocity can be calculated as well, though by different methods (as exemplified at the end of Section 10.11). The heterodyne detection (radar, lidar) is linked to the so called range-velocity ambiguity (Section 10.15).

10.7 The advent of sodar

In the 1930s the radar was introduced and the atmosphere found another group of scientists interested in it. In an attempt to discover the cause of a severe fading, in some atmospheric conditions, of an experimental trans-horizon microwave radio link (over the 64 km distance) between New York and Neshanic, New Jersey, G.W. Gilman and collaborators—working for the Bell Telephone Laboratories in New York—built a device acting like an acoustic radar. They named it sodar. The device was rather rudimentary, with a detection range a little over 100 m. In spite of that intense acoustic echoes taking place in presence of low-level temperature inversions were detected. Since calculations showed that even a strong inversion could not account for the strength of the signal, Gilman and collaborators stated: "it maybe that slight turbulence and eddies even in seemingly calm air present significant discontinuities by virtue of variations in temperature, humidity, air velocity and static pressure" (Gilman et al. [1946]).

Interested in Gilman's new creature—the sodar—Linday McAllister [1968], working in Adelaide for the Australian Defense Scientific Service, built a greatly improved one—that he called "acoustic sounder"—with the same antenna to transmit and receive the signal. To record the signal McAllister introduced a key innovation: he adopted the facsimile recorder, already in use since the 1920s in ocean sonars to

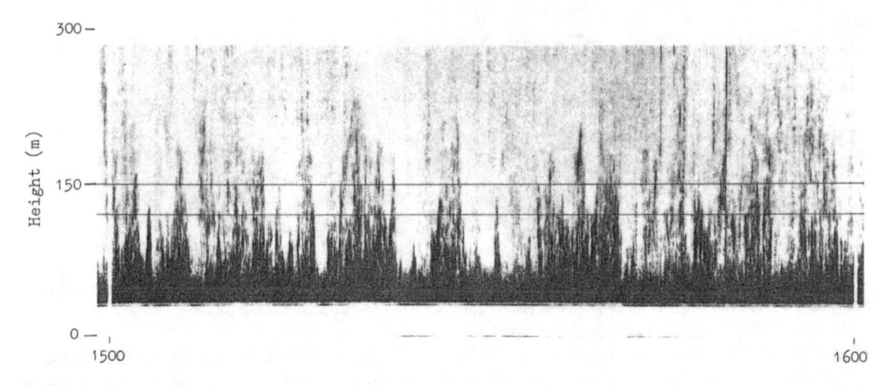

Figure 10.11 Acoustic sounder record of thermal plumes, 15:00–16:00 C.S.T., April 2, 1971. Height of range gate shown by parallel lines on record. (McAllister [1971])

image the bottom of the sea or fish schools crossing the acoustic beam. In this way each pulse is recorded as a vertical trace whose darkness represent the strength of the signal received from the atmospheric height marked on the ordinate of the chart. Along the abscissa the time of successive pulses is recorded. The resulting picture shows the evolution of atmospheric structures in time (Fig. 10.11). In McAllister short and elegant paper the echoes were attributed to temperature discontinuities: "The velocity of propagation of sound waves through the atmosphere is, to a first approximation, a function of the absolute temperature of the air. Therefore, given an acoustic sounding system (Gilman et al. [1946]) of sufficient sensitivity and an integrated recorder, it is possible to obtain echoes from temperature discontinuities in the lower atmosphere, and so to determine the structural characteristics of one of the key atmospheric parameters as a continuous function of height and time." Subsequent meteorological records showed that a clearly detectable echo occurred in presence of temperature fluctuations with a range of more than 0.05 °C at a frequency greater than 1 Hz (McAllister et al. [1969]).

Meanwhile in Boulder, Colorado, at the Wave Propagation Laboratory of the National Oceanic and Atmospheric Administration (NOAA), Gordon Little had started a program for the study of optical propagation and remote sensing of the atmosphere by means of laser beams. In the process Little became acquainted with the work of the Russian school on the scattering of waves by turbulence, such as the monograph by Tatarski [1961], as well as acoustic studies by Kallistratova [1961] and Monin [1962]. In 1968, during a visit to Australia, Little shared his knowledge of the results of the Russian school. Then, using turbulent acoustic scattering theory, Little was first to obtain quantitative estimates of the strength of the received signal, later confirmed by experimental results (Little [1969]). He foresaw the great potential of sodar and stated that "the acoustic sounding technique could be developed to monitor to heights of at least 1500 meters: 1) the vertical profile of wind speed and direction (by utilizing the Doppler technique); 2) the vertical profile of humidity; 3) the location and intensity of temperature inversions; 4) the three-dimensional spectrum of mechanical turbulence; 5) the three-dimensional spectrum of temperature inhomogeneities." Later research has followed up all these suggestions.

Next to Little's paper there appeared the coordinated paper (McAllister et al. [1969]), also adopting turbulent scattering theory as a mechanism for the production of the returned signal and presenting as well facsimile recordings of various atmospheric structures like thermal plumes (Fig. 10.11), inversions, and breaking waves (Section 10.10). In his paper Little had acknowledged the "stimulation provided by McAllister experimental work."

The first sodar by McAllister was followed by others: in 1969 by Little in U.S., in 1974 by G. Fiocco and G. Mastrantonio in Italy and by A. Spizzichino in France, and then in other nations.

10.8 The refractive index of air to acoustic, radio, and light waves

Refraction is the change of direction of a wave, due to its change in speed, as the transmission medium changes (Fig. 10.12). In case of two different isotropic media Snell's law states that

$$\frac{\sin\theta_1}{\sin\theta_2} = \frac{n_2}{n_1} = \frac{v_1}{v_2} \tag{10.6}$$

where θ_1 is the angle of incidence, θ_2 the angle of refraction, n_1 and n_2 the corresponding indices of refraction, v_1 and v_2 the velocities of the wave in the two media.

Variations of the refractive index lie at the origin of the scattering of acoustic waves as well as electromagnetic waves. The *acoustic refractive index* n_a is defined as the ratio of two velocities

$$n_a = \frac{C_a}{c_a}$$

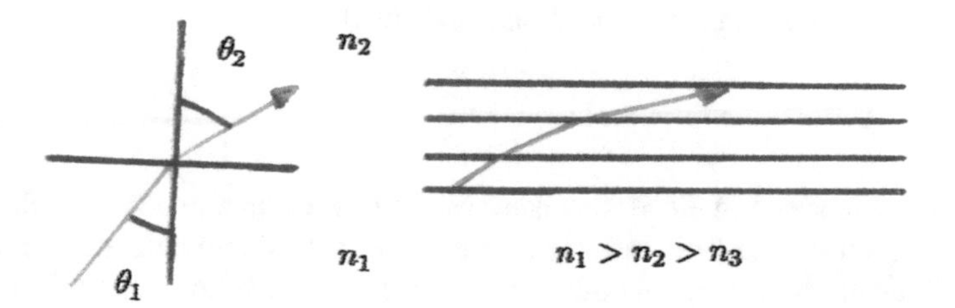

Figure 10.12 Left: Snell's law of refraction with $n_1 > n_2$. Right: Since the atmospheric refractive index typically falls with height, the refracted ray moves away from the vertical as it moves to higher altitudes making possible trans-horizon radio transmissions.

Table 10.1 Sound speed in dry air. (Galati [2002])

T (°C)	C_a (m/s)
0	331.5
10	337.8
20	343.8
30	349.6
40	355.3

where C_a is the reference velocity of sound in a dry and stationary atmosphere at a reference absolute temperature T_0 (in kelvin) and c_a is the velocity of the probing wave. By the formula

$$C_a = 20.05\,T_0{}^{1/2}[m/s]$$

with $T_0 = 288\,°K$, corresponding to 15 °C, it is obtained $C_a = 340$ m/s (Tab. 10.1).

In moist stationary atmosphere the velocity of sound is slightly higher (the more so in water where is of the order of 1 km/s). To a good approximation it can be calculated by the following formula

$$c_a = 20.05\left(1 + 0.14\frac{e}{p}\right)T^{1/2}[m/s] \tag{10.7}$$

where T denotes absolute temperature and e/p the ratio of water vapor pressure e to total pressure p. So the velocity of acoustic waves is independent of the wavelength, but depends on air temperature and, to a lesser degree, on humidity since the total contribution of the atmospheric water vapor to the velocity of acoustic waves is typically no more than 1 m/s. The same dependence holds for the refractive index. Therefore variations of the above two parameters — to which wind must be added— result in the scattering of acoustic waves. (For completeness it is mentioned that in case of frequencies above 25 kHz—and so above the audible range (Chapter 4) — the presence of CO_2 makes the velocity of acoustic waves wavelength dependent, thus changing sound during propagation).

The radio refractive index n_r is similarly defined

$$n_r = \frac{c_0}{c} \tag{10.8}$$

where c_0 is the speed of electromagnetic waves in vacuum and c is the velocity of the probing wave. The radio refractive index of air varies with height and normally falls with height in a standard atmosphere. Hence trans-horizon radio transmission is made possible, since the refracted ray moves away from the vertical (Fig. 10.12) by Snell's law.

Typically $(n_r - 1)$ is very close to zero. Multiplied by 10^6 it gives the more convenient refractive index $N_r = (n_r - 1)10^6$ "expressed in N units". To a good approximation it holds

$$N_r = (n_r - 1)\,10^6 = 77.6\mathrm{T}^{-1}(p + 4810e\mathrm{T}^{-1}) \tag{10.9}$$

with T absolute temperature, p total pressure and e water vapor pressure expressed in millibar.[2] So at radio wavelengths, the velocity c is strongly affected by water vapor but it is not significantly wavelength dependent.

At optical wavelengths the opposite is true: c is not appreciably affected by water vapor, but is strongly wavelength dependent. The formula for the refractive optical index N_{opt}, expressed in N units, reads (Valley [1965])

$$N_{opt} = \frac{77.6p}{\mathrm{T}} + \frac{0.584p}{\mathrm{T}\lambda^2} \tag{10.10}$$

where λ is measured in microns. Typical example of refractive indices n_{opt} for yellow light ($\lambda = 5893 \cdot 10^{-10}$ m) are: 1.00029 for air; 1.333 for water; 1.00045 for carbon dioxide; 1.00013 for hydrogen; 2.417 for diamond at normal temperature and pressure (Morgan [1953]).

Similarly the acoustic refractive index n_a is "expressed in N units" as follows $N_a = (n_a - 1)10^6$. Variations of 1 °C in temperature or 1 mb in water vapor give rise, respectively, to variations of N_r of 1 and 4 and to variations of N_a of 1700 and 140. Fluctuations of 1 m/s in wind velocity, assumed to be in the direction orthogonal to the wave front, result in variations of N_r and N_a, respectively, of 0 and 3000 (Ottersten et al. [1973]). These estimates show acoustic waves to be much more sensitive to fluctuations of atmospheric parameters than radio waves. In the boundary layer the refractive index of acoustic waves is about 1000 times more intense. Since the scattered power is proportional to the square of the refractive index fluctuations, the scatter of acoustic waves may be expected to be roughly a million times stronger than for radio waves, as already observed in Little [1969]. Hence acoustic probes are especially well suited for atmospheric studies.

Yet the strong attenuation of acoustic waves limits their scope to the height of 1.5 km. Indeed the intensity of an acoustic wave, measured in watt per square meter, decays exponentially with distance: if I_0 is the reference intensity that crosses a reference plane, then the intensity I at a distance x is given by

$$I = I_0 e^{-\alpha x} \tag{10.11}$$

where α, that denotes the *attenuation coefficient,* has the dimension of m^{-1} and is often measured in dB/m. The attenuation coefficient depends on temperature, pressure, humidity and frequency. Figure 10.13 shows the strong dependence of attenuation from air humidity as well as from frequency. Data at different temperatures can be found on Table 1 at https://en.wikibooks.org/wiki/Engineering_Acoustics/Outdoor_Sound_Propagation.

---

[2]The relation 1 atm= $1.013 \cdot 10^3$ mb makes the millibar slightly less than one-thousandth of a standard atmosphere.

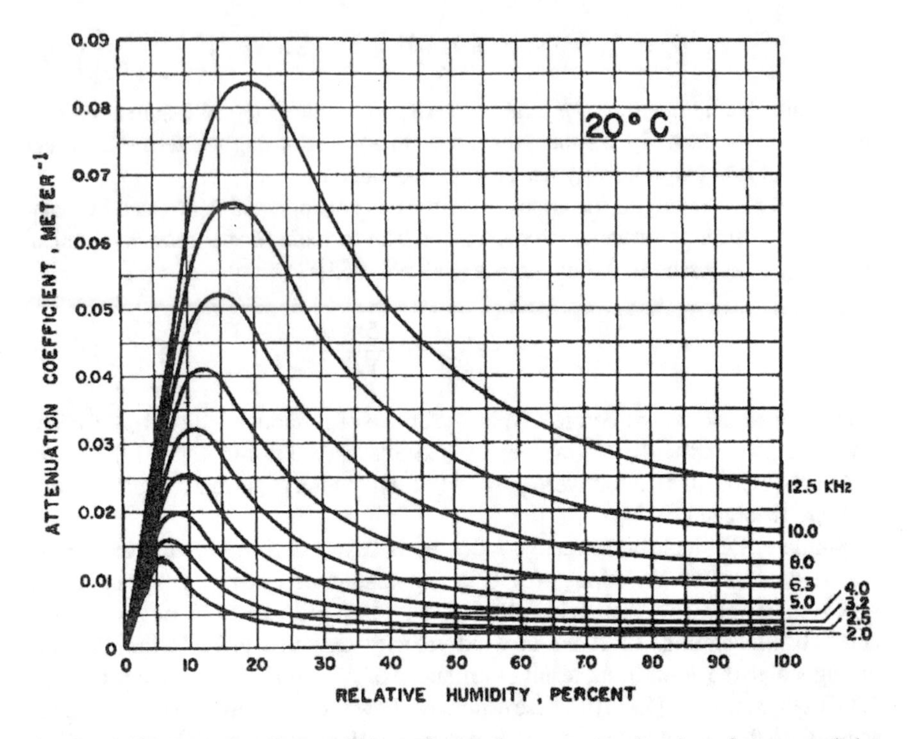

Figure 10.13 The attenuation coefficient of acoustic waves as a function of percent relative humidity—with air at 20 °C and standard atmospheric pressure—for frequencies between 2.0 and 12.5 kHz at 1/3 octave interval. (Reproduced from Harris C.M. [1966], "Absorption of sound in air versus humidity and temperature," *J. Acoust. Soc. Amer.* 40, 148–159, with the permission of Acoustical Society of America)

10.9 Acoustic scattering theory and the role of the Fourier transform

The physical process of the interaction of acoustic waves with turbulence depends on the fluctuations of the refractive index which in turn depend on the fluctuations of temperature, water vapor pressure, and wind. Measurements indicate that the isotropic turbulence *limiting microscale* l_m—the smallest scales at which turbulence retains inertial range characteristics — varies from few millimeters to about 1 cm, whereas the outer scale L_0—the largest scale for nonnegligible correlation of isotropic turbulence—lies between few tens of a meter to about 10 m (Section 10.5). So, given the wavelengths involved, the corresponding wave numbers can be assumed to belong to the inertial range of the turbulence spectrum. (However, in case of anisotropic eddies, correlations may extend up to several hundred meters).

The scattered power (watt)—a fraction of the unit incident acoustic power—depends on several parameters and in particular on the *effective cross section* σ_0 (cm^{-1}) which stands for the cross section per unit volume and per unit solid angle.

Let θ denote the scattering angle—the angle between the direction of the incident wave and the direction of the scattered wave (2θ is the different notation for the scattering angle in Figure 6.6). In a dry atmosphere Margarita Kallistratova [1961] and Monin [1962] stated $\sigma_0(\theta)$ to be the sum of two terms, one involving $\Phi_T(\mathbf{K})$ and the other $E(\mathbf{K})$ which are, respectively, the intensity of spatial Fourier transform of the temperature and of the wind velocity mean square fluctuations. The formula reads

$$d\sigma_0 = 2\pi k^4 \cos^2\theta \Big[\frac{1}{c_a{}^2} E(\mathbf{K}) \cos^2\frac{\theta}{2} + \frac{1}{4\mathrm{T}^2}\Phi_\mathrm{T}(\mathbf{K})\Big] d\Omega \tag{10.12}$$

where c_a is the mean velocity of sound, T the mean absolute temperature of the scattering volume, $k = 2\pi/\lambda$ the wave number of the acoustic wave, $\mathbf{K}$ a three-dimensional wave number such that $||\mathbf{K}|| = 2k \sin(\theta/2)$, and where θ belongs to a cone of solid angle $d\Omega$. In particular no energy is scattered at $\theta = 90^\circ$, a peculiarity of acoustics-turbulence interaction. The above equation is not complete since it disregards the effect of humidity. Nevertheless it is adequate for a first-order presentation and calculation of the scattering by atmospheric irregularities. For a more general theory the reader is referred to Brown and Hall [1978].

In case of locally homogeneous and locally isotropic turbulence (which can usually be assumed for acoustic, radio, and light scattering) only $\mathrm{K} = ||\mathbf{K}||$ matters. Then in Kallistratova's formula just $\Phi_\mathrm{T}(\mathrm{K})$ and $E(\mathrm{K})$ appear with $\mathrm{K} = 2k \sin(\theta/2)$. Hence $\sigma_0(\theta)$ depends on the spectral intensities of temperature and wind velocity fluctuations at scale $l = 2\pi/\mathrm{K} = \lambda/[2\sin(\theta/2)]$, with $\theta \neq 0$. This is Bragg's diffraction law for periodic structures (6.1). So, once the turbulence field is decomposed by a spatial Fourier transform then, for any scattering angle θ, turbulence can be interpreted as originating from a regular crystal lattice, with lattice spacing $l = l(\theta)$ satisfying Bragg's law. Hence turbulence scattering can be regarded somehow as a "coherent" scattering or as a result of constructive interference. It is for this reason that Fourier transforms, via the 3D spectrum of refractivity fluctuations, "play such a major role in the theoretical analysis of the interaction of sound with turbulence" (Brown, Hall [1978]).

In case of backscattering, that is $\theta = 180^\circ$ so that the term $\cos^2(\theta/2)$is zero, no echo is received from "mechanical" turbulence and Kallistratova's formula reads

$$\sigma_0(180^\circ) = 2\pi k^4 \frac{\Phi_\mathrm{T}(2k)}{4\mathrm{T}^2} \tag{10.13}$$

Only temperature fluctuations matter. The scale is $l = \lambda/2$, half the incident acoustic wavelength, and the refractivity fluctuations can be dealt with as if they were originated by a set of partially reflecting parallel mirrors with spacing equal to $\lambda/2$. Confirming Kallistratova's experiments of 1959, McAllister et al. [1969] remarked: "This is consistent with observations of the temperature records which showed the strong echoes at a carrier frequency of 950 Hz ($\lambda = 0, 36$ m) were associated with the existence of perceptible fluctuations with a scale $l \cong 0.2$m within the scattering region."

From Kolmogorov theory of turbulence Tatarski [1961] derived

$$\Phi_{\mathrm{T}}(\mathrm{K}) = 0.033\mathrm{C}_{\mathrm{T}}^{2}\mathrm{K}^{-11/3} \tag{10.14}$$

with $k_0 < \mathrm{K} < k_m$ (cm^{-1}), where $k_0 = 2\pi/L_0$ and $k_m = 2\pi/l_m$. The *structure parameter* C_{T} in (10.14), expressed in deg $\cdot$ cm$^{-1/3}$, can be obtained from measurements of the *structure function* of the temperature field

$$D_{\mathrm{T}}(r) =< |\mathrm{T}(r_1 + r) - \mathrm{T}(r_1)|^2 > \tag{10.15}$$

Here $\mathrm{T}(r_1 + r)$ and $\mathrm{T}(r_1)$ are the temperature at two points—where sensors are located—separated by a distance r, and $<>$ denotes average over the scattering volume. Within the inertial subrange the structure function depends only on the distance between the two points of measurements, not on their position in space, and obeys the 2/3 *law*

$$D_{\mathrm{T}}(r) = C_{\mathrm{T}}^{2} r^{2/3} \tag{10.16}$$

for $l_m < r < L_0$. With C_{T}^2 so determined, substituting (10.14) into (10.13) gives

$$\sigma(180\,^\circ) = 0.008\frac{C_{\mathrm{T}}^{2}}{\mathrm{T}^{2}}\lambda^{-1/3} \tag{10.17}$$

showing a relative weak dependence on wavelength λ.Taking $\lambda = 2\pi 10^{-2}$m, $C_T = 4.6 \cdot 10^{-2}$ (expressed in deg $\cdot$ m$^{-1/3}$) and T = 300 °K, then

$$\sigma_0 = 4.4 \cdot 10^{-10} \cdot \mathrm{m}^{-1}$$

Using σ_0, the received power can be estimated and compared with the interfering noise level existing at the input to the preamplifier (Little [1969]).

10.10 How the sodar works: thermal plumes, laminated structures, and breaking waves

The sodar in its simplest *monostatic* configuration — colocated transmitter and receiver—works even nowadays similarly to the one built by McAllister in 1968, with a single antenna emitting strong acoustic tones (via a loudspeaker) and receiving the echoes. Thus the angle between the direction of transmission and the direction of detection measures 180 °. The frequency of the audible tones belongs to the interval [1.5–5] kHz, equivalently the wavelength λ belongs to the interval [0.06–0.2] m. The *time length* τ of the transmitted pulse is usually fixed in between [0.01–0.1] seconds. The *repetition time* depends on the height to be probed. If echoes up to 1000 m are of interest then the optimal repetition time is 6 s, for in about 3 s the probing wave reaches height 1000 m and in about 3 s is back, assuming the velocity of sound to be 340 m/s. Higher frequencies undergo a stronger attenuation (Fig. 10.13) and so only lower strata can be probed: up to 200 m in case of [4-5] kHz. Then the repetition time is [1–2] seconds and the resolution in radial wind retrieving may be higher.

After a pulse emission the antenna is set in reception, hence sodars spend most of their time in reception. The backscattered signal (echo), a function of time hence of the height of provenance, is sampled (Fig. 10.14) and a single vertical trace is obtained whose darkness represents the intensity of the echo from the different heights (Fig. 10.16). For the correlation between the echo and the heights of provenance, observe that at any fixed time t after emission ($t > \tau$) the antenna receives echoes from all heights in the interval $[(c_a/2)/(t-\tau), (c_a/2)/t]$, the round trip taken into account. Thus the heights of provenance make up an interval of size $c_a\tau/2$. Now assume that at time t_1 the strata involved are those between heights A and B, and at time t_2 those between C and D. Then in the time interval $[t_1, t_2]$ all the strata between A and D are involved, but only the strata between B and C are crossed by the whole transmitted signal (Fig. 10.15). So the results of the analysis of the echo received from time t_1 to time t_2 are associated to the interval centered at Z and thickness ΔZ (*range gate*).

The most characteristic of returns recorded by the sodar are those obtained on days of light wind and cloudless sky (Fig. 10.16). Plumes are observed during the day from 11:00 to 16:00 involving a layer thicker than 500 m, and reaching a maximum rate of occurrence of 30 to 40 per hour during the hottest part of the day (their number depends on the speed of the wind carrying them over the sodar). On those hours the air in contact with the ground heats up, reaching a temperature higher than in the surrounding atmosphere. This increases internal turbulence and leads to the formation of thermal plumes, that is columns of air that rise being hotter and more turbulent than the surrounding air. Between plumes the air that descend, closing the *convective cell,* experiences a thermal turbulence much less intense. Thus, in the facsimile recording, plumes are characterized by dark vertical lines separated by a lighter region. In these conditions turbulence reaches the high strata where usually wind is stronger (and capable of dispersing pollution eventually present). Plumes rapidly disappear as

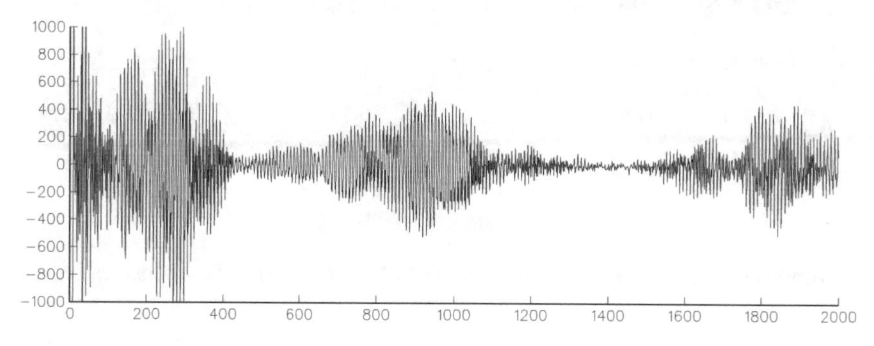

Figure 10.14 First 2000 samples of an echo, sampled at 1600 Hz, after analog-to-digital conversion in the ordinate. (Mastrantonio [2002])

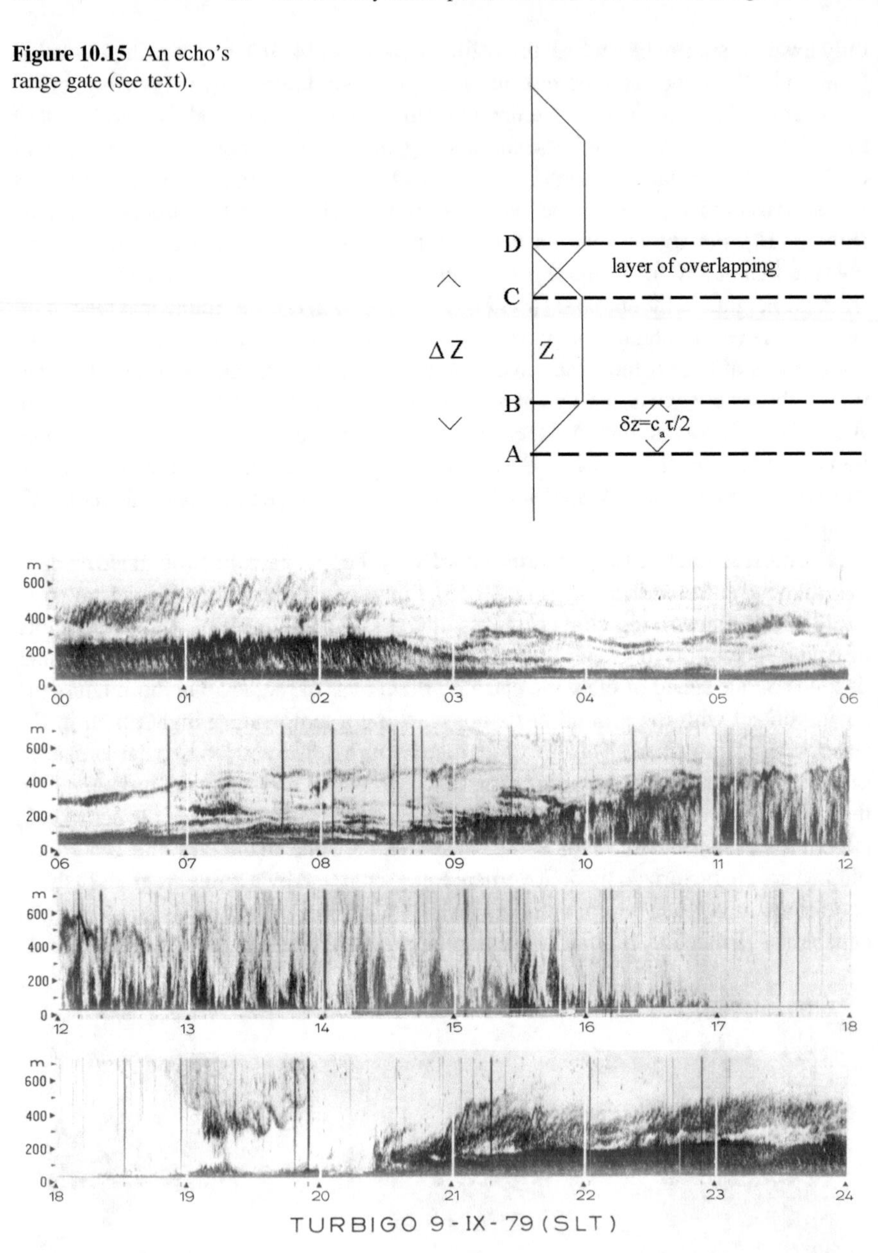

Figure 10.15 An echo's range gate (see text).

Figure 10.16 A typical facsimile recording over 24 hours. (*Courtesy of G. Mastrantonio, ISAC-CNR*)

ground temperature begins to decline toward the evening. Indeed usually half an hour before sunset, convection rapidly ceases.

After sunset the ground starts to lose heat by radiation and the air in close contact becomes cooler leading to a nocturnal *inversion* (warm air above, cold air below) and to the formation of stable strata that appear as horizontal structures whose height may vary in time. They are due to air currents that shear on one another. Usually different in temperature, speed and direction, such currents generate in the contact region thermal turbulence and the resulting echo. Between 00:45 and 01:30, in presence of stratified stability, a breaking wave is visible in Figure 10.16. Breaking waves may occur in stable conditions (temperature increasing with height) when wind shear is present. In this case every push upward or downward of air bubbles is followed by a restoring force that tends to bring the air bubbles toward equilibrium. This process leads to wavelike oscillations that sometimes may break. Characterized by the sawtoothed line shape of their echo, they are the analogue of sea waves breaking on the beach.

By a monostatic sodar with three antennas (Fig. 10.17) it is possible to determine the wind profile in three dimensions. An example, of great complexity, of the wind horizontal component is shown in Figure 10.18. Two air masses flowing one over another, both coming from the Antarctic plateau, are seen to follow different paths: from West and from North (Argentini, Mastrantonio et al. [1992]).

The continuous remote measurements obtained by sodars, describing the dynamics of the lower strata of the troposphere, are of great significance to meteorolo-

Figure 10.17 A configuration of a sodar with three (synchronized) monostatic antennas in Antartica. The penguins had to be gently moved away when the sodar was in operation, their noise affecting sodar performances. (*Courtesy of G. Mastrantonio, ISAC-CNR*)

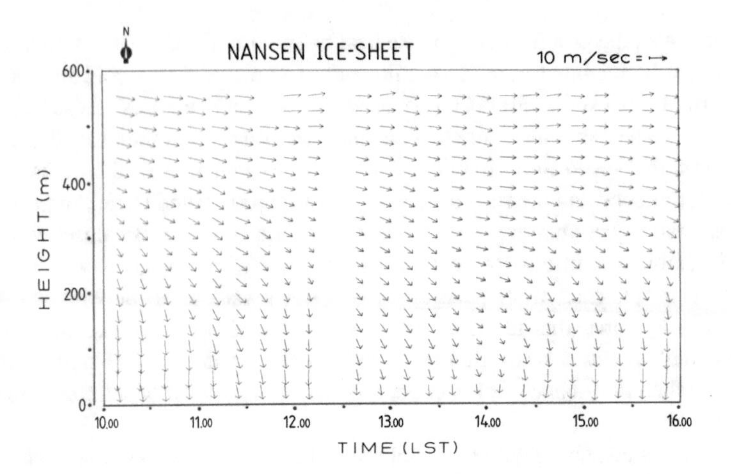

Figure 10.18 Wind profiles for a 6 hours period on January 9, 1989, as observed by a 2 kHz sodar on the Nansen Ice Sheet in Antarctica. Wind vector estimates correspond to 15 minutes averages. (*Courtesy of G. Mastrantonio, ISAC-CNR*)

gists and to those concerned with atmospheric turbulence, diffusion, and pollution. Besides, sodar techniques find application in the location of sites apt to the production of wind energy.

10.11 The sodar, the wind, and the Doppler shift

The Doppler effect—named after the Austrian physicist who discovered it in 1842—is the frequency shift that a wave undergoes in case source and/or observer are in motion with respect to each other. It is a common experience that the sound of a siren is higher in pitch as the siren approaches and is lower when it recedes, even though the siren keeps emitting at a constant frequency. The motion of the approaching siren gives rise to a compression of successive wave crests hence to an increase of the detected frequency. The opposite occurs if the siren recedes from the observer. The following formula

$$v_r = -\frac{\lambda \Delta\omega}{2} \tag{10.18}$$

where $\Delta\omega = \omega_d - \omega_0$ is the Doppler shift, ω_0 the transmitted frequency and ω_d the detected frequency, will be proved in Section 10.14. It holds under the assumption that v_r, the radial velocity of the target with respect to the receiver (i.e., the component of the velocity vector along the line of sight), is much lower than the speed of the probing wave. It is seen that v_r turns out to be negative if the target is approaching and positive otherwise. If the antenna is pointing vertically the vertical component of the wind can be determined. With other two antennas, pointing in different directions, the horizontal components of the wind velocity vector can be determined.

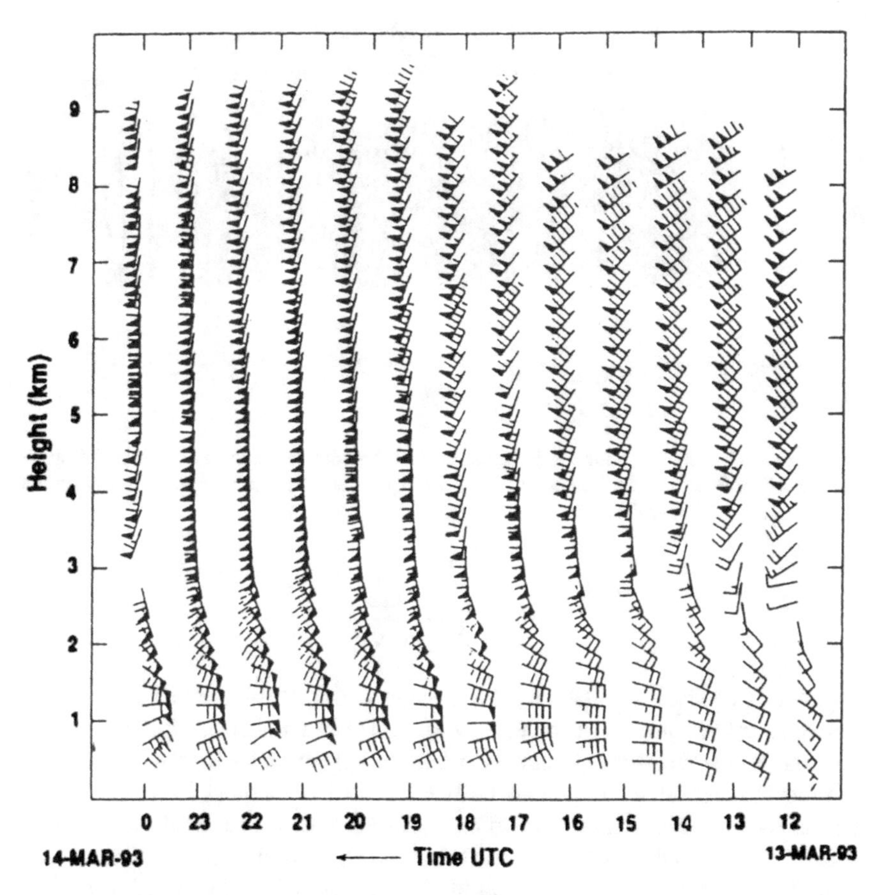

Figure 10.19 Time/height radar wind profile for a 12-h period as observed by the NOAA 404-MHz network wind profiler located in Bloomfield, CT. The wind velocities correspond to 1-h averages. The barbs and solid triangles represent 5 m/s and 25 m/s wind speed increments, respectively. (NOAA)

Figure 10.18 is an example of an horizontal wind profile showing a large variation of directions depending on height. Note the relatively low minimum height (about 25 m) and the good height resolution (about 25 m) compared to radar (Fig. 10.19) and lidar (Fig. 10.20). More on Clifford et al. [1994].

The method to calculate the Doppler shift for sodar, is different from that for radar (Section 10.15). In case of sodar the Fourier transform of the echo is involved, in case of radar it is the phase of the echo to undergo Fourier analysis. Here an example will be given in case of sodar, and it will be seen why the same method is inadequate in case of radar. Formula (10.18) requires the "center" frequency ω_d of the echo to be determined and compared to the known "center" frequency ω_0 of the transmitted signal. In absence of noise, ω_d is defined via the spectral power density $W(\omega)$—the square magnitude of the fast Fourier transform of the echo—as follows

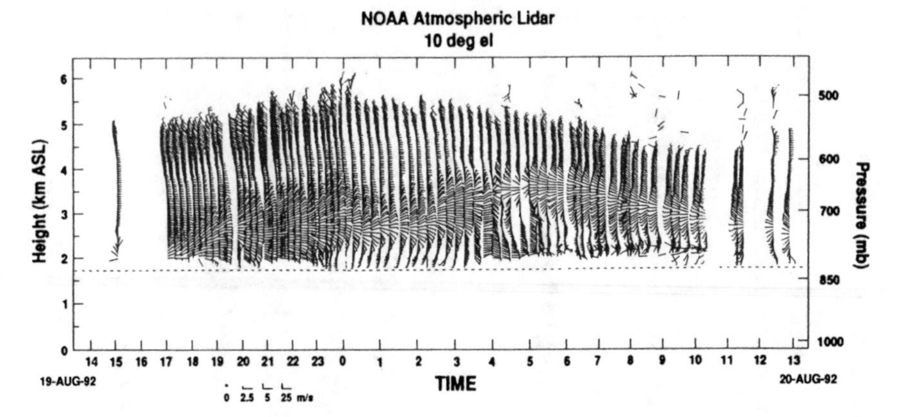

Figure 10.20 Time/height lidar wind profile for a 24-h period as observed by NOAA's Doppler lidar located in Boulder, CO. The wind values correspond to a 15-min average. The dashed line identifies ground level. (NOAA)

$$\omega_d = \frac{\int W(\omega)\omega d\omega}{\int W(\omega) d\omega} \tag{10.19}$$

To provide frequency ω_d relatively to a range gate via (10.19), the FFT is calculated over consecutive samples of the echo (Fig. 10.14) in number of 128, 256, or 512 according to the range-gate thickness and to the spectral resolution required. Then the process is repeated over the next 128 (256 or 512) consecutive samples of the echo and so on. (Hence at the edge of each range gate there is a small overlap of the order of $c_a t/4$ with the contiguous one (Fig. 10.15)). Formula (10.19), that involves the first moment of $W(\omega)$, is recognized to be the formula of the coordinate of the "center of gravity" of $W(\omega)$. It may also be pointed out that the detected signal $s_d(t)$ is always echo plus ambient and/or instrument noise. In Mastrantonio and Fiocco [1982] a detailed procedure, to compute ω_d with improved accuracy in case of a poor signal-to-noise ratio, is presented.

To show that the above procedure is adequate observe that (10.18) becomes

$$v_r \cong -\frac{c_a}{2}\frac{\Delta\omega}{\omega_0} \tag{10.20}$$

being $\lambda = c_a/\omega_0$, where c_a stands for the speed of the probing sound. Now let $\Delta\omega$ be the Doppler shift due to a radial wind of velocity $v_r = 10$ m/s (a rather strong wind). Then by (10.20)

$$\Delta\omega = \omega_d - \omega_0 = -\frac{2}{c_a} v_r \omega_0 \tag{10.21}$$

Assuming $\omega_0 = 2000$ Hz and $c_a = 340$ m/s, then $|\Delta\omega| \cong 117$ Hz by (10.21). A spectral resolution that leads to detect such a Doppler shift is easily obtained. It suffices to receive echoes lasting 0.1 s, whose inverse gives the spectral resolution of 10 Hz. In 0.1 s the transmitted signal reaches 34 m height. The echo, detected at time $t = 0.1$s from the beginning of transmission, comes from the height of 17 m: indeed having started at time $t = 0$ it has reached a height of 17 m and it has come back to the sensor after traveling 17 m on its way back, for a total of 34 m. Similarly the echo that reaches the sensor at time $t = 0.2$s had traveled a total of 78 m and so it comes from height 34 m and so on. Hence the measurements of v_r are averages over atmospheric layers of thickness roughly 17 m. For instance, to probe the layer from 78 to 95 meters the reception time should be from 0.4 to 0.5 seconds.

In case of radar it is easy to see that the above direct method does not work. Indeed for the same v_r, with $\omega_0 = 3\,\text{GHz}$ and the speed of radio waves $c = 3 \cdot 10^8$m/s, it turns out $|\Delta\omega| = 200$ Hz by (10.21). A spectral resolution that leads to detect such a Doppler shift requires to receive echoes lasting at least 1/200 = 0.005 second. In 5 ms though the radar signal travels round trip 1500 km. The related atmospheric thickness to which such a measurement would refer is 750 km: completely out of scale, for the entire troposphere is only 13 km thick in the average.

10.12 Tracing the origin of radar meteorology

Aimed at detecting submarines, acoustic techniques were vigorously pursued mostly after World War I. Radio detection took more time: the design of practical equipment, such as pulsed transmitters and receivers, required considerable efforts.

The first patent for radio *detection* was issued both in Germany and England to the German engineer Christian Hulsmeyer and the first public demonstration took place on May 1904 in Cologne, Germany, where river boats were detected as they crossed the beam of continuous waves (Fig. 2.1) of about 40–50 cm wavelength (Swords [1986]). Hulsmeyer's technique did not receive much interest from the German Navy on the ground that it was only a little better than a visual observer. In 1922 Guglielmo Marconi (1874–1937) himself advanced the idea of radio detection in a speech delivered at the U.S. Institute of Radio Engineers (Marconi [1922]):

"It seems to me that it should be possible to design apparatus by means of which a ship could radiate or project a divergent beam of these rays in any desired direction, which rays, if coming across a metallic object, such as another steamer or ship, would be reflected back to a receiver screened from the local transmitter on the sending ship, and thereby, immediately reveal the presence and bearing of the other ship in fog or thick weather."

Although Marconi had predicted and successfully demonstrated radio communications between continents raising a sensation in every part of the civilized world, this idea of him did not gain support. Nonetheless it apparently stimulated A.H. Taylor and L.C. Young of the U.S. Naval Research Laboratory to seek an experimental confirmation. In the autumn of 1922 they detected a wooden ship using a (CW) continuous wave of 5 m wavelength, with separated transmitter and receiver. And there the idea rested since no more funds were allocated to Taylor and Young's project (Skolnik [1962]).

Atmospheric studies provided the first demonstration of detection and *ranging*. It took place in December 1924 in England, when E.W. Appleton of King's College in London and M.A.F. Barnett of Cambridge University measured the height (or *range*) of a ionospheric layer from the beat frequency they observed using a (FM-CW) frequency-modulated radio wave (Fig. 10.21). The first ranging with *pulsed* radio waves (Fig. 10.29), commonly associated with radars, is due to G. Breit and M.A. Tuve of the Department of Terrestrial Magnetism of the Carnegie Institution in collaboration with the radio engineers of the U.S. Naval Research Laboratory. In July 1925 they detected echoes from an atmospheric layer about 150 km above the Earth where ion and free electron concentration reaches its maximum.

The true impetus for the development of radars though came about in the late 1930s and 1940s. It was due to the more urgent need to locate aircraft. To inquire about the use of radio waves for detecting enemy aircraft, in January 1935 the British Committee for the Scientific Survey of Air Defense approached Robert Watson-Watt of the British Meteorological Office who had done work to provide a system for timely thunderstorm warnings to World War I aviators. At the time engineers of the British Post Office had already detected aircraft that crossed the beam of the Postal radio transmitters (Sword [1986]) and published reports about it. Swiftly, in a memo dated February 1935, Watson-Watt and colleague A.F. Wilkins proposed a radar system to detect and locate aircrafts that, after substantial modifications and improvements, led to the Chain Home (Section 9.3). Watson-Watt later acknowledged the work of atmospheric scientists: "... without Breit and Tuve and the bloodstream of the living organism of international science open literature, I might not have been privileged to become ... the Father of Radar" (Watson-Watt [1957]). The looming of a global conflict led to secret independent parallel efforts in radar development in Germany, Italy, Japan, France, Holland, and Hungary.

The engulfing secrecy makes it difficult to trace the first radar detection of precipitation. The beginning of radar meteorology may be placed with the work of Ryde of the General Electric Corporation Research Laboratory in Wembly, England. There, beginning in July 1940, a 10 cm radar system was operating, likely out of concern for

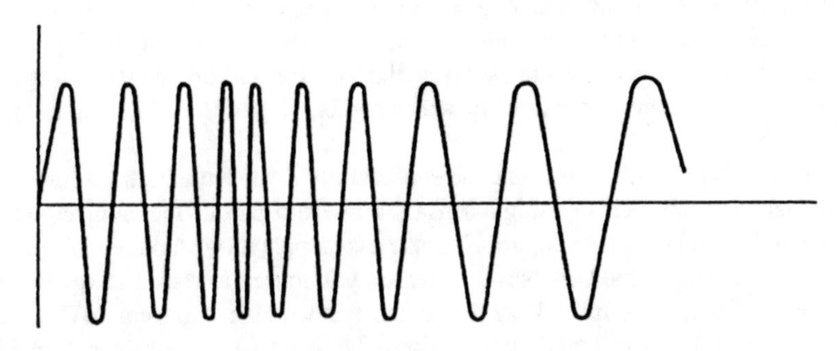

Figure 10.21 A frequency-modulated wave.

the effects of precipitation on aircraft detection. In Ryde [1947] attenuation and echoing properties of clouds and rain are estimated. Detection of echoes from the *clear air* troposphere, by vertically pointing radio beams, dates back to 1935 (Colwell, Friend [1936]). At first the echoes were thought to originate from ionized layers. That this was not the case was proved both experimentally and theoretically by Englund et al. [1938] working with wavelengths of about 5 m at Bell Laboratories: they showed that the echoes were due to dielectric boundaries of different air masses. Shortly afterwards this was confirmed by *in situ* measurements made aboard an aircraft (Friend [1939]).

Doppler effects are observed in radio receivers when echoes from moving objects are received *simultaneously* with the direct radiation of the transmitter, or with radiation scattered from fixed objects. The mixing of the two signals produces a *beat* or fluctuation of the echo intensity at a frequency equal to the Doppler shift. The early observations mentioned above, that provided the incentive for the development of the modern pulse radar, were based on the Doppler shift. Indeed the continuous wave (CW) radar can only measure velocity, via the Doppler shift. The frequency-modulated continuous-wave (FM-CW) radar can measure both range and velocity but with some limitations and by a procedure more complicated with respect to the pulsed radar in case of a distributed target (i.e., not a point target).

The first application of *pulsed* Doppler radar principles to meteorological measurements is due to P. Barratt and I.C. Browne [1953]. They showed that the shape of the Doppler spectrum, they had detected with a radar beam vertically pointing into a rain shower, was in agreement with the spectrum expected from raindrops of different sizes falling at different speed except for a displacement consistent with a downdraft of about 2 m/s (Rogers [1990]). The development into the modern pulsed Doppler radar required a formidable amount of signal processing to extract and process data of the Doppler shift at each of the thousands or more *range locations* that a radar surveys. This took place only in the late 1960s and early 1970s. Radars devoted to measurements of winds in all weather conditions are called *profilers.* They operate with a vertical beam to measure vertical velocity and with two or more beam fixed positions to measure the wind horizontal components.

10.13 The radar for stormy and clear air observations

Weather radars can "see" through conditions such as darkness, haze, fog, rain, snow. By penetrating the storm clouds, radars can reveal in real time the storm internal structure where tornados, hurricanes, or other hazards can be developing in case of severe storms. Precipitation particles are common scatterers. Radar can determine intensity and even type (rain, snow, hail,...) by use of dual polarization. In successive scans, radar detects and continuously tracks the movement of a storm providing range, angular position, and velocity. Short term forecast of weather phenomena, made familiar by television broadcasts, is then obtained by dedicated software.

The weather radar emits a pulse of microwave radiation (Fig. 10.29) lasting about a microsecond (*pulse width*) which is repeated with repetition time of the order

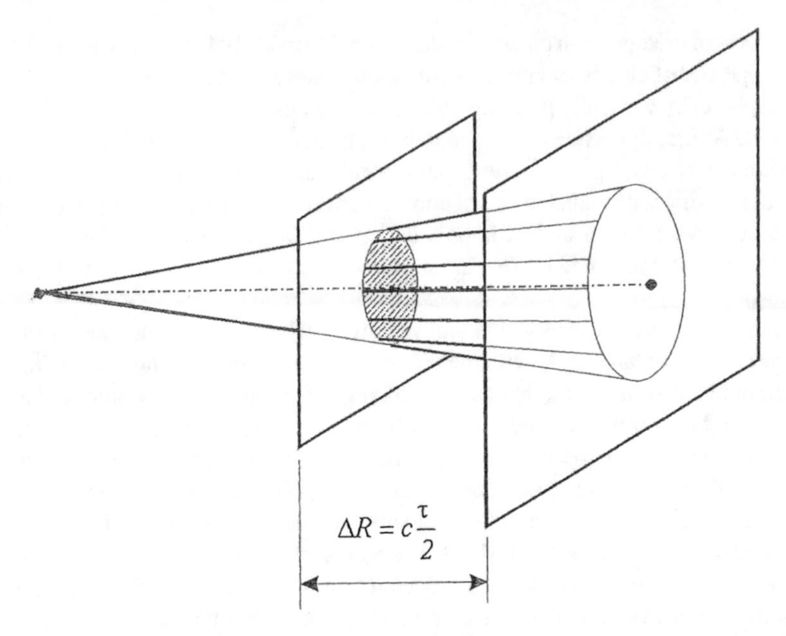

Figure 10.22 The radar conical beam and a space resolution cell of thickness $c\tau/2$. (Galati [2002], modified)

of a millisecond (*pulse repetition time*). Right after each pulse emission, the radar switches to detect the echo from target. Hence radar spends most of its time in detection. Radar continuously scans in all directions. After each scanning rotation the antenna elevation is changed and the same process of scanned rotation is repeated. The elevation starts at around 0 ° to reach 25 ° or more. The lowest elevation angle allows to sense precipitation as close as possible to ground in order to obtain estimates of precipitation that can be referred to ground. However, to avoid obstacles such as mountains or nearby buildings, higher elevation angles are needed. Also higher angles are needed to identify the inner structure of clouds. The maximum range—determined by the pulse repetition time — could be several hundred kilometers. The beam width could be 1 °. As the beam propagates in space, the volume covered during a pulse width—the so-called *resolution volume*—grows larger (Fig. 10.22). The echo, that reaches back the radar, is a superposition of returns from individual scatterers within each resolution volume. The time delay between the transmitted pulse and the echo determines the distance, hence the location of the volume. Distances are measured in discrete increments which are referred to as *range gates*. The resolution volume, growing larger and larger, at 150–200 km could be of the order of a cubic kilometer. Thus the radar spatial resolution decreases with distance.

Each radar operates at a fixed wavelength λ. The term radio waves is a generic term that refers to electromagnetic radiation of wavelength between a fraction of

Table 10.2 Radar frequency bands according to IEEE standards. (https://en.wikipedia.org/wiki/Radio_spectrum)

Band designation	Frequency range
HF	0.003 to 0.03 GHz
VHF	0.03 to 0.3 GHz
UHF	0.3 to 1 GHz
L	1 to 2 GHz
S	2 to 4 GHz
C	4 to 8 GHz
X	8 to 12 GHz
Ku	12 to 18 GHz
K	18 to 27 GHz
Ka	27 to 40 GHz
V	40 to 75 GHz
W	75 to 110 GHz
mm	110 to 300 GHz

a millimeter to 100 km. The corresponding frequencies are divided in bands (Tab. 10.2) but do not cover the entire frequency region indicated in the table, rather they are grouped into separate bands for historical reasons. In case of weather radars, wavelength λ falls into the interval 3 mm–10 cm. The S band (10 cm) and the C band (5 cm) are the most common and are used for detecting rain drops or ice particles. The X band (3 cm) has emerged as a standard for cheaper systems to be used for short-range (50–60 km) monitoring of storms. The higher frequencies of the W band (3 mm) and Ka band (1 cm) are used to detect smaller particles such as cloud droplets by the so-called *cloud radars*. Typically they do not scan, but operate at fixed orientation such as the vertical to reveal the detailed vertical structure of clouds.

When an object is illuminated by an electromagnetic wave, such as radio or light waves, a portion of the incident energy is absorbed as heat and the remainder is scattered in all directions. The backscattered portion is of chief interest. If the energy scattered in other directions is important then a bistatic radar — receiver and transmitter not colocated—is used. Radar measures different characteristics of the echo: first of all *power*, then *phase* involved in detecting velocity (Section 10.15), and *polarization* for shape and orientation. Power depends on the precipitation intensity which is determined by the meteorological target concentration (number of drops/volume), diameter, density (mass/volume) measured in g/cm^3 and approximately equal to 0.99 for water, 0.92 for ice, while for ice hydrometeors it can range from 0.90 (hail) to 0.10 (snowflakes).

The *radar fundamental equation* has the power of the backscattered echo expressed in terms of several radar parameters (transmitted power, wavelength and others) and of the important *radar cross section* σ (cm^2) that represents the scattering properties of target (Fig. 10.23, Tab. 10.3). Targets of interest can be complex objects like

Figure 10.23 The C-54, a military aircraft of the 1940s, was known for its long range and large cargo capacity. Its cross section is in between 10 and 1000 m^2 depending on the viewing angle. (Wikipedia)

ground or sea surfaces in case of meteorological radar, ships and aircraft in case of surveillance radar. Their cross sections are complicated functions that do not have a simple relationship to the physical area of the target. Scatterers such as birds and insects have sufficiently large backscattering cross section and therefore are visible by radar as Figure 10.28 shows. For hydrometeors, with thousands of droplets in a cubic meter, the cross section is defined per unit volume (cm^{-1}) and is called *reflectivity* in general radar terminology.

In case of a conductive sphere of diameter D, small compared to λ (i.e., $D < \lambda/10$), and refractive index n the following formula provides a good approximation of the cross section

$$\sigma \cong \frac{\pi^5}{\lambda^4}\left(\frac{n^2-1}{n^2+2}\right)^2 D^6 \tag{10.22}$$

This relation is called the *Rayleigh approximation,* after Lord Rayleigh who in the early 1870s was the first to study the scattering of *light* by small particles. Indeed (10.22) holds as well for the cross section of atmospheric molecules with diameter small compared to optical wavelengths (Section 10.17). Relation (10.22) shows that waves of shorter λ are more strongly scattered, a fact that Rayleigh used to explain

Table 10.3 Backscattering cross sections at $\lambda = 10\,\text{cm}$. (*Courtesy of R. Doviak and D. Zrnić*)

Object	σ (m^2)
C-54 aircraft	10 to 1000
Man	0.14 to 1.05
Weather baloon, sea gull	10^{-2}
Small birds	10^{-3}
Wingless hawkmoth	10^{-5}
Bee, dragonfly	3×10^{-6} to 10^{-7}
Water sphere (D = 2mm)	1.8×10^{-10}
Free electron	8×10^{-30}

why the is blue. Indeed the factor λ^{-4} implies that the scattering of blue light ($\lambda = 0.47\mu\text{m}$) is about six times larger than that of red light ($\lambda = 0.72\mu\text{m}$). This accounts for the blue color of the sky at midday in clear air. At sunset instead, due to the longer path in the atmosphere the ray has to cross, only the longer wavelengths (red) reach the site where the observer is located.

In case of a simple object of small dimensions with respect to the resolution cell, the *radar cross section*—also called *equivalent area*—is defined to be the area that, intercepting the transmitted power density at the target and reradiating isotropically in all directions, produces at the radar an echo equal to that produced by the target. In case of backscattering it is given by the formula

$$\sigma = \lim_{R\to\infty} 4\pi R^2 \frac{|E(R)|^2}{|E_i|^2}$$

where R is the distance between radar and target, $E(R)$ the reflected field strength, E_i the strength of the incident field, and where the term $4\pi R^2$ accounts for the decrease in power density due to distance from the scatterer.

The radar cross section of the most simple target shape, namely the sphere, is given in Figure 10.24 as a function of its circumference measured in wavelengths. The region where $2\pi a/\lambda << 1$ is called the *Rayleigh region,* with associated *Rayleigh scattering.* The cross section of raindrops and other meteorological particles falls within this region at the usual weather radar frequencies. Since the cross section of objects within the Rayleigh region varies as λ^{-4}, to observe raindrops shorter wavelengths are preferable. Indeed rain and clouds are essentially invisible to radars operating at longer wavelengths. Yet raindrop diameters can be as large as 8 mm, in which case the simple formula (10.22) can be applied only for wavelengths about 10 cm or longer. On the other extreme is the *optical region* where the radius of the sphere is much larger than wavelength: $2\pi a/\lambda >> 1$. Here the radar cross section approaches the optical cross section πa^2. In between the Rayleigh and the optical region lies the *Mie region* or *resonance region* with associated *Mie scattering.* Here

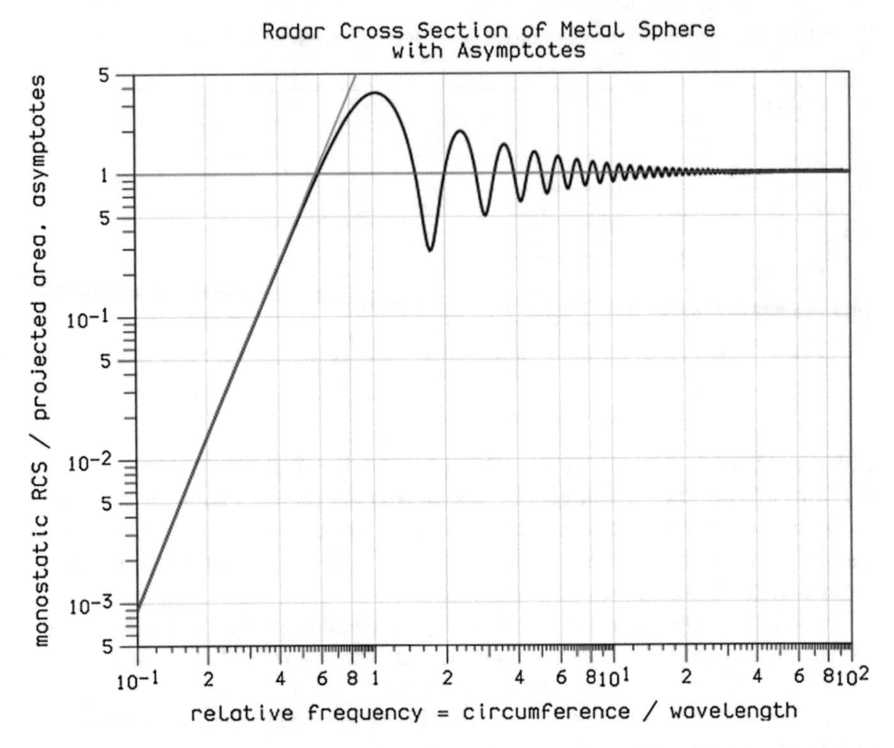

Figure 10.24 Radar cross section of a sphere of radius a, normalized by πa^2, as a function of circumference/ wavelength (with asymptotes). (https://en.wikipedia.org/wiki/Mie_scattering)

the sphere cross section oscillates with λ, internal resonance being responsible for the fluctuations (Probert-Jones [1984]).

Concerning shape, observe that large droplets may be far from spherical. It depends on the terminal velocity: the higher the velocity the more flat and oblate the shape. For example, drops of 8.00 and 2.70 mm diameter have terminal velocity 9.2 and 7.70 m/s (Prupacker and Beard [1970]). Ice particles pose a much more challenging task to radar experts as the about 80 types of snowflakes in the *Magono and Lee Classification* of 1966 shows (Fig. 10.25). Besides it may be difficult to distinguish ice from liquid phase precipitation even with the help of temperature. Indeed in convective storms liquid water can be present at temperature colder than 0 °C and ice can be found at temperatures warmer than 0 °C. Nevertheless to accurately quantify rain-, snow-, or hail-fall rates requires detailed knowledge of the precipitation particle size distribution and shape. Radar may work jointly with raingauges for a better rainfall estimate and with satellite instruments for a better discrimination of hail from rain or other forms of frozen precipitation. Other improvements are needed to sense, more accurately and reliably, events not so unusual such as flash floods, large hail, or heavy snow fall.

Magono and Lee Snowflake Classification System

N1a	Elementary needle
N1b	Bundle of elementary needles
N1c	Elementary sheath
N1d	Bundle of elementary sheaths
N1e	Long solid column
N2a	Combination of needles
N2b	Combination of sheaths
N2c	Combination of long solid columns
C1a	Pyramid
C1b	Cup
C1c	Solid bullet
C1d	Hollow bullet
C1e	Solid column
C1f	Hollow column
C1g	Solid thick plate
C1h	Thick plate of skeleton form
C1i	Scroll
C2a	Combination of bullets
C2b	Combination of columns
P1a	Hexagonal plate
P1b	Crystal with sectorlike branches
P1c	Crystal with broad branches
P1d	Stellar crystal
P1e	Ordinary dendritic crystal
P1f	Fernlike crystal
P2a	Stellar crystal with plates at ends
P2b	Stellar crystal with sectorlike ends
P2c	Dendritic crystal with plates at ends
P2d	Dendritic crystal with sectorlike ends
P2e	Plate with simple extensions
P2f	Plate with sectorlike extensions
P2g	Plate with dendritic extensions
P3a	Two-branched crystal
P3b	Three-branched crystal
P3c	Four-branched crystal
P4a	Broad branch crystal with 12 branches
P4b	Dendritic crystal with 12 branches
P5	Malformed crystal
P6a	Plate with spatial plates
P6b	Plate with spatial dendrites
P6c	Stellar crystal with spatial plates
P6d	Stellar crystal with spatial dendrites
P7a	Radiating assemblage of plates
P7b	Radiating assemblage of dendrites
CP1a	Column with plates
CP1b	Column with dendrites
CP1c	Multiple capped column
CP2a	Bullet with plates
CP2b	Bullet with dendrites
CP3a	Stellar crystal with needles
CP3b	Stellar crystal with columns
CP3c	Stellar crystal with scrolls at ends
CP3d	Plate with scrolls at ends
S1	Side planes
S2	Scalelike side planes
S3	Combination of side planes, bullets, and columns
R1a	Rimed needle crystal
R1b	Rimed columnar crystal
R1c	Rimed plate or sector
R1d	Rimed stellar crystal
R2a	Densely rimed plate or sector
R2b	Densely rimed stellar crystal
R2c	Stellar crystal with rimed spatial branches
R3a	Graupel-like snow of hexagonal type
R3b	Graupel-like snow of lump type
R3c	Graupel-like snow with nonrimed extensions
R4a	Hexagonal graupel
R4b	Lump graupel
R4c	Conelike graupel
I1	Ice particle
I2	Rimed particle
I3a	Broken branch
I3b	Rimed broken branch
I4	Miscellaneous
G1	Minute column
G2	Germ of skeleton form
G3	Minute hexagonal plate
G4	Minute stellar crystal
G5	Minute assemblage of plates
G6	Irregular germ

5. The meteorological classification of snow crystals according to the scheme of Magono and Lee. This scheme permits much more detailed classification than the International one. It also applies to falling snow.

Figure 10.25 The Magono and Lee snowflakes classification of 1966. (NASA) (www.nasa.gov/pdf/183515main_Magono_and_Lee_Snowflake.jpg)

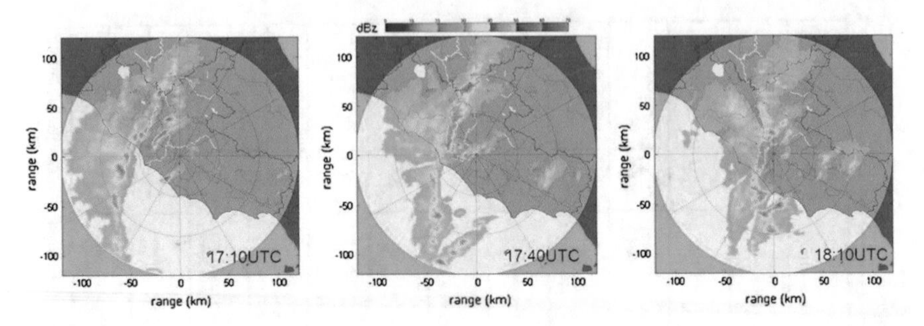

Figure 10.26 Reflectivity images at different times (17:10, 17:40, and 18:10 UTC) collected by Polar 55C radar (ISAC-CNR) in Rome during October 15, 2012 (fixed elevation at 1.6 °). (*Courtesy of L. Baldini, ISAC-CNR)*

Figures 10.26 and 10.27 refer to a convective event that occurred in Rome on October 15, 2012 from 17:00 to 19:00 UTC with a remarkable maximum rainfall rate of 35 mm/h at 18:00 UTC. The spatial distribution and the evolution of precipitation can be inferred from the radar reflectivity pattern (Fig. 10.26). Two intense convective cells were present between 10 and 12 km and between 20 and 23 km from the radar at 17:56 UTC (Fig. 10.27 left). The closest one is characterized by the presence of graupel from ground to a height of 6 km, while the second one is characterized by graupel up to 4 km as well as hail mixed with rain in proximity of the ground (Fig. 10.27 right). Images of the same event at different times can be found in Ferretti, Baldini et al. [2014] and in Roberto, Baldini et al. [2016].

Although *clear air* does not contain extreme phenomena like tornados, hurricanes or baseball-sized hail found in severe storms, nevertheless its structures can lead to the

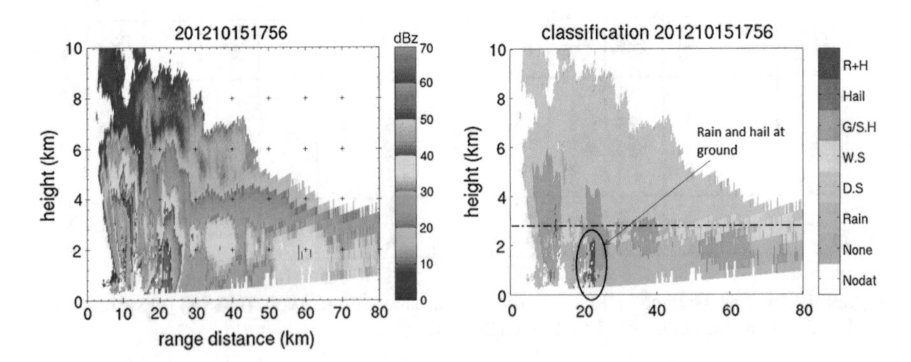

Figure 10.27 Left: fixed azimuth at 293 ° and variable elevation to detect the vertical structure of the hydrometeor observed in Rome on October 15, 2012 at 17:56 UTC. Right: color bar of the hydrometeor classes identified by multisensor analysis: white denotes no data, gray denotes data that are not classified, light blue denotes rain, green denotes dry snow, yellow denotes wet snow, orange denotes graupel and small hail, red denotes hail, and dark red denotes a hail/rain mix. (*Courtesy of L. Baldini, ISAC-CNR)*

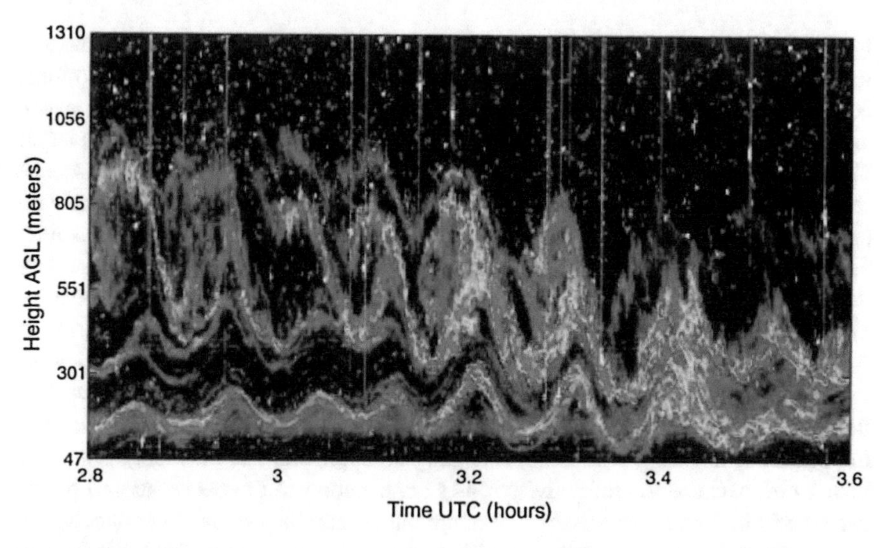

Figure 10.28 FM-CW radar record showing low-level waves on October 20, 1992 (lighter shading indicates a larger reflectivity). The continuous shading depicts waves features resulting from Bragg scattering, the points echoes from insects, and the vertical streaks from bats and birds whose reflectivity saturate the receiver. (Eaton et al. [1995])

development of these storms. Example are convective cells that spring from heated surfaces and breaking waves generated by shear instability at higher altitude zones. Bragg scattering (Section 10.9) provides the theoretical model for the backscattered signal interpretation. With a wavelength of the order of the decimeter, structures of the troposphere of comparable size are observed at different heights and over distances approaching 100 km. The first studies in clear air were mainly focused on determining whether radar returns were due to variations of the refractive index or were caused by swarms of insects and flocks of birds. Then starting in 1966 the three high-powered pulse radars with large steerable antennas at Wallops Islands, Virginia, were the first to be used intensively to sense and resolve turbulent structures in clear air up to high altitudes ($<$35 km) and out to long ranges ($\sim$100 km). In Eaton et al. [1995] other images besides the one in Figure 10.28—taken by the FW-CW radar at the White Sands Missile Range, New Mexico — can be found. Quoting this paper "The FM-CW radar technique, with its ultrasensitivity and unequaled range resolution, is ideal for examining fine details of wave activity in the atmospheric boundary layer."

The basic equations for calculations of Bragg scattered signal returns at 180° has been treated by Tatarski [1961] and apply to sodar, radar and lidar.[3] There are differences though. For instance, electromagnetic waves are unaffected by wind. On the contrary the velocity of sound waves is increased by the component of the wind veloc-

[3] For scatter through other than 180° formulas for electromagnetic and acoustic scatter are slightly different (Derr, Little [1969]).

ity in the direction of the wave normal. Also, under normal atmospheric conditions, the acoustic power scattered from the humidity field is about two orders of magnitude below that from the temperature field so that the influence of water vapor may be usually neglected. For radar waves instead the scattered power from humidity is larger than that from temperature fluctuations by about one order of magnitude. Hence radar scattering depends primarily on fluctuations in water vapor and secondly on temperature fluctuations (but in higher and drier parts of the atmosphere temperature variations may dominate) while acoustic scattering depends mainly on temperature and wind variability. In common the different waves have the feature to be sensitive only to the three-dimensional spectral frequencies **K** such that $l = 2\pi/||\mathbf{K}||$ meets Bragg's law of diffraction. So the scales of atmospheric structures, that interact most strongly with the probing radiation from a remote sensor, are approximately one-half the wavelength of the radiation itself, in case of backscattering. The clear air scattering mechanism (Bragg scattering) is usually not significant at wavelengths smaller than 1 cm, because the refractivity scales contributing to the backscattered power can be within the viscous dissipative range rather than the inertial subrange. Moreover at these short wavelengths particulate and molecular scatter (Rayleigh scatter) dominate.

10.14 Target localization and radial velocity

Pulsed radars are the great majority of existing radars, being capable to determine easily the target range. The pulse parameters are τ the *pulse length* (or pulse width), and T the *pulse repetition time* (or pulse period). In Figure 10.29 three consecutive pulses are represented.

The distance R of a target follows from the formula of the round trip—of length $2R$—of the signal

$$2R = ct.$$

Here t is the delay from the pulse transmission and echo detection and c is the velocity of the wave which is, to a good approximation, the velocity of light in vacuum in case

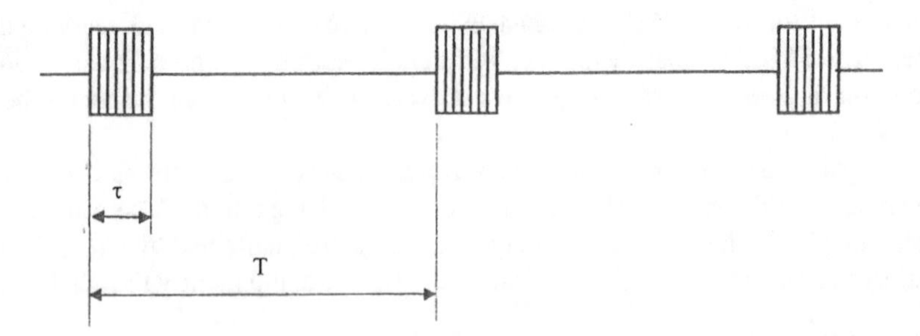

Figure 10.29 Three consecutive pulses of a radio wave. Each pulse oscillates about a thousand times.

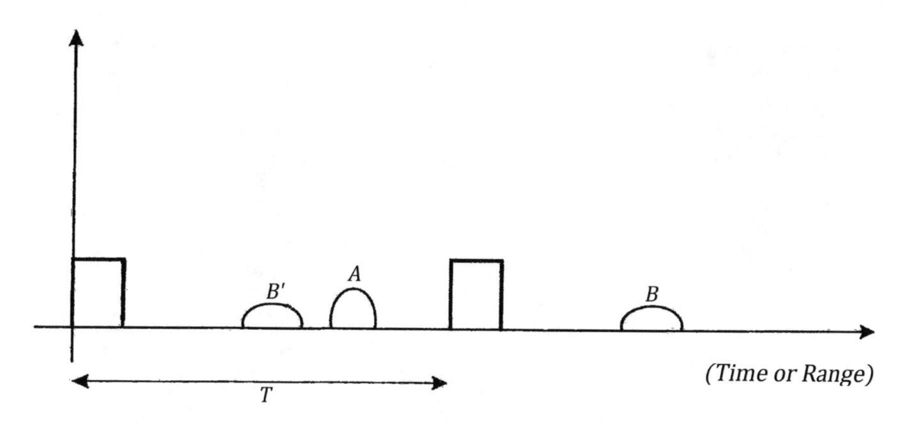

Figure 10.30 A false alarm: target B is incorrectly "seen" in B' (see text).

of a radio wave traveling in the atmosphere. To determine the maximum distance R_{max}, that can be measured without ambiguity, observe that time t starts from zero at the beginning of each pulse, hence $0 \leq t \leq T$. Therefore

$$R_{max} = \frac{cT}{2} \tag{10.23}$$

To illustrate this point a "false alarm" of air control traffic will be considered. In relatively rare cases it is possible indeed for echoes from more distant targets—i.e., $R > R_{max}$—of high reflectivity to be detected. In Figure 10.30 target A is located within R_{max}, the maximum unambiguous range. Target B is at a distance greater than R_{max}, but less than $2R_{max}$. Only the range measured for target A is correct. The echo from target B is associated to time $t_2' = t_2 - T$. Therefore target B is incorrectly "seen" in B'. Similarly, in weather radar, intense cells farther than R_{max} can cause weak echoes close to the radar.

As for *radar resolution* it is clear that two targets, close to each other of $\Delta R = c\tau/2$ or less, give rise to a unique echo (Fig. 10.15) usually of length bigger than τ. For instance with an impulse of length $\tau = 1\mu s$ the corresponding $\Delta R = 150$ m. So the radar cannot resolve — "see" as distinct—two targets less than 150 m apart. (Similarly in case of sodar to $\tau = 300$ ms there corresponds $\Delta R = 4.5$ m). So τ must be short to discriminate between echoes from different targets, yet τ must be long for the energy of the echo to be high enough to give rise to detection. A suitable compromise has to be worked out.

Angular localization requires spherical coordinates centered at the radar location O as shown in Figure 10.31. Since two targets falling inside the radar beam cannot be resolved, the antenna beam width is used to measure the *angular resolution* (Section 9.5). For radar it is typically 1 °.

Next the formula of the radial velocity (10.13.1) will be proved. It will be seen that T is a fundamental parameter not only for R_{max} (10.23), but for $|v_r|_{max}$ as well (10.29). Formula (10.18) will be derived under the usual assumption $v_r << c$ and in

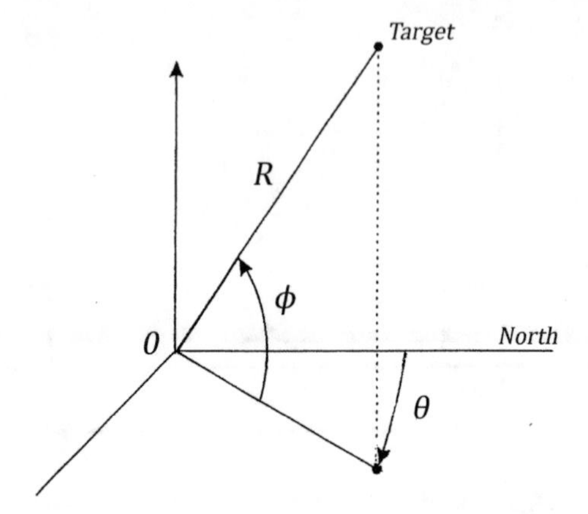

Figure 10.31 Spherical coordinates: R=distance of the target, ϕ=elevation, θ= azimuth, measured clockwise from North.

a simplified setting namely for a point target located at distance R_0 at time $t = 0$ and moving at constant radial velocity v_r (assumed to be positive if the target is moving away and negative if approaching). The transmitted signal is assumed to be

$$s(t) = \sin 2\pi\omega_0 t \tag{10.24}$$

where ω_0 is the transmitted frequency. This is the simple case of a CW radar. The detected signal (echo) $s_d(t)$—aside from a scaling factor (attenuation)—is a delayed replica of the transmitted signal and is given by

$$s_d(t) = s\left(t - \frac{2R(t)}{c}\right) = \sin\left[2\pi\omega_0\left(t - \frac{2R(t)}{c}\right)\right]$$

where $R(t) = R_0 + v_r t$. Hence, with $\alpha = 1 - 2v_r/c$ and $t_0 = 2R_0/c$,

$$s_d(t) = s\left[\left(1 - \frac{2v_r}{c}\right)t - \frac{2R_0}{c}\right] = s(\alpha t - t_0) = \sin\left[2\pi\omega_0\left(\alpha t - t_0\right)\right] \tag{10.25}$$

It is seen that frequency $\omega_d = \omega_0\alpha$. For a more general $s(t)$, the argument goes as follows. By (2.17) and (2.18) the relation between the Fourier transforms $S_d(\omega)$ and $S(\omega)$ is the following one

$$S_d(\omega) = \frac{1}{\alpha}S\left(\frac{\omega}{\alpha}\right)e^{-2\pi i\frac{\omega}{\alpha}t_0}$$

To illustrate this formula, first assume $v_r = 0$. Then $\alpha = 1$ and the echo $s_d(t) = s(t - t_0)$ shows a delay equal to t_0. Correspondingly the Fourier transform undergoes a phase shift t_0 by (2.18): $S_d(\omega) = S(\omega)e^{-2\pi i\omega t_0}$. Next assume that target is approaching: $v_r < 0$ and so $\alpha > 1$. Then $s_d(t)$ undergoes a contraction by α, followed by a translation as (10.25) shows. Correspondingly $S_d(\omega)$, besides the phase shift t_0, undergoes a dilation by (2.17). So its frequency ω_d satisfies the relation $\omega_d/\alpha = \omega_0$ and so it is higher then the emitted frequency ω_0. Instead if the target is moving away then $\alpha < 1$. In this case $S_d(\omega)$, besides the phase shift t_0, undergoes a contraction and its frequency $\omega_d/\alpha = \omega_0$ turns out to be smaller than ω_0.

Either way for the frequency shift $\Delta\omega$, also called *Doppler frequency* and denoted ω_D, it holds

$$\omega_D = \Delta\omega = \omega_d - \omega_0 = \omega_0(\alpha - 1) = \omega_0 \left(-\frac{2v_r}{c}\right) = -\frac{2v_r}{\lambda} \tag{10.26}$$

and (10.18) proved.

10.15 The radar and the wind: Nyquist frequency and range-velocity ambiguity

In Section 10.11 formula (10.18), proved above, has been applied in case of sodar. At the same time it was pointed out that the direct method there does not work for radar: the Doppler shift of meteorological phenomena is of the order of 100 Hz and so it is about 1/10 of the carrier frequency (1 kHz) of sodar, but only 0.1 parts per million that of radar (1 GHz). In this last case it is the *phase shift* $\Delta\Phi$ to be sampled in order to calculate, via the fast Fourier transform, its frequency to be denoted by $1/T_0$. Then v_r is found since it holds

$$|v_r| = \frac{\lambda}{2T_0} \tag{10.27}$$

as proved below. The transmitted signal (wave$_1$) and the echo signal (wave$_2$) are *heterodyned* at the mixer stage of the receiver. The mixed signal contains the sum and the difference of the frequencies of the two waves. The sum-frequency is a high frequency and is filtered out. The difference-frequency is a low frequency, so the corresponding signal can be detected with great accuracy. Now to prove (10.27) observe that in complex notations the transmitted wave of amplitude 1 is represented, as a function of space and time, as follows

$$e^{i\Phi(x,t)} = e^{2\pi i(\frac{x}{\lambda} - \omega_0 t)}$$

where λ is wavelength and ω_0 frequency. If wave$_1$ and wave$_2$ are *simultaneously* detected at the receiver at time t_0, then their phases $\Phi_1(x_1, t_0)$ and $\Phi_2(x_2, t_0)$ will be different only because of the different distances the two waves have traveled. Precisely

$$\Delta\Phi = \Phi_2 - \Phi_1 = 2\pi\frac{x_2 - x_1}{\lambda} = 2\pi\frac{2d}{\lambda} \tag{10.28}$$

where d is the distance of target and $2d$ the round trip. If target is stationary then $\Delta\Phi$ is constant. If target is nonstationary then $d = d(t)$ and $\Delta\Phi$ varies in time. In case of weather radar, scatterer velocities typically are of the order of meters per second and changes in the trigonometric functions of the phase are extremely small during a pulse width $\tau \cong 10^{-6}$s. Thus the phase is sampled with sampling rate equal to the pulse repetition time $T \cong 10^{-3}$s. Now $|\Delta\Phi| \leq 2\pi$ since Φ takes values in the interval $[-\pi, \pi]$. Assume $\Delta\Phi = 0$ at time τ_s, the *range time*, i.e., the time delay between the transmitted pulse and its echo, that specifies in units of time the scatterer's location. With v_r constant, target travels equal spaces in equal times: if the values of $\Delta\Phi$ are acquired at times $\tau_s + (n-1)T$, with $n = 1, 2, ...$ then $\Delta\Phi$ will be seen increasing (or decreasing) by a fixed small amount. After time T_0 it will be found $\Delta\Phi = 2\pi$ (or $\Delta\Phi = -2\pi$). This will keep repeating, since $\Delta\Phi$ is a periodic function of time with period denoted by T_0. When $\Delta\Phi = 2\pi$, target has moved away by $\lambda/2$ as (10.28) shows. Target, moving at constant speed v_r, travels distance $\lambda/2$ in time T_0. Then $\lambda/2 = |v_r|T_0$, and (10.27) is proved.

To determine frequency $1/T_0$ of $\Delta\Phi$, the FFT of $\Delta\Phi$ is computed. Sampling time is T, as motivated above. The Nyquist frequency—the maximum frequency that can be determined without ambiguity—is $1/2T$. Hence

$$\left|\frac{1}{T_0}\right|_{max} = \frac{1}{2T}$$

and by (10.27)

$$|v_r|_{(\max)} = \frac{\lambda}{4T} \tag{10.29}$$

It may be interesting to mention how the phase Φ is determined. Given a real valued signal (of amplitude 1) $I(t) = \cos\Phi(\mathrm{t})$ called *in-phase signal*, the *quadrature signal* $Q(t) = \cos(\Phi(\mathrm{t}) + 90^\circ) = \sin\Phi(\mathrm{t})$ is provided by the instruments (quadrature phase shifter). Then the phase $\Phi(t) = \tan^{-1}(\mathrm{Q(t)/I(t)})$ is known. In Figure 10.32 the amplitude of in-phase and quadrature components of three successive echoes of a moving scatter are represented superimposed to make more evident the relative changes. Observe that echoes from fixed targets remain constant from sweep to sweep, while echoes from moving targets vary and produce, once superimposed, a "butterfly effect". The transmitted pulse width is $\tau = 1.2\,\mu$s. Since the moving scatterer gives rise to an echo that almost replicates the transmitted one, it may be inferred that there is only one moving scatterer (at about 700 m) in the total range interval of 1500 m, displayed in Figure 10.32, that corresponds to 10μs. (The echoes of the stationary scatterers instead are not a replica of the transmitted pulse because more than one scatterer lies in the range interval of size $c\tau/2$). Now trace 1 has I_1 small and negative, Q_1 big and positive. Placing I_1 and Q_1 in the complex plane it is seen that ϕ_1 is slightly bigger than 90°. Similarly locating ϕ_2 and ϕ_3 it is seen that $\phi_1 < \phi_2 < \phi_3 < 180^\circ$. The increasing phase signals that the Doppler shift of the moving scatterer is positive, corresponding to a negative radial velocity. Hence target is approaching.

Finally by (10.29) the *range/velocity ambiguity* follows. It is expressed by the relation

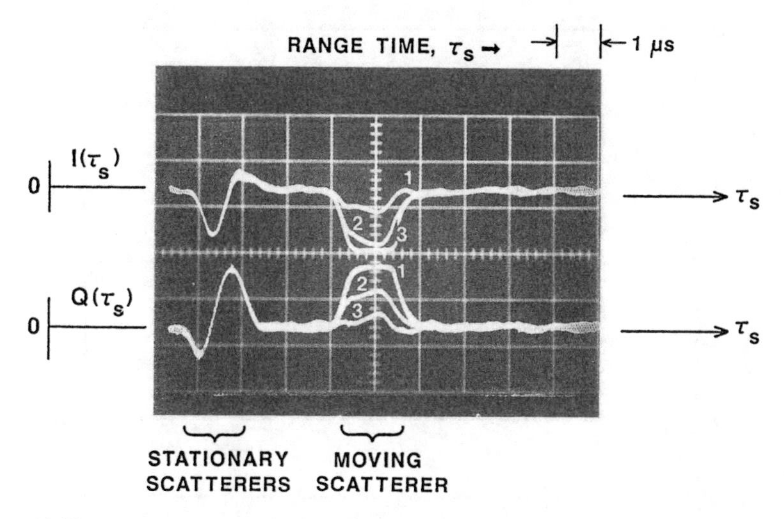

Figure 10.32 Amplitude of in-phase and quadrature components, of three successive echoes of a moving scatterer, are represented superimposed to show the relative change. (*Courtesy of R. Doviak and D. Zrnić*)

$$R_{max} \cdot |v_r|_{max} = \frac{cT}{2} \cdot \frac{\lambda}{4T} = \frac{\lambda c}{8} \tag{10.30}$$

So $Rv_r \leq \lambda c/8$ is required. This condition depends only on wavelength λ. With λ of the order of the centimeter, for targets at distance of hundreds of kilometers, ambiguity turns out to be unavoidable in case v_r is larger than some tens of m/s. This may even be the case in natural phenomena like clouds or rain and it is known as the "Doppler dilemma". If a pulse repetition interval is chosen to achieve a large unambiguous range, then a small unambiguous velocity range will turn out. Methods have been developed to extend the unambiguous velocity (Doviak and Zrnić [1993]).

In Figure 10.33 the wind is assumed to be uniform, blowing from West at 36 m/s at a particular height in the atmosphere. As the radar antenna rotates 360°. in azimuth, the Doppler velocity at a specified range-gate location varies as the sketched sinusoidal curve. In this case spatial continuity permits to visually identify the presence of aliasing, but in general it is not possible to determine whether an individual Doppler velocity data point has been aliased without additional information. Clearly condition (10.30) becomes harder to satisfy by C band with respect to S band radar. It may be interesting to point out that the prevailing type of weather radar in Europe is the C band, while in the U.S. and tropical regions it is the S band, better suited to locate and track extreme meteorological phenomena like hurricanes and tornadoes.

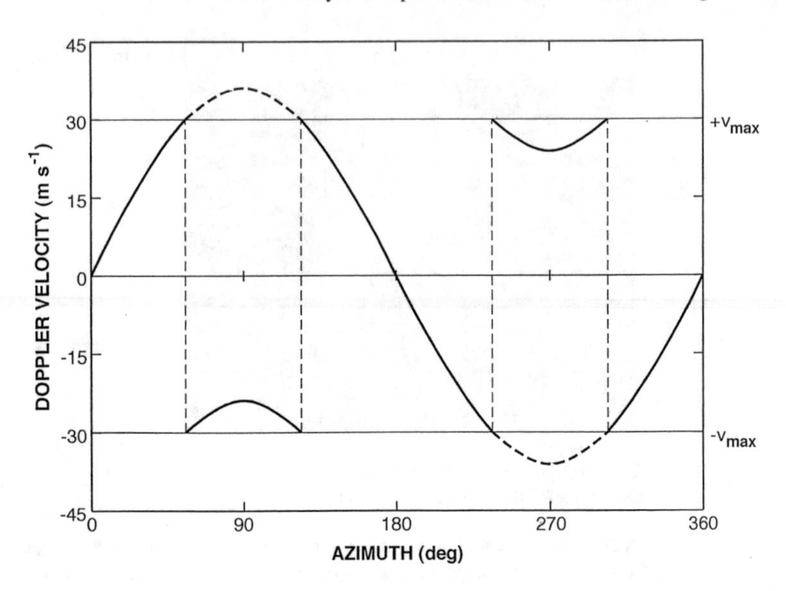

Figure 10.33 Display of calculated Doppler velocity at a given range gate as a function of azimuth when the antenna rotates at a fixed elevation angle. The sinusoidal curve represents true Doppler velocity measurements. The thick solid curve represents Doppler velocity values within the Nyquist interval [-30, 30] m/s. (NOAA) (www.nssl.noaa.gov/publications/dopplerguide)

10.16 After laser came lidar

Lidar principles, going back as far as the 1930s, predate laser times. This holds as well for the acronym lidar that dates 1953 (Middleton, Spilhaus [1953]). Then in 1960 at the Laboratory of the Hughes Aircraft Company in Malibu, California, the first laser was built (Maiman [1960]) and a technique for producing intense pulses of optical radiation was introduced in 1962 (McClung, Hellarth [1962]). Applications quickly followed. The lidar, initially called optical radar, is one of them.

Lidar was developed at the beginning of the 1960s at the Massachusetts Institute of Technology by Louis Smullin and the Italian engineer Giorgio Fiocco to measure the Earth–Moon distance, for the first time via the echo of a pulsed laser light (Smullin, Fiocco [1962]). The difficulty to identify and detect the minuscule number of photons of the backscattered laser beam did not escape the two researchers. With sense of humor they called the project "Project Luna See" (Luna is Moon in Italian). Their successful experiment was reported in newspapers worldwide. Soon after the lidar was used to observe the upper atmosphere (Fiocco, Smullin [1963]) and the stratospheric layer of aerosols that had originated from the eruption in 1963 of Mount Agung in Bali, Indonesia (Fiocco, Grams [1964]). Due to difficulties related to laser technology as well as gaps in the theory of optical interactions with the atmosphere, years went by before the lidar could become a reliable and accurate instrument. Yet the richness of such interactions raised considerable interest in the lidar's potential and about a decade later all basic lidar techniques had been advanced and demonstrated (Hinkley [1976]). At present lidar is largely used in atmospheric studies to determine aerosol concentration, wind profile, temperature profile, pressure profile, water vapor content, and concentration of gases (ozone, methane, ...). The height of the boundary layer itself can be measured. Indeed at the upper limit of the boundary layer a discontinuity in aerosol concentration and water vapor content—that had lifted up from the Earth surface and mixed—is detected. As a technology to make high resolution maps, lidar is employed in many fields such as agriculture, archeology, forestry, geomorphology. Lidar is used as well in the laboratory for the diagnostics of flames (Fig. 10.37).

Lidar operates in the visible as well as in the near infrared and ultraviolet (Fig. 10.34) by means of lasers which are light sources of outstanding properties. A beam of laser light has a very narrow bandwidth (single frequency or single wavelength, also named monochromaticity) which allows discrimination against background noise by use of narrow-band filters. Moreover it is coherent (constant phase) over long path lengths, and maintains its collimation (very low divergence angles). This last characteristics makes it possible to lower the direction of the beam and measure radial winds without the beam or parts of it interacting with obstacles present in the immediate vicinity of the beam path. Wavelengths, depending on application, belong to the interval [0.25–11]μm, *pulse width* could be as small as 10^{-8}s, *pulse repetition time* as small a fraction of 1 ms, *range resolution* as small as few meters and *beam width* as small as few hundred μrad. The majority of lidars use mirror telescopes. Lenses can be used only for small aperture receivers.

Bragg scattering, already described for acoustic and radio waves, applies to light waves as well. Rayleigh and Mie scattering, at optical wavelengths, deliver information on the presence and location of molecules, aerosol and cloud layers. These two types of interaction, based on the wave model that holds for light as well as radio waves (Section 10.13), are *elastic scatterings*: no energy exchange between the incident light and the target takes place. Thus the incident and scattered light have the same frequency. No energy exchange implies that the scattered signal is not specific to any particular species and cannot be employed for measurements of individual species concentrations, but only total density measurements. At optical wavelengths a further type of interaction takes place, namely Raman scattering. This is an *inelastic scattering*. It requires light and discriminates between species.

10.17 Rayleigh, Mie, and Raman scattering

Rayleigh scattering refers to particles much smaller than wavelength λ, typically molecules whose dimension D is of the order of the angstrom (10^{-10}m), so $D/\lambda << 1$ (Fig. 10.24). Indeed, in the context of lidar, Rayleigh scattering is used as a synonym for molecular scattering. Nitrogen (N_2) and oxygen (O_2) make up about 99% of the Earth atmosphere. Hence these two gases are main sources of Rayleigh backscattered radiation which, having intensity proportional to λ^{-4}, is more intense for shorter λ.

Figure 10.34 A lidar beam at the Institute of Atmospheric Sciences and Climate in Frascati, Italy. The transmitter operates at two wavelengths 532 nm (green) and 355 nm (ultraviolet). (*Courtesy of F. Congeduti, ISAC-CNR*)

Mie scattering is usually associated to particles of diameter bigger or comparable to λ, typically aerosol. It consists of atmospheric particles, liquid or solid, whose dimensions range from a fraction of micron to some hundreds micron (Fig. 10.10). Chemical composition and shape vary. Scattering from particles very large with respect to λ does not depend on λ. For particles of magnitude similar to λ, i.e., in the radius range from 0.5μ to few μ, scattering intensity varies strongly with wavelength (Fig. 10.24). In this last case Mie scattering can be used to obtain information on the size of the atmospheric particles involved. The theory, developed by Gustav Mie in 1908 (Mie [1908], Van der Hulst [1957]), gives the analytical solution for the radiation scattered by a sphere of arbitrary radius and refractive index, so it includes Rayleigh scattering from small particles.

Particles in the atmosphere have many different *shapes*, not only spherical. If dimensions are small compared to λ then the actual shape does not play a significant role in scattering, but for large particles shape matters. In case of large nonspherical particles—like ice crystal, mineral dust, sea salt—Mie theory does not apply and more elaborate substitutes are used. The presence of such particles can be easily detected by lidar: while spherical scatterers do not change the polarization state of light (if scattered at $180\,°$) nonspherical scatterers lead to depolarization. Polarization-sensitive light detection is particularly useful for investigating cirrus clouds (made of ice crystals) (Fig. 10.35) and dust layers.

Elastic scattering lidar is used in the troposphere to measure the emission of fine particulate, like PM 10 and PM 2.5. It is also widely used for an accurate detection of the cloud base height, vertical extent and optical density. Indeed clouds have a significant influence on the models of planetary atmospheric circulation, but their characteristics and role are not known in detail. Finally, elastic backscattered lidars are routinely used in traffic control at airports to measure visibility and cloud heights.

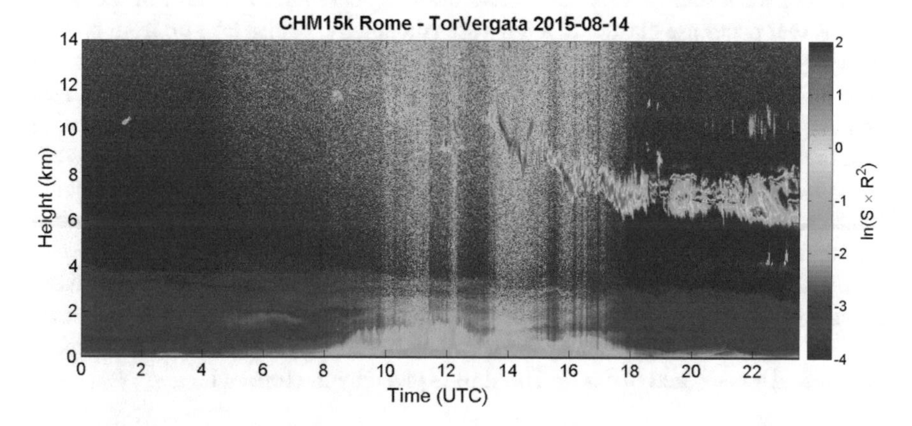

Figure 10.35 Cirrus (ice clouds) at 8–10 km are in red, mixed-phase clouds (ice and liquid water) at 6–8 km are in yellow/red, particulate matter from 0 to 4 km are green-light blue. Vertical strips are caused by increasing instrument and/or background noise due to thin clouds (August 14, 2015). (*Courtesy of G.P. Gobbi, ISAC-CNR*)

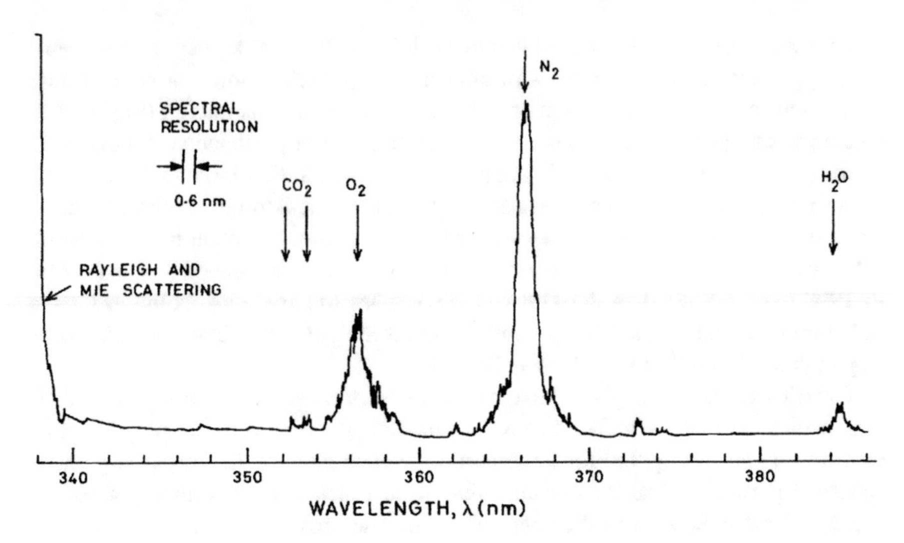

Figure 10.36 Atmospheric backscattered signal as a function of wavelength between 337 nm (excitation wavelength) and 386 nm. (Inaba and Kobayashi [1972])

Raman scattering, an inelastic scattering that requires light, leads to a frequency shift in the backscattered signal (Fig. 10.36). The usefulness of Raman scattering is that this frequency shift does not depend on the excitation frequency and is characteristic of the scattering molecule. It is named after the physicist Chandrasekhara V. Raman (1888–1970) working in Calcutta, India, who discovered it in 1928 and for it received the Nobel Prize for Physics in 1930. In Raman scattering, light is modeled as consisting of photons of energy proportional to the light frequency via the Planck's constant. As photons strike the sample's molecules most of the interactions that occur are elastic. In a few cases though molecules take up from or give up energy (vibrational and/or rotational) to the photons which are then scattered with diminished or increased energy, i.e., with lower or higher frequency. With ω_0 denoting the frequency of the incident photon and ω_s that of the scattered photon, in case the molecule absorbs energy the frequency of the scattered photon is decreased: $\omega_s = \omega_0 - |\Delta\omega|$ and the wavelength is redshifted. On the contrary if the molecule transfers energy to the scattered photon then $\omega_s = \omega_0 + |\Delta\omega|$ and the wavelength is blueshifted.

Usually frequency $\omega = c_0/\lambda$ and frequency shift $\Delta\omega$ are measured in Hz. Nevertheless in Raman spectroscopy, for consistency with the spectroscopic literature, they are respectively replaced by the wavenumber $\nu = 1/\lambda$ (still called frequency) and shift $\Delta\nu$ measured in cm^{-1}. The shift is given by the formula

$$\Delta\nu = \frac{1}{\lambda_0} - \frac{1}{\lambda_s} = \nu_0 - \nu_s$$

For instance, for N_2 the shift equals 2331 cm^{-1} and for O_2 equals 1556 cm^{-1}.

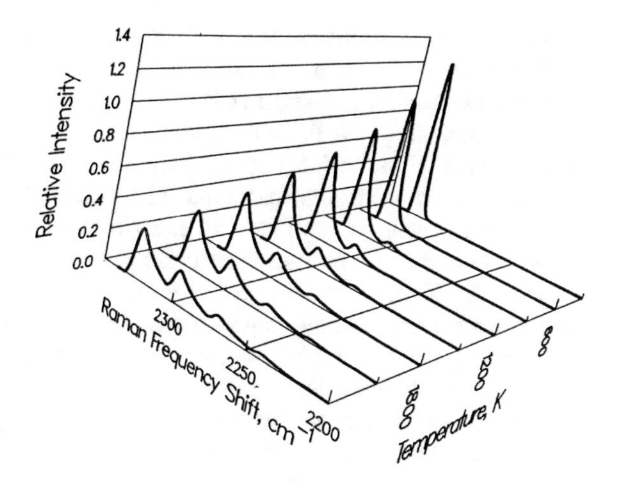

Figure 10.37 N_2 vibrational Raman spectrum as a function of temperature, at constant density, in flames. Spectral resolution is $5\,\mathrm{cm}^{-1}$. (NASA)

The overall Raman spectrum consists of two parts: the vibrational spectrum and of the rotational spectrum (Long [2002]). The *vibrational Raman spectrum*, corresponding to vibration–rotation transitions, has a band centered at ν_s—the frequency shifted by $\Delta\nu$ with respect to the excitation frequency ν_0—and of sidebands. The sidebands above ν_s are called anti-Stokes, while those below ν_s are called Stokes. Stokes bands—such as those appearing in Figure 10.37 — are more intense than anti-Stokes and more easily measured. Raman vibrational spectrum can be used to identify the scatterer and measure its *density* (number of molecules per volume). The density of atmospheric species—N_2, O_2, and H_2O to mention some—as a function of height (profile) can be obtained from the intensity of the echoes. Raman profiles of water vapor (humidity profiles), very important for the dynamics of the atmosphere, were first obtained in 1969 up to 2 km (Cooney [1969], Melfi et al. [1969]), in 1985 up to 5 km, and in 1988 for the whole troposphere. Temperature affects the populations of the molecular energy levels so that Raman spectrum as a whole depends on temperature, but not the shift. In Figure 10.37 the peak at ν_s decreases as an increased temperature makes higher energy levels more populated. Correspondingly additional peaks appear in the spectrum. This remarkable picture has to do with the Space Shuttle (J.A. Shirley [1986], Eckbreth [1987]). Temperatures in the Space Shuttle Main Engine were measured with diagnostics designed at UTRC (at the time the center of excellence in combustion diagnostics) to verify experimentally the temperatures predicted by the NASA thermochemical equilibrium code CEC-73. The accuracy of the code had been questioned and, to set to rest the issue, gas temperature inside the engine preburner was measured conveying the Raman scattering signal out of the combustion chamber by means of cooled optical fibers. The measurements showed that the NASA code produced indeed realistic temperature predictions.

Another application of Raman lidar is for atmospheric temperature profiles at low altitude. They are obtained from the temperature dependent spectral bands in the rotational spectrum of N_2 and O_2. *Rotational Raman spectra*, corresponding to pure rotation transitions, have shift close to zero and bands more separated than those of the vibrational spectrum. At altitude Raman signal is too weak and a different method, based on Rayleigh scattering and involving the gas equation, is followed.

The cross section in Raman scattering is at most 2% of the cross section for Rayleigh scattering and the intensity of the backscattered signal might be only 1/100,000 that of the incident beam. So for a long time Raman lidars were mainly used at night to avoid daylight background noise, until this effect was sufficiently suppressed by improved transmitters and detection systems. Moreover since the wavelength dependence of the Raman scattering cross section goes as λ^{-4}, the shorter wavelengths of the ultraviolet are preferred. Yet the low Raman scattering cross section requires comparably high concentrations of the atmospheric components under investigation. Nitrogen, oxygen, and water vapor, being the main constituent gases in air, are of prime interest (Fig. 10.36).

Other major branches of lidar's techniques are the *high spectral resolution lidar* that, by employing an extremely narrow filter, removes the aerosol particles component from the molecular component in the elastic backscattered signal; the *differential absorption lidar* used for high-sensitivity detection of atmospheric gases; the *resonance scattering lidar* used for detection of the remote mesopause atmospheric region ([80–110] km in height) where layers containing metallic atoms and ions such as Na, K, Ca, Ca+, Li, and Fe are found; and the *Doppler wind lidar* (Weitkamp [2005]).

10.18 The lidar and the wind

Back to elastic scattering and to frequencies for wind velocity measurements. A well known application of the optical Doppler effect has been the determination of the shifts of light coming from distant stars. They were all toward longer wavelengths (redshifted), signaling an expanding universe (Chapter 9). These measurements were comparatively easy to make. Indeed the relative shift $\Delta\omega/\omega_0$ is proportional to v/c and distant stars move away at high velocity v. Velocities of air masses in the Earth's atmosphere—that preemptively need to be illuminated in order to emit radiation—range from 0.1 to 100 m/s. The corresponding ratios to the speed of light range from 3 parts in 10^{10} to 3 parts in 10^7, thus a highly sophisticated equipment is required for detection.

With a monochromatic transmitted signal—i.e., a delta function on the Fourier transform side—and in absence of wind, the backscattered signal would appear as the solid line in Figure 10.38. The broad, bell-shaped spectrum is a Gaussian whose width at half-height is approximately 2.5 GHz, in case $\lambda = 532\,\text{nm}$ and absolute temperature T=280 K (Di Sarra [2002]). Frequently called the Rayleigh component, it has a molecular origin. It is mainly due to thermal motions, but it is also the effect of molecular collisions. Typically the random movement of molecules is much faster than wind, the more so the higher the temperature. From the broad width of the molecular spectrum it is possible to measure temperatures and obtain temperature profiles as mentioned in the previous section. The first experiments were due to Fiocco et al. [1971]. At the same temperature aerosol particles, because of their higher mass, move more slowly than molecules.

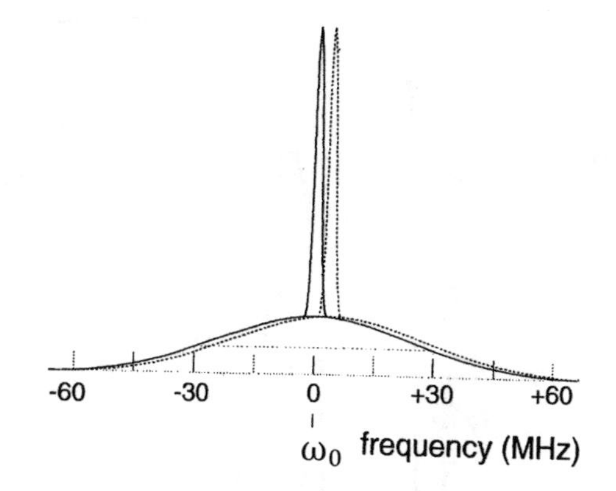

Figure 10.38 Schematic representation of the echo's frequency distribution in absence of wind (solid line) and in presence of wind (dotted line) for a monochromatic signal of frequency ω_0, set at 0. The broad spectrum has a molecular origin, the narrow spike is due to aerosols. (Werner [2005])

If aerosol is present then it appears, superimposed on the broad spectrum, a narrow spike often called the Mie component. Its width at half-height is 3.8 MHz, in case 0.2 μm is the particles diameter and $\lambda = 532\,\text{nm}$.

In presence of wind the echo is shifted and it would appear as the dotted line in Figure 10.38. Since the transmitted frequency ω_0 is shifted toward higher values $\Delta\omega$ is positive, so $v_r < 0$ and the wind comes toward the lidar (10.18). The shift is the same for molecules and aerosol but "by looking at the aerosol peak, it is seen that the spectral range to be explored is smaller than in case of molecules, and it is possible to increase the spectral resolution of the spectrum analyzer" (Benedetti-Michelangeli et al. [1972]). Even minor aerosol concentrations suffice to assure an accurate measurements. Nevertheless at high elevation, above 35-40 km, aerosol is absent and only the echo of molecular origin is available.

The radial velocity of 3.18 m/s (Fig. 10.39) was obtained by formula (10.18) (at $\lambda = 0.4880\,\mu$m and for $v_r = 1$ m/s, shift $\Delta\omega = 4.098$ MHz). The very detailed frequency analysis required was performed by passing the return signal through narrow-band optical filters (*direct detection Doppler lidar*). The echoes obtained were from layers some hundred meters thick around a height of approximately 750 m above ground. To measure wind velocity though the workhorse is the *heterodyne detection Doppler lidar*, historically the first to offer accurate and dependable results on a routine basis. It involves mixing the return signal with the signal from a local optical oscillator (Section 10.15). The range-velocity ambiguity $Rv_r \leq \lambda c/8$ (10.30) is more stringent here since the wavelengths employed by lidar are much smaller.

An example of a profile obtained from a Doppler wind lidar is in Figure 10.20. Examples of wind flows from the mountains and back on an area covering a half circle of about 8 km, obtained by a Doppler lidar at NOAA Laboratory in Boulder, Colorado, are given in Banta et al. [2004]. To measure air speed, not in the free atmosphere but in the laboratory or wind tunnels, other methods are available.

A certain degree of overlapping exists between the Doppler lidar and the Doppler radar (wind profiler), yet there are differences. The Doppler radar, whose wavelength range between 6–30 cm, detects the signal backscattered from atmospheric inhomogeneities. The need of a large antenna makes difficult to change the direction

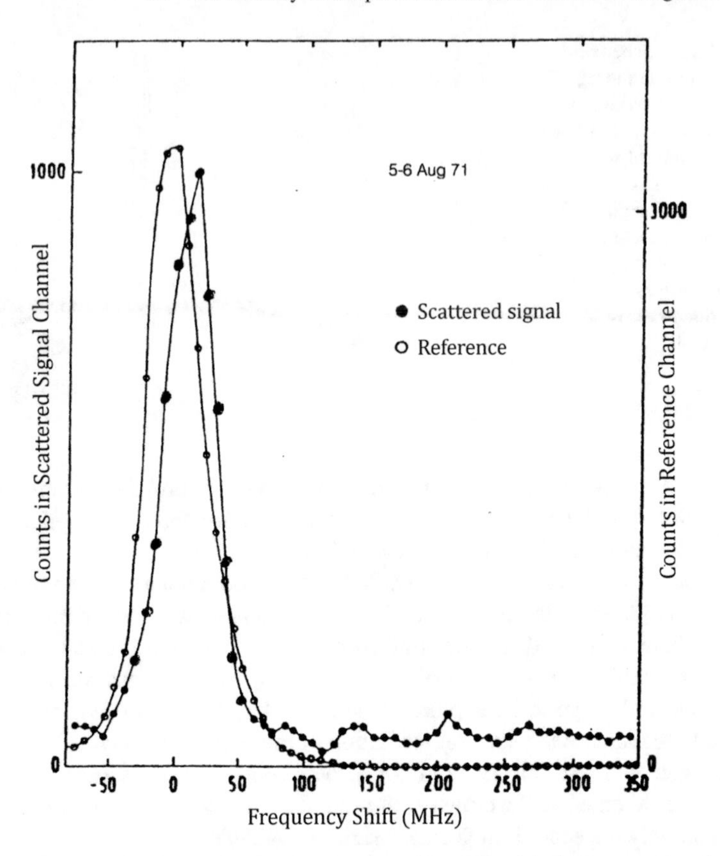

Figure 10.39 Doppler shift of the signal backscattered by aerosol (only narrower band peak) as detected by the lidar in Frascati (Italy). Wavelength is 488 nm and the shift corresponds to a radial velocity of 3.18 m/s. Observe the different scales of the photons count for reference (right) and for the backscattered signal (left). (Reproduced from Benedetti-Michelangeli G., Congeduti F., Fiocco G. [1972], "Measurement of aerosol motion and wind velocity in the lower troposphere by Doppler optical radar," *J. Atmos. Sci.* 29, 906-910. © American Meteorological Society. Used with permission.)

of orientation. Lidar employs mirror telescopes of smaller dimensions that make selection of orientation easier. Moreover radar cannot target winds at low elevations while lidar, with its narrow beam, is more apt to measure winds in regions close to obstacles. Lidar operates only in presence of adequate optical transmission. Its beam does not penetrate far into dense clouds and cannot provide meaningful data from regions at a distance of many times the visual range.

To summarize, the choice of the remote sensor to detect natural atmospheric tracers goes as follows: for molecules (water vapor, ozone,...) and particles less than 1μm in diameter, short-wavelength lidar; for larger particles, from 1 to 20μm in diameter, infrared lidar or very short wavelength radar; for rain and cloud droplets, radar; for temperature fluctuations, sodar; for insects and birds, radar. Temporal resolution matters as well. An example for long range radar may be hurricanes. Having a spatial scale of the order of 1000 km and temporal scale of the order of 1 day, they require a horizontal scale of spacing of the order of 100 km and temporal scale of the order of 1 hour or less.

10.19 Milankovitch theory of climate and the temperature of the Earth

Back to the temperature of the Earth that aroused Fourier interest and made him write in [1827]: "The problem of terrestrial temperatures, one of the most important and difficult of the whole natural philosophy, is composed by rather different elements that ought to be considered from a unique general point of view" (Section 1.1).

The gravitational forces exerted by the other planets induce slow variations of the Earth's astronomical parameters. The first to compute the long-term variations of the Earth's elliptical orbit was Lagrange in [1781] and [1782]. Interest in the ice ages was alive as well (Section 10.2). In *Études sur les Glaciers* of 1840 the biologist and geologist of Swiss origin Louis Agassiz showed geological evidence of the ice ages in the relatively recent past of the Earth. The search for a correlation between geological evidence of large climatic changes and variations of the Earth's astronomical parameters could start. It led to the theory of the Earth's insolation by Milutin Milankovitch (1879–1958). Born of Serbian parents in the village of Dalj, then in the Austria-Hungarian Empire and now in modern Croatia, he studied and became doctor of Technical Sciences in Vienna. After working as chief engineer for a large construction company, in 1909 was elected professor at the University of Belgrade where he lectured on rational mechanics, theoretical physics, and celestial mechanics. His papers, written in Serbian, were printed in Cyrillic characters by the Serbian Academy of Sciences, but some he personally translated in German and French (Milankovitch [1920]). He devoted his life to develop a mathematical theory of climate (Milankovitch [1941]). Having recognized that the solar energy received on the Earth surface (*insolation*) — and therefore the position of the Earth with respect to the Sun—is at the origin of climatic changes, he stated the main factors involved, namely

- elliptical orbit described by the Earth in its motion around the Sun (*eccentricity*);
- angle that the Earth's axis of rotation makes with a perpendicular to the ecliptic (*tilt*);
- spinning of the Earth's axis of rotation (*precession*).

With their cyclical variations (Fig. 10.40) they create peaks in the amount of solar energy striking the Earth.

Calculated long-term variations over million years in the past as well as in the future can be found in Laskar [2004], who states: "The understanding of the climate response to the orbital forcing has evolved, but all the necessary ingredients for the insolation computations were present in Milankovitch work." Variations of the above

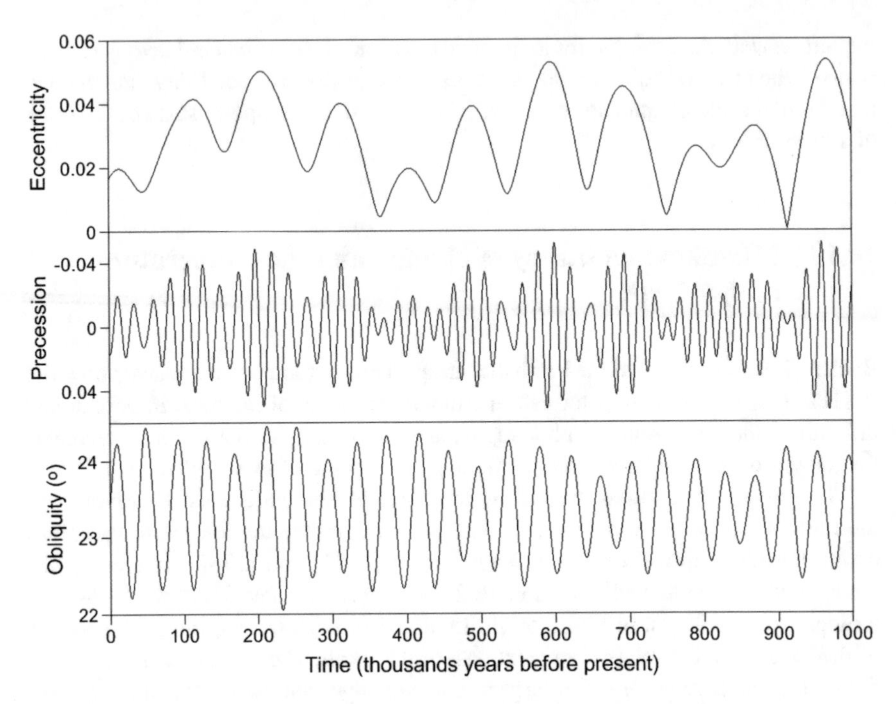

Figure 10.40 Calculated long-term variations of eccentricity, precession, and tilt. (*Courtesy of A. Berger and Q. Z. Yin.* Data from Berger [1978])

three parameters, in conjunction with insolation and data from ocean sediments and Antarctic ice cores, can be found at Wikipedia (Milankovitch Cycles). Observe that the eccentricity is defined by $e = \sqrt{1 - (b^2/a^2)}$ with $a \geq b$, where a and b are the semiaxes of the ellipse. So e=0 corresponds to the circle. The eccentricity of the Earth's orbit—at present 0.017 and decreasing—is always rather close to zero so that the orbit varies from nearly circular to mildly elliptical. In the middle of Figure 10.40 is the precession index, defined by $e \sin\bar{\omega}$, where $\bar{\omega}$ is the longitude of the perihelion. This is the main parameter for calculating insolation.

Since the pioneering work by Milankovitch the cronologies of the above orbital elements and insolation have been recalculated several times. They point to peaks of solar energy, striking the Earth, occurring at intervals of 100,000 years, 41,000 years, and 23,000 years, respectively, since astronomers believe that patterns of orbital variations did not change in any appreciable way over the last several million years. From 1941, when Milankovitch published the final version of his theory, until the early 1970s attempts were made to compare in *time domain* geological records with climate curves predicted by this theory. Indeed deep-sea cores, transferring unknown astronomical data of solar insolation into sedimentological data, are a geological thermometer recording changes going back million years. These attempts clearly presented substantial difficulties but in the mid 1970s, with improved geological data and improved time scale for the Pleistocene marine deposits, this theory was revived by the seminal paper Hays et al. [1976]. (Pleistocene, from 2.59 million years ago to 12,000 years ago, spans the world's recent period of repeated glaciations. Holocene, the latest interglacial period, follows the Pleistocene and continues to the present days). Tests in the *frequency domain*, comparing orbital and climatic spectra, confirmed unambiguously that orbital variations significantly influenced the Earth's climate during the Pleistocene (Imbrie [1982]).

Milankovitch theory suggests that the Earth should begin to head into its next ice age cycle, calculated to take place in about 10,000 years and to last for about 80,000 years. Indeed the dominant cycle over the past million years has been approximately 100,000 years long (Abe-Ouchi et al. [2013], where several examples of tests in the frequency domain can also be found). Records from 1940 to 1970 showed a decrease of about 0.3 °C. In spite of that in the 1960s Mikhail Budyko (1920-2001)—director in Leningrad of one of the oldest Russian geophysical observatory and Lenin Prize in 1958—by calculations of radiative balance predicted that cooling would soon switch to warming and the Arctic Ocean's ice cap melted away as early as 2050 (Lindzen and Farrel [1977], Walsh and Rackauckas [2015]).

The Earth temperature and climate have been shaped by natural phenomena, independent of human activity, until the times — about 7,000 years ago—of early man-made greenhouse gas emissions linked to forest clearance and related land use. The cumulative effect of man-made greenhouse gases could become dominant with respect to natural effects, or already is according to the paleoclimatologist Bill Ruddiman [2013].

A question under much debate is whether or not human emissions of greenhouse gases are helpful to prevent or postpone the next ice age. A study by Berger et al. [2002], [2003] was pointing to an exceptional length of the present interglacial period. Glacial inception represents a bifurcation between interglacial and glacial climate states and, under the current low orbital eccentricity, the Earth is very close to such a bifurcation. A recent model by Ganopolski et al. [2016], that replicates the beginning of the past eight glacial cycles, "might anticipate future periods of glacial inception." This study shows that a glacial inception was narrowly missed *before* the Industrial Revolution of the eighteenth century and also that, given the present levels of CO_2 concentration, a glacial inception will be postponed by at least 100,000 years.

Even though there is still uncertainty with regard to location and timing of early glaciation within the North Atlantic, Arctic marine cores suggest that some North Atlantic ice sheets could date back as early 5.4 million years ago and were fully established about 2 million years ago. So the formation of the Arctic ice cap took at least 2 million years with glaciers advancing to the sea during several intervals (Larsen et al. [1994]). Climate had fluctuated greatly in the short term too, nevertheless even these short trends were long, lasting thousand years, so the effect on climate and vegetation could have been gradual: animals could have moved with trends, competition arising where resources became scarce. Instead in case of severe climatic changes—living spaces and food supply drastically reduced—a tremendous stress must have been experienced by animals and plants alike, accompanied by extinction of species (Muzzolini [1993], Reader [1998]).

Table 10.4 Global mean annual temperature averaged per decade. (Data by the Goddard Institute for Space Studies (GISS) of NASA based in New York City.) (https://www.currentresults.com/Environment-Facts/change-in-earth-temperature.php)

Decade	°C	°F
1880s	13.73	56.71
1890s	13.75	56.74
1900s	13.74	56.73
1910s	13.72	56.70
1920s	13.83	56.89
1930s	13.96	57.12
1940s	14.04	57.26
1950s	13.98	57.16
1960s	13.99	57.18
1970s	14.00	57.20
1980s	14.18	57.52
1990s	14.31	57.76
2000s	14.51	58.12

From 1880 the Earth mean global temperature has been calculated each year on data gathered around the world at weather stations, on ships and by satellites for the sea surface (Tab. 10.4). Most of the increase since the 1880s took place in the last three decades, with the very last one—the first decade of the twenty-first century—the warmest. This brought the Earth temperature to rise, during the last century, by 0.8 °C (1.5 °F) which is a lot for vegetation. It might be pointed out that the rise has not been uniform. Land warms faster than the oceans, yet the global average is dominated by the sea surface temperature since water covers 70% of the planet. The Northern Hemisphere has warmed faster than the Southern Hemisphere, and the Arctic faster than anywhere else in the world. Comprehensive data can be found in the Intergovernmental Panel on Climate Change (IPCC) 5th Assessment Report (www.climatechange2013).

Following Paul Crutzen [2006], Nobel Prize for Chemistry in 1995 — who relabeled Antropocene the present era—some are looking for an escape route from a global warming they foresee turning severe (Keith [2013]). Meanwhile, to curb emissions of greenhouse gases, international conferences are held. The last one of December 2015 in Paris saw the participation of the representatives of 195 nations who agreed on actions to be taken in order to contain the rise in temperatures since the 1880s below 2 °C. Maybe we are witnessing the beginning of a new era of awareness, restrain, and science.

Appendix

Wavelets: a new frontier

The Fourier transform is not the most efficient tool to analyze transients like the brief sound of a consonant or the sound of a gong (Chapter 4), or a detail in the corner of a picture, to give an example in two dimensions. Indeed if f is supported on a very small interval, then $\hat{f}$ has a large spread by the uncertainty principle (2.17).

The Hungarian-born electrical engineer Dennis Gabor (1900–1979) inventor of holography and recipient of the Nobel prize for physics in 1971 — proposed in Gabor [1946] to decompose signals using elementary waveforms well localized in time and frequency. In Gabor [1947] he related such a decomposition to our sensitivity to sound.

The human ear is a good frequency analyzer and does so by segmenting the sound while it is in progress. When high frequency resolution is needed — as for a drawn out note of a violin — long time segments are attended to. When high time resolution is needed — as for recognition of the plosive consonant "p" — attention shifts to shorter time segments.

Gabor used as a "window" the Gaussian $g(t) = e^{-t^2}$. Even though nonzero for every t it is well localized in space as well as in frequency (Fig. 3.7). By a translation both in time and frequency domains he obtained "atoms"

$$g_{u,\xi}(t) = e^{-(t-u)^2} e^{i\xi t}$$

concentrated in time in a neighborhood of u. In Gabor's approach they were to replace the oscillating term $e^{i\xi t}$ of constant amplitude that appears in the definition of the Fourier transform. Note that the Fourier transform of Gabor's atoms

$$\hat{g}_{u,\xi}(\omega) = \hat{g}(\omega - \xi)e^{-iu(\omega-\xi)}$$

is localized near the frequency ξ. Gabor proposed to analyze a function f by integrating $f(t)$ against $g_{u,\xi}(t)$, thus obtaining "quanta of information" over time-frequency

E. Prestini, *The Evolution of Applied Harmonic Analysis*,
Applied and Numerical Harmonic Analysis,
DOI 10.1007/978-1-4899-7989-6

rectangles in the $(u,\ \xi)$-plane. The resulting Gabor transform, which will evolve in the so-called *windowed Fourier analysis*, while customary for analysis, is not suited for synthesis.

More recently the engineer Jean Morlet, working in reflection seismology for oil prospect surveys (Elf-Aquitaine), used short waveforms of high frequency obtained by scaling a single function called a "wavelet," in order to analyze the subsurface structure. Then in collaboration with the theoretical physicist Alex Grossman he introduced the *wavelet transform* (Grossman, Morlet [1984]).

A wavelet ψ is a smooth function supported on an interval, or else rapidly decaying at infinity, of average zero,

$$\int_{-\infty}^{\infty} \psi(t)dt = 0.$$

This last condition means that ψ is somewhat oscillating. The wavelet is no longer multiplied by an oscillating factor like $e^{i\xi t}$ but translated at u and dilated by a scaling factor s,

$$\psi_{u,s}(t) = s^{-1/2}\psi(\frac{t-u}{s}).$$

The Fourier transform is

$$\hat{\psi}_{u,s}(\omega) = e^{-iu\omega}\sqrt{s}\hat{\psi}(s\omega)$$

and the wavelet transform is defined by

$$Wf(u,\ s) = \int_{-\infty}^{\infty} f(t)\bar{\psi}_{u,s}(t)\,dt$$

where $\bar{\psi}$ denotes the complex conjugate of ψ. One of the usual choices for ψ is the second derivative of the Gaussian, also called *Mexican hat*.

The Grossman, Morlet paper acted as a catalyst and researchers from different fields joined to pursue the subject. Some of the fields involved are acoustic and psychoacoustic, visual psychophysics, signal and image processing, and turbulence.

The literature on wavelets grew rapidly. The discrete wavelet transform was introduced as well as several fast wavelet algorithms and extensions to several dimensions. We mention Daubechies [1992], Meyer [1993], Benedetto, Frazier [1994], Hernández, Weiss [1996], Mallat [1998], Bracewell [2000], and Gröchenig [2001] for further readings and references.

References

Abell, G. O. [1982], *Exploration of the Universe*. Philadelphia: CBS College Publishing.
Abe-Ouchi, A., et al. [2013], "Insolation driven 100,000-year glacial cycles and hysteresis of ice-sheet volume," *Nature*, *500*, 190–193.
Abrikosov, A. A., Gorkov, L. P., & Dzyaloshinski, I. E. [1975], *Methods of Quantum Field Theory in Statistical Physics*. New York: Dover.
Alfvén, H., & Herlofson, N. [1950], "Cosmic radiation and radio stars," *Phys. Rev.*, *78*, 616.
Altarelli M., Schlachter F., Cross J. [1998], "Making ultrabright X-rays," *Scientific American*, (December), pp. 35–43.
Andrews E. R. [1988], "NMR in medicine: a historical review," article in *Magnetic Resonance Imaging*, 2nd ed., Partain et al. (Eds.), Philadelphia: Saunders.
Andrews, H. C., & Hunt, B. R. [1977], *Digital Image Restoration*. Englewood Cliffs, New Jersey: Prentice-Hall.
Argentini, S., Mastrantonio, G., Fiocco, G., Ocone, R. [1992], "Complexity of the wind field as observed by a sodar system and by automatic weather stations on Nansen Ice Sheet, Antartica, during summer 1988–1989: two case studies," *Tellus,* vol. 44B, 422–429.
Arnold, J. T., Darmatti, S. S., & Packard, M. E. [1951], "Chemical effects on nuclear induction signals from organic compounds," *J. Chem. Phys.*, *19*, 507.
Balzarotti, A. [1975], *Luce di sincrotrone*. Istituto della Enciclopedia Italiana, Roma: Enciclopedia delle Scienze Fisiche.
Banta, R.M., Darby, L.S., Fast, J.D., et al. [2004], "Nocturnal low-level jet in mountain basin complex. Part I: Evolution and effects on local flows," *J. Appl. Meteor.,* 43, 1348–1365.
Banwell, C. N. [1966], *Molecular Spectroscopy*. London: McGraw-Hill.
Barratt, P., Browne, I.C. [1953], "A new method of measuring vertical air currents," *Q. J. R. Meteorol. Soc.,* 79, 550.
Beauchamp, K., & Yuen, C. [1979], *Digital Methods in Signal Analysis*. London: Allen & Unwin.
Bell, E. T. [1937], *Men of Mathematics*. New York: Simon & Schuster.
Bell, R. J. [1972], *Introductory Fourier Transform Spectroscopy*. New York: Academic Press.
Bell Burnell J. [1983], "The discovery of pulsars" in *Serendipitous Discoveries in Radio Astronomy*, Kelleman K., Sheets B. (Eds.), National Radio Astronomy Observatory, p. 169.
Belliveau, J. W., et al. [1990], "Functional cerebral imaging by susceptibility contrast NMR," *Magn. Reson. Med.*, *14*, 538–546.
Benade, A. H. [1990], *Fundamental of Musical Acoustics*. New York: Dover.

E. Prestini, *The Evolution of Applied Harmonic Analysis*, Applied and Numerical Harmonic Analysis, DOI 10.1007/978-1-4899-7989-6

Benedetto, J. J., & Frazier, M. W. (Eds.). [1994], *Wavelets*. Mathematics and Applications. Boca Raton, Florida: CRC Press.
Benedetto, J. J. [1996], *Harmonic Analysis and Applications*. Boca Raton, Florida: CRC Press.
Benedetti-Michelangeli, G., Congeduti, F., Fiocco, G. [1972], "Measurement of aerosol motion and wind velocity in the lower troposphere by Doppler optical radar," *J. of Atmospheric Sciences, 29*, 906–910.
Bengtsson I., Gabrielsson A. [1983], "Analysis and synthesis of musical rhythm," in: *Studies of Musical Performance*, Sundberg J. (Ed.), pp. 27–60.
Berger, A. [1978], "Long-term variations of daily insolation and Quaternary climatic changes," *J. Atmos. Science*, 35(12), 2362–2367.
Berger, A., & Loutre, M. F. [2002], "An exceptionally long interglacial ahead?" *Science*, 297, 1287–1288.
Berger, A., Loutre, M. F., & Crucifix, M. [2003], "The Earth's climate in the next hundred thousand years," *Survey in Geophysics*, 24, 117–138.
Berisio, R., et al. [1999], "Protein titration in the crystal state," *J. Mol. Biol.*, *292*, 845–854.
Berisio, R., Vitagliano, L., Sorrentino, G., Carotenuto, L., Piccolo, C., Mazzarella, L., et al. [2000], "Effects of microgravity on thecrystal quality of a collagen-like polypeptide," Acta Crys. *D*, *56*, 55–61.
Bhagavantam, S., & Venkatarayudu, T. [1951], *Theory of Groups and Its Applications to Physical Problems* (2nd ed.). Waltair: Andhra University.
Bhagavantam, S. [1966], *Crystal Symmetry and Physical Properties*. London: Academic Press.
Bignami, G. F., & Hermsen, W. [1983], "Galactic gamma ray sources," *Ann. Rev. Astron. Astroph.*, *21*, 67–108.
Blackham E.D. [1965], "The physics of the piano," *Scientific American*, (December), pp. 88–99.
Bloch, F. [1928], "Über Die Quantenmechanik der Elektronen in Kristallgittern," *Z. Phys.*, *52*, 555–600.
Bloch, F. [1946], "Nuclear induction," *Phys. Rev.*, *70*, 460–474.
Bloch, F., et al. [1946], "The nuclear induction experiment," *Phys. Rev.*, *79*, 474–485.
Bloembergen, N., Purcell, E. M., & Pound, V. [1948], "Relaxation effects in nuclear magnetic absorption," *Phys. Rev.*, *73*, 679–712.
Bloembergen, N. [1961], *Nuclear Magnetic Relaxation*. New York: Benjamin.
Bôcher M. [1905–1906], "Introduction to the theory of Fourier series," *Ann. of Math.* 7, pp. 81–152.
Bochner, S., & Chandrasekharan, K. [1949], *Fourier Transforms*. Princeton, New Jersey: Princeton Univ. Press.
Bode, H. [1984], "History of electronic sound modification," *Journal of Audio Engineering Society*, *32*(10), 730–739.
Bom, M., & Wolf, E. [1980], *Principles of Optics*. Oxford: Pergamon.
Bourbaki N. [1981], *Groupes et Algèbres de Lie*. Paris: Masson.
Boyce, W. E., & DiPrima, R. C. [1969], *Elementary Differential Equations and Boundary Value Problems*. New York: Wiley.
Bracewell, R. N. [1956], "Strip integration in radio astronomy," *Austr. J. Phys.*, *9*, 198–217.
Bracewell R. N. [1958], "Restoration in presence of errors," *Proc. I.R.E.* 46, pp. 106–111.
Bracewell, R. N. [1960], "Communications from superior galactic communities," *Nature*, *186*(4726), 670–671.
Bracewell R. N. [1979], "Image reconstruction in radio astronomy," article in *Image Reconstruction from Projections*, G.T. Herman (Ed.), Berlin: Springer.
Bracewell, R. N. [1986], *The Hartley Transform*. Oxford: Oxford University Press.
Bracewell R. N. [1989], "The Fourier transform," *Scientific American*, (June), pp. 62–69.
Bracewell, R. N. [1995], *Two-Dimensional Imaging*. Englewood Cliffs, New Jersey: Prentice-Hall.
Bracewell, R. N. [2000], *The Fourier Transform and Its Applications* (3rd ed.). New York: McGraw-Hill.
Bracewell, R. N., & Riddle, A. C. [1967], "Inversion of fan-beam scans in radioastronomy," Astro Phys. *J.*, *150*, 427–434.

Brigham, E. O. [1988], *The Fast Fourier Transform*. Englewood Cliffs, New Jersey: Prentice-Hall.
Brown, E.H., & Hall, F.H., Jr [1978], "Advances in atmospheric acostics," *Rev. Geo. Space Physics*, 16 (1), 47–110.
Brown, M. A., & Semelka, R. C. [1999], *MRI Basic Principles and Applications*. New York: Wiley.
Burkhardt H. [1908], *Entwicklungen nach oscillierenden Functionen und integration der Differentialgleichungen der mathematischen Physik*, Leipzig.
Cagnet, M., Franąon, M., & Thrierr, J. C. [1962], *Atlas of Optical Phenomena*. Berlin: Springer.
Callaghan, P. T. [1991], *Principles of Nuclear Magnetic Resonance*. Clarendon Press, Oxford: Oxford Science Publications.
Cappellini, V. [1985], *Elaborazione numerica delle immagini*. Torino: Boringhieri.
Carleson, L. [1966], "On the convergence and growth of partial sums of Fourier series," *Acta Math.*, *116*, 135–157.
Carr, H. Y. [1993], "Letter to the Editor," *Physics Today*, (January), p. 94.
Carslaw, H. S. [1930], *Introduction to the Theory of Fourier's Series and Integrals* (3rd ed.). New York: Dover.
Carslaw, H. S. [1925], "A historical note on Gibbs' phenomenon in Fourier's series and integrals," *Bull. Amer. Math. Soc.*, *31*, 420–424.
Cavaliere, S., & Piccialli, A. [1997], "Granular synthesis of musical signals," article in *Musical Signal Processing*, Roads C. et al. (Eds), Lisse: Swets & Zeitlinger.
Cellai, L., Cerrini, S., & Lamba, D. [1995], "30-Dechloro-30-methoxy-25-Omethyl-N-methylnapthomicin A," *Acta Crystallographica*, Sect. C, pp. 2060–2064.
Cercignani C. [1972], *Teoria ed applicazioni delle serie* di *Fourier*. Milano: Tamburini.
Champeney, D. C. [1973], *Fourier Transforms and their Physical Applications*. London: Academic Press.
Chang-Díaz F. R. [2000], "The VASIMR Rocket," *Scientific American*, (November), pp. 72–79.
Chayen N. E. [1997], "The role of oil in macromolecular crystallisation," *Structure* 5, pp. 1269–1274.
Chiarotti, G., & Giuliotto, L. [1954], "Proton relaxation in water," *Phys. Rev.*, *93*(6), 1241.
Cho, Z. H., et al. [1982], "Fourier transform nuclear magnetic resonance tomographic imaging," Proc. *IEEE*, *70*, 1152–1173.
Chowning, J. [1963], "The synthesis of complex audio spectra by means of frequency modulation," *J. of the Audio Engineering Society*, *21*, 526–534.
Chowning, J., & Bristow, D. [1980], *FM Theory and Applications*. Tokyo: Yamaha Music Foundation.
Christiansen, W. N., & Högbom, J. A. [1985], *Radiotelescopes*. Cambridge: Cambridge University Press.
Christiansen, W. N., & Warburton, J. A. [1957], "The Sun in 2-dimensions at 21 cm," *The Observatory*, *75*, 9–10.
Clairaut A. C. [1754], "Mémoire sur l'Orbite Apparente du Soleil Autour de la Terre, en Ayant égard aux Perturbations Produites par les Actions de la Lune et des Planètes Principales," *Mémoires de Mathématique et de Physique de l' Académie Royale des Sciences*, no. 9, pp. 801–870.
Clifford, S.F., et al. [1994], "Ground-based remote profiling in atmospheric studies: an overview," *Proc. IEEE*, 82 (3), 313–355.
Cocconi, G., & Morrison, P. [1959], "Searching for interstellar communications," *Nature*, *184*, 844–846.
Cochran, W., Crick, F. H. C., & Vand, V. [1952], "The structure of synthetic polypeptides. I. The transform of atoms on a helix," *Acta Cryst.*, *5*, 581–586.
Coles, B. A. [1976], "Dual frequency proton spin relaxation measurements on tissues from normal and tumor bearing mice," *J. Nat. Cancer Inst.*, *57*, 389.
Colwell, R.C., & Friend, A.W. [1936], "The D region of the ionosphere," *Nature*, 137, p. 782.
Conklin, E. K., & Bracewell, R. N. [1967a], "Isotropy of Cosmic Background Radiation at 10690 MHz," *Phys. Rev. Lett.*, *18*, 614.

Conklin, E. K., & Bracewell, R. N. [1967b], "Limits on small scale variations in Cosmic Background Radiation," *Nature*, *216*, 670.

Connor, F. R. [1982], *Signals*. London: Arnold.

Connor, F. R. [1982], *Noise*. London: Arnold.

Cooley, J. W., & Tukey, J. W. [1965], "An algorithm for machine calculation of complex Fourier transform," Math. *Computation*, *19*, 297–301.

Cooley J. W., Lewis P. A. W., Welch P. D. [1967], "Historical notes on the Fast Fourier transform," *Proc. IEEE*, Vol. 55, no. 10 (October), pp. 1675–1677.

Coppel, W. A. [1969], J.B. Fourier - "On the occasion of his two hundredth birthday," Amer. *Math. Monthly*, *76*, 468–483.

Corkum P. [2000], "Attosecond pulses at last," *Nature*, (February), pp. 845–846.

Cormack, A. M. [1963], "Representation of a function by its line integrals, with some radiological applications," *J. Applied Phys.*, *34*, 2722–2727.

Cormack, A. M. [1964], "Representation of a function by its line integrals, with some radiological applications II," *J. Applied Phys.*, *35*, 2908–2912.

Cowley, J. M. [1981], *Diffraction Physics*. Amsterdam: North-Holland.

Crawford I. [2000], "Where are they?" *Scientific American*, (June), pp. 28–33.

Crowther, R. A., & Klug, A. [1974], "Three-dimensional image reconstruction on an extended field - a fast stable algorithm," *Nature*, *251*, 490–492.

Crutzen, P.J. [2006], "Albedo enhancement by stratospheric sulfur injections: a contribution to resolve a policy dilemma?" *Clim. Change*, 77, 211-220.

Cullity, B. D. [1978], *Elements of X-Ray Diffraction*. London: Addison-Wesley.

Damadian, R. V. [1971], "Tumor detection by nuclear magnetic resonance," *Science*, *171*(March), 1151–1153.

Damadian, R., Zaner, K., Hor, D., & Di Maio, T. [1971], "Human tumors detected by nuclear magnetic resonance," *Proc. Nat. Acad. Sc. USA*, *71*, 1471–1473.

Danielson G. C., Lanczos C. [1942], "Some improvements in practical Fourier analysis and their application to X-ray scattering from liquids," *J. Franklin Inst.*, Vol. 233, pp. 365–380 and 435–452.

Darboux G. (Ed.) [1888–90], *Oevres de Fourier*, 2 vols., Paris: Gauthier-Villars.

Daubechies, I. [1992], *Ten Lectures on Wavelets*. Philadelphia: SIAM.

David E., Mathews M., McDonald H. [1958], "Description and results of experiments with speech using digital computer simulation," *Proceedings of the National Electronics Conference*, Vol. IV, Chicago, pp. 766–775.

Davis, P. J., & Hersh, R. [1981], *The Mathematical Experience*. Boston: Birkhäuser.

de Bernardis, P., et al. [2000], "A flat Universe from high resolution maps of the cosmic microwave background radiation," *Nature*, *404*, 955–959.

de la Vallée-Poussin, C. [1950], *Intégrales de Lebesgue* (2nd ed.). Paris: GauthierVillars.

Deans, S. R. [1983], *The Radon Transform and Some of its Applications*. New York: Wiley.

Derham, W. [1708], "Experimenta et observationes de soni motu," *Phil. Trans. Roy. Soc. London* 26, 2–35; (translated in English: Google Books - From 1703 to 1712).

Derr, V.E., & Little, C.G. [1970], " A comparison of remote sensing of the clear atmosphere by optical, radio, and acoustic radar techniques," *Appl. Optics,* vol. 9, No. 9, 1976–1992.

Dirac, P. A. M. [1947], *The Principles of Quantum Mechanics*. Oxford: Oxford University Press.

Di Sarra, A. [2002], "Tecniche lidar e sensori attivi ottici," article in *Tecniche e Strumenti per il Telerilevamento Ambientale,* (G. Galati and A. Gilardini Eds.) vol. 3, 587–664, Consiglio Nazionale delle Ricerche, Roma, Italy.

Doviak, R.J., Zrníc, D.S. [1993], *Doppler Radar and Weather Observations,* New York: Academic Press.

Dunnington, G. W. [1955], *Carl Friedrich Gauss: Titan of Science*. New York: Exposition Press.

Dym, H., & McKean, H. P. [1972], *Fourier Series and Integrals*. San Diego, California: Academic Press.

Eaton, F.D., McLaughlin, S.A., Hines, J.R. [1995], "A new frequency-modulated continuous wave radar for studying planetary boundary layer morphology," *Radio Sc.* 30, 75–88.

Eckbreth, A.C. [1987], *Laser Diagnostics for Combustion Temperature and Species.* Cambridge, MA: Abacus.

Edwards, R. E. [1967], *Fourier Series, a Modern Introduction*. New York: Holt Rinehart and Winston.

Einstein, A. [1905], "An heuristic viewpoint concerned with the generation and transformation of light," *Ann. Phys.*, *322*, 132–148.

Elder, F. R., Gurewitsch, A. M., Langmuir, R. V., & Pollock, H. C. [1947], "Radiation from electrons in a synchrotron," *Phys. Rev.*, *71*, 829–830.

Elliott, R. S. [1966], *Electromagnetics*. New York: McGraw-Hill.

Emsley J. W., Feeney J., Sutcliffe L. H. [1965], *High Resolution Nuclear Magnetic Resonance Spectroscopy*, 2 vols., Oxford: Pergamon Press.

Engel, P. [1986], *Geometric Crystallography*. Dordrecht (Holland): Reidel.

Engl, H. W., Hanke, M., & Neubauer, A. [1996], *Regularisation of Inverse Problems*. Dordrecht: Kluwer.

Englund, C.R., Crawford, D.R., Mumford, W.W. [1938], "Ultra-short-wave transmission and atmospheric irregularities," *Bell Sys. Tech. J.*, 17, 489–519.

Ernst, R. R. [1965], "Sensitivity enhancement in magnetic resonance. I. Analysis of the method of time averaging," *Review of Scientlffic Instruments*, *36*, 1689–1695.

Ernst, R. R., & Anderson, W. A. [1966], "Application of Fourier transform spectroscopy to magnetic resonance," *Review of Scientific Instruments*, *37*, 93–102.

Ernst, R. R., Bodenhausen, G., & Wokaun, A. [1987], *Principles of Nuclear Magnetic Resonance in One and Two Dimensions*. Oxford: Clarendon Press.

Euler, L. [1748], *Introductio in Analysis Infinitorum*, Bousquet et Comp., Lausanne. Reprinted in *Leonhardi Euleri Opera Omnia*, Series I, Vol. 8.

Euler, L. [1750], "De Propagatione Pulsuum per Medium Elasticum," *Novi Commentarii Academiae Scientiarum Petropolitanae*, Vol. 1, pp. 67–105. Reprinted in *Leonhardi Euleri Opera Omnia*, Series II, Vol. 10, pp. 98–131.

Euler, L. [1753], "De Serierum Determinatione seu Nova Methodus Inveniendi Terminos Generales Seriemm," *Novi CommentariiAcademiae Scientiarum Petropolitanae*, Vol. 3, pp. 36–85. Reprinted in *Leonhardi Euleri Opera Omnia*, Series I, Vol. 14, pp. 463–515.

Euler, L. [1793], "Observationes Generales Circa Series Quarum Termini Secundum Sinus vel Cosinus Angulorum Multiplorum Progrediuntur," *Nova Acta Academiae Scientiarum Petropolitanae*, Vol. 7, pp. 87–98. Reprinted in *Leonhardi Euleri Opera Omnia*, Series I, Vol. 16, pp. 163–177.

Euler, L. [1798], "Methodus facilis Inveniendi Series per Sinus Cisinusve Angulorum Multiplorum Proceddentes Quarum Usus in Universa Theoria Astronomiae est Amplissimus," *Nova Acta Academiae Scientiarum Petropolitanae*, vol. 11, pp. 94–113. Reprinted in *Leonhardi Euleri Opera Omnia*, Series I, Vol. 16, pp. 311–332.

Ewald, P. P. (Ed.). [1962], *Fifty Years of X-ray Diffraction*. Utrecht: Oosthoek.

Faridani, A. [2003], "Introduction to the mathematics of computed tomography," In G. Uhlmann (Ed.), *Inside Out*. Cambridge: Cambridge University Press.

Faridani, A., Ritman, E. L., & Smith, K. T. [1992a], "Local tomography," *SIAM J. Appl. Math.*, *52*(2), 459–484.

Faridani, A., Ritman, E. L., Smith, K. T. [1992b], "Examples of local tomography," *SIAM J. Appl. Math.*, 52 (2), 1193–1198.

Faridani, A., Finch, D. V., Ritman, E. L., & Smith, K. T. [1997], "Local tomography II," *SIAM. J. Appl. Math.*, *57*, 1095–1127.

Fedorov, von E. S. [1885], *An Introduction to the Theory of Figures*, (in Russian), Zipiski Imperator St. Petersburg Miner, Obscestra [2], 21, pp. 1–279. Reprinted: Leningrad (1953).

Fefferman, C. [1971], "The multiplier problem for the ball," *Ann. of Math.*, 94 (2), 330–336.

Ferretti, R., Pichelli, E., Gentile, S., Maiello, I., Cimini, D., Davolio, S., Miglietta, M.M., Panegrossi, G., Baldini, L., et al. [2014], " Overview of the first HyMeX Special Observation Period over Italy: observations and model results," *Hydrol. Earth Syst. Sci.*, 18, 1953–1977.

Feynman, R. P. [1963–1965], *Lectures in Physics*, 3 vols., Reading, Massachusetts: Addison-Wesley.

Feynman, R. P. [1989], *What Do You Care What Other People Think?*. New York: Bantam.
Field, G. B., & Chaisson, E. J. [1985], *The Invisible Universe*. Boston: Birkhäuser.
Fiocco, G., & Smullin, L.D. [1963], "Detection of scattering layers in the upper atmosphere (60-140 km) by optical radar," *Nature*, 199, 1275–1276.
Fiocco, G., & Grams, G.W. [1964], "Observations of the aerosol layer at 20 km by optical radar," *J. Atmos. Sciences*, 21, p.322
Fiocco, G., Benedetti-Michelangeli, G., et al. [1971], "Measurement of temperature and aerosol to molecule ratio in the troposphere by optical radar," *Nature*, 229, 78–79.
Fleury, P., & Mathieu, J. P. [1970], *Immagini ottiche*. Bologna: Zanichelli.
Födermayr F., Deutch W.A. [1993], "Parmi veder le lacrime. One aria, three interpretations," *Proc. Music Acoustic Conference*, Royal Swedish Academy of Music (Ed.), pp. 96–99.
Fourier, J. [1978], *The Analytical Theory of Heat*. New York, London: Cambridge University Press. Reprinted by Dover.
Franklin, R. E., & Gosling, R. G. [1953], "Molecular configuration in sodium thymonucleate," *Nature*, *171*, 740–741.
Friedrich, W., Knipping, P., & von Laue, M. [1912], 1913. *Interferenzerscheinungen bei Rontgenstrahlen. Ann. Physik*, *41*(5), 971–988.
Friend, A.W. [1939], "Continuous determination of air-mass boundaries by radio," *Bull. Amer. Meteorol. Soc.*, 20, 202–205.
Frisch, U. [1995], *Turbulence: The Legacy of A. N. Kolmogorov*. Cambridge (UK): Cambridge Univ. Press.
Gabor D. [1946], "Theory of communication," *Journal of the IEE* 93 (III), pp. 429–457.
Gabor, D. [1947], "Acoustical quanta and the theory of hearing," *Nature*, *4044*, 591–594.
Galati, G. (Ed.). [1993], *Advanced Radar Techniques and Systems*. London: Institute of Electrical Engineers.
Galati, G., & Naldi, M. [2002], "Sensori attivi a microonde," article in *Tecniche e Strumenti per il Telerilevamento Ambientale,* (G. Galati and A. Gilardini Eds.) vol. 3, 883–955, Consiglio Nazionale delle Ricerche, Roma, Italy.
Ganopolski, A., Winkelmann, R., Schellnhuber, H.J. [2016], "Critical insolation-CO_2 relation for diagnosing past and future glacial inception," *Nature*, 529, 200–203.
Garcia, F. L., Zahn, R., Riek, R., & Wüthrich, K. [2000], "NMR structure of the bovine prion protein," *Proc. Natl. Acad. Sci. U.S.A.*, *97*, 8334–8339.
Gargett, A.E., Osborn, T.R., Nasmyth, T.W. [1984], "Local isotropy and decay of turbulence in a stratified fluid," *J. Fluid Mech.,* 144, 231–280.
Gauss, C. F. [1866], "Nachlass, Theoria Interpolationis Methodo Nova Tractata," *Carl Friedrich Gauss Werke* (Vol. 3, pp. 265–330). Göttingen: Königlichen Gesellschaft der Wissenschaften.
Gelfand, I. M., & Shilov, G. E. [1964], *Generalized Functions*. New York: Academic Press.
Gerlach, W., & Stern, O. [1922], "The experimental detection of quantized orientations in a magnetic field," *Z. Phys.*, *9*, 349–352.
Giacovazzo, C. [1980], *DirectMethods in Crystallography*. London: Academic Press.
Giacovazzo, C., et al. [1992], *Fundamentals of Crystallography*. Oxford: Oxford University Press.
Gibbs, J. W. [1899], *Nature*, *59*, 606.
Gillespie, C. C. (Ed.). [1981], *Scientific Biography*. New York: Scribners's Sons.
Gilman, G.W., et al. [1946], "Reflection of sound signals in the troposphere," *J. Acoust. Soc. Amer.*, 18 (2), 274–283.
Givi, P. [1989], "Model-Free Simulations of Turbulent Reactive Flows," *Progress in Tubulence and Combustion Science,* vol. 15, No. 1, 1–107.
Glover, G. H. [1997], *Functional Neurologic MR Imaging* (pp. 145–152). RSNA Categorical Course in Physics: The Basic Physics of MR Imaging.
Glusker, J. P., & Trueblood, K. N. [1985], *Crystal Structure Analysis*. Oxford: Oxford University Press.
Goldberg, R. R. [1961], *Fourier Transforms*. London: Cambridge University Press.

Goldman, M. [1988], *Quantum Description of High-Resolution NMR in Liquids*. Oxford: Oxford Science Publications.
Goldstine, H. H. [1977], *A History of Numerical Analysis from the 16th Through the 19th Century*. Berlin: Springer-Verlag.
Gonzales, R. C., & Wintz, P. [1987], *Digital Image Processing*. London: Addison-Wesley.
Goodman, J. W. [1968], *Introduction to Fourier Optics*. San Francisco: McGraw-Hill.
Gordon, R., Bender, R., & Herman, G. T. [1970], "Algebraic reconstruction techniques (ART) for three-dimensional electron microscopy and X-ray photography," *J. Theor. Biol.*, *29*, 471–481.
Gorter C.J. [1936], "Negative result of an attempt to detect nuclear magnetic spins," *Physica* 3, pp. 995–998.
Gowan, E.H. [1929], "Low frequency sound waves and the upper atmosphere," *Nature*, 124, 452–454.
Grattan, Guinness I. [1972], *Joseph Fourier 1768–1830*. Cambridge, Massachusetts: MIT Press.
Grimson, W. E., et al. [1999], "Image-guided surgery," *Scientific American*, (June), pp. 55–61.
Gröchenig, K. [2001], *Foundations of Time-Frequency Analysis*. Boston: Birkhäuser.
Grossman, A., & Morlet, J. [1984], "Decomposition of Hardy functions into square integrable wavelets of constant shape," *SIAM J. of Appl. Math.*, *15*(4), 723–736.
Gunsteren van W. F., Berendsen H. J. C. [1990], "Computer simulation of molecular dynamics: methodology, applications and perspective in chemistry," *Angew. Chem. Int. Ed. Engl.* 29, pp. 992–1023.
Güttinger, P. [1991], "Das Verhalten von Atomen in magnetischen Drehfeld," *Z. Phys.* 73, pp. 169–184.
Hanany, S., et al. [2000], "A measurement of the cosmic microwave background anysotropy on angular scales of 10 arcminutes to 5 degrees," *J. Astro-Phys.*, *L5*, 545.
Harburn, G., Taylor, C. A., & Welberry, T. R. [1979], *Atlas of Optical Transforms*. London: Bell.
Hardy, G. H., & Rogosinski, W. [1944], *Fourier Series*. London: Cambridge University Press.
Harford, J. [1997], *Korolev*. New York: Wiley.
Harris, C.M. [1966], "Absorption of sound in air versus humidity and temperature," *J. Acoust. Soc. Amer.*, 40, 148–159.
Hauptman, H. A. [1972], *Crystal Structure Determination: the Role of the Cosine Invariants*. New York: Plenum.
Hauptman, H. A. [1990], *"A minimal principle in direct methods of X-ray crystallography," Advances in Fourier Analysis and its Applications*. Dordrecht: Kluwer.
Hausser, K. H., & Kalbitzer, H. R. [1991], *NMR in Medicine and Biology*. Berlin: Springer-Verlag.
Haüy M. l'Abbé [1784], "Essai d'une théorie sur la structure des cristaux, appliquée à plusieurs genres de substances cristallisées," *Gogué et Née de la Rochelle*, Libraires, Paris.
Hays, J.D., Imbrie, J., Schackleton, N.J. [1976], "Variations in the Earth's orbit: pacemaker of the Ice Ages," *Science*, 194, 1121–1132.
Hazard, C., Mackey, M. B., & Shimmins, A. J. [1963], "Investigations of the radio source 3C273 by the method of lunar occultations," *Nature*, *197*, 1037–1039.
Hecht, E., & Zajac, A. [1987], *Optics*. Reading, Massachusetts: Addison-Wesley.
Heideman, M. T., Johnson, D. H., & Burrus, C. S. [1985], "Gauss and the History of the Fast Fourier Transform," *Archive for History of Exact Science*, *34*, 265–277.
Heitler W. [1945], *Elementary Wave Mechanics*. London: Oxford University Press (Clarendon).
Helgason, S. [1980], *The Radon Transform*. Boston: Birkhäuser.
von Helmholtz, H. [1954], *On the Sensation of Tone*. New York: Dover.
Hendee W. R. [1988], *The Imaging Process. Textbook of Diagnostic Imaging*, Putman C. E., Ravin C. E. (Eds.), Philadelphia: Saunders Company.
Hérivel, J. [1975], *Joseph Fourier - The Man and the Physicist*. Oxford: Clarendon.
Herman G. T. (Ed.) [1983], "Special issue in computerized tomography," *Proc. IEEE*, pp. 291–435.
Herman, G. T. (Ed.). [1991], *Mathematical Methods in Tomography*. Berlin: Springer-Verlag.
Herman, G. T., & Rowland, S. W. [1973], "Three methods for reconstructing objects from X-rays: a comparative study," *Comput. Graphic Process*, *2*, 151–178.

Herman G.T., Lewitt R.M. [1979], "Overview of image reconstruction from projections," article in *Overview of Image Reconstruction from Projections*, Herman G. T. (Ed.), Vol. 32, pp. 1–8, New York: Springer-Verlag.

Hernández, E., & Weiss, G. [1996], *A First Course on Wavelets*. Boca Raton, Florida: CRC Press.

Hertz H. [1893], *Electric Waves*, reprinted in 1962 by Dover, New York.

Hewish, A., Bell, S. J., Pilkington, J. D. H., Scott, P. F., & Collins, R. A. [1968], "Observation of a rapidly pulsating radio source," *Nature*, *217*, 709–713.

Hinkley, E. D. (Ed.) [1976], *Laser Monitoring of the Atmosphere*. Berlin: Springer.

Hinshaw, W. S., et al. [1983], "An introduction to NMR Imaging: from the Bloch equation to the Imaging equation," *Proc. IEEE*, *71*(3), 338–350.

Hobson, E. W. [1957], *The Theory of Functions of a Real Variable and the Theory of Fourier's Series*. New York: Dover.

Holden, A., & Morrison, P. [1999], *Crystals and Crystal Growing*. Cambridge, Massachusetts: MIT Press.

Horowitz, A. L. [1995], *MRI Physics for Radiologists*. New York: Springer-Verlag.

Hounsfield, G. N. [1972], "A method of and apparatus for examination of a body by radiation such as X-ray or gamma radiation," British Patent No. 1283915, London.

Hounsfield, G. N. [1973], "Computed transverse axial scanning (tomography)": I. *Description of system. Brit. J. Radiol.*, *46*, 1016–1022.

Hu, W. [2000], "Ringing in the new cosmology," *Nature*, *404*, 939–940.

Huygens C. [1690], *Traité de la Lumière*, Leiden.

Iisuka, K. [1985], *Engineering Optics*. New York: Springer-Verlag.

Inaba, H., & Kobayashi, T. [1972], "Laser-Raman radar," *Opto-Electron.* 4, 101–123.

Inouye H., Kirschner D. A. [1997], "X-ray diffraction analysis of scrapie prion: intermediate and folded structures in a peptide containing two putative alpha-helices," *J. Mol. Biol.* 268, pp. 375–389.

International Tables for X-ray Crystallography [1983], Mol. A, Space-Group Symmetry, Dordrecht (Holland): Reidel.

Imbrie, J. [1982], "Astronomical theory of the Pleistocene Ice Ages: a brief historical review," *Icarus* vol. 50, p. 411.

Ivanenko, D., & Pomeranchuk, J. [1944], "On a maximal energy attainable in a betatron," *Phys. Rev.*, *65*, 343.

Jacobi C. G. J. [1846, 1851], *Mathematische Werke*, 2 vols., Berlin: Reimer.

Jackson, J. D. [1975], *Classical Electrodynamics*. New York: Wiley.

Jahnke, E., & Emde, F. [1945], *Tables of Functions*. New York: Dover.

Jansky, K. G. (1933], "Electrical disturbances apparently of extraterrestrial origin," Proc. *IRE*, *21*, 1387–1398.

Jeffery, G. A., & Saenger, W. [1981], *Hydrogen Bonding in Biological Structures*. Berlin: Springer-Verlag.

John F. [1955], *Plane Waves and Spherical Means Applied to Partial Differential Equations*. New York: Wiley (Interscience).

Jolesz F. A. [1995], "MRI-guided interventions," article in *Progress in RMN*, Cammisa M., Scarabino T. (Eds.), Gnocchi: Napoli.

Judson, H. F. [1979]. *The Eight Day of Creation*. New York: Simon & Schuster.

Kalender, W. A., Seissler, W., et al. [1990], "Spiral volumetric CT with single-breath-hold technique, continuous transport, and continuous scanner rotation," *Radiology*, *176*, 181–183.

Kallistratova, M.A. [1961], "Experimental investigation of sound wave scattering in the atmosphere," *Trudy Inst. Fiz. Atmos., Atmosfernaya Turbulentnost.*, 4, 203–256.

Karle J., Hauptman H. [1953], Monograph no. 3, American Crystallography Association, The Letter Shop, Delaware: Wilmington.

Katakura, T., Kimura, K., et al. [1989], "Fundamental research in CT report 9 -Trials in helical scanning," *Journal of the Tomographic Imaging Association*, *16*, 247–250.

Katznelson, Y. [1968], *An Introduction to Harmonic Analysis*. New York: Wiley.

Kaufmann, W. J. [1993], *Discovering the Universe*. New York: Freeman.

Keith, D. [2013], *A Case for Climate Engineering,* A Boston Review Book, Boston, MA: MIT Press.

Kiepenheuer, K. O. [1950], "Cosmic rays as a source of the general galactic radio emission," *Phys. Rev., 79*, 738–739.

Kimura, K., & Koga, S. (Eds.). [1993], *Basic Principles and Clinical Applications of Helical Scan*. Tokyo: Iryokagakusha.

Kittel, C. [1971], *Solid State Physics*. New York: Wiley.

Klein, A. H. [1974], *The Science of Measurement: A Historical Survey*. New York: Dover.

Kline, M. [1967], *Mathematics for the Nonmathematician*. New York: Dover.

Kline, M. [1991], *Storia del pensiero matematico*. Torino: Einaudi.

Klug, H. P., & Alexander, L. E. [1974], *X-Ray Diffraction Procedures*. New York: Wiley.

Knight, W. D. [1949], "Nuclear magnetic resonance shift in metals," *Phys. Rev., 76*, 1259–1260.

Kolmogorov, A. N. [1926], "Une série de Fourier-Lebesgue divergente partout," C.R. *Acad. Sci. Paris Sér. A-B, 183*, 1327–1328.

Kolmogorov, A.N. [1941], "The local structure of turbulence in incompressible viscous fluid for very large Reynolds number," *Doklady Acad. Nauk. SSSR,* 30, 301–305; "On degeneration (decay) of isotropic turbulence in an incompressible viscous fluid for very large Reynolds number," *Doklady Acad. Nauk. SSSR* 31, 538–540; "Energy dissipation in locally isotropic turbulence," *Doklady Acad. Nauk. SSSR* 32, No.1, 16–18.

Kolmogorov, A.N. [1962], "A refinement of previous hypotheses concerning the local structure of tubulence in a viscous incompressible fluid at high Reynolds number," *J. Fluid Mech.,* 12, 82–85.

Körner, T. W. [1988], *Fourier Analysis*. Cambridge: Cambridge University Press.

Kraus, J. D. [1966], *Radio Astronomy*. New York: McGraw-Hill.

Kuhl D. E., Edwards R. Q. [1963], "Image separation radioisotope scanning," *Radiology* 80, pp. 653–661.

Kumar A., Welti D., Ernst R. R. [1975], "NMR Fourier zeumatography," *J. Magn. Res.* 18, pp. 69–83.

Kuo, K.K., Acharya, R. [2012], *Fundamentals of Turbulent and Multiphase Combustion*. Hoboken (USA): Wiley.

Ladd, M. F. C., & Palmer, R. A. [1977], *Structure Determination by X-Ray Crystallography*. New York: Plenum.

Lagrange, J. L. [1759], "Recherches sur la Nature et la Propagation du Son," *Miscellanea Taurinensia*, Vol. I, no. I-X, pp. 1–112. Reprinted in *Oevres de Lagrange*, Vol. 1, pp. 39–148, Paris: Gauthier-Villars.

Lagrange, J. L. [1762–1765], "Solution de Différents Problèmes de Calcul Intégral," *Miscellanea Taurinensia*, Vol. III. Reprinted in *Oevres de Lagrange*, Vol. 1, pp. 469–668, Paris: Gauthier-Villars.

Lagrange, J.L. [1781], *Oevres Completes*, t. V, (1870), p. 125. Paris: Gauthier-Villars.

Lagrange, J.L. [1782], *Oevres Completes*, t. V, (1870), p. 211. Paris : Gauthier-Villars.

Lanczos, C. [1956], *AppliedAnalysis*. Englewood Cliffs, New Jersey: Prentice Hall.

Langer, R. E. [1947], "Fourier Series -The Genesis and Evolution of a Theory," *American Mathematical Monthly, 54*(supplement), 1–81.

Larmor, J. [1897], "On the Theory of the Magnetic Influence on Spectra; and on the Radiation from moving Ions," *Phylosophical Magazine, 44*, 503–512.

Larsen, H.C., et al. [1994], "Seven million years of glaciation in Greenland," *Science*, 264, 952–955.

Laskar, J., et al. [2004], "A long-term numerical solution for the insolation quantities of the Earth," *Astronomy and Astrophysics*, 428 (1), 261–285.

Lauterbur, P. C. [1973], "Image formation by induced local interactions: examples employing nuclear magnetic resonance," *Nature, 242*(March), 190–192.

Lauterbur P. C., Lai C. M. [1980], "Zeugmatography by reconstruction from projections," *IEEE Trans. Nucl. Sci.*, Vol. NS-2, pp. 1227–1231.

Lebesque, H. [1906], *Leçons sur les series trigonometriques*. Paris: Gauthier-Villars.

Lesieur, M. [1987], *Turbulence in Fluids*. Dordrecht (Holland): Martinus Nijhoff.

Les Cahiers de Science et Vie [1994], *Naissance de la Radioastronomie*, no. 8, Paris: Excelsior Publications.

Lipson G. H. [1995], *Optical Transforms*, Cambridge University Press.

Lipson, H., & Taylor, C. A. [1958], *Fourier Transform and X-Ray Diffraction*. London: Bell.

Lipson, H., & Cochran, W. [1966], *The Determination of Crystal Structures*. London: Bell.

Little, C.G. [1969], "Acoustic methods for the remote probing of the lower atmosphere," *Proc. IEEE,* 57, 571–578.

Liu, H., Farr-Jones, S., Ulyanov, N. B., Llinas, M., Marqusee, S., Groth, D., et al. [1999], "Solution structure of syrian hamster prion protein rPrP(90–231)," *Biochemistry*, *38*, 5362–5377.

Lloyd, L. S. [1937], *Music and Sound*. London: Oxford University Press.

Logan, B. F. [1975], "The uncertainty principle in reconstructing functions from projections," Duke Math. *J.*, *42*, 661–706.

Long, D.A. [2002], *The Raman Effect.* New York: Wiley.

Lorenz E.N. [1963], "Deterministic nonperiodic flow," *J. Atmos. Sci.* 20, 130–141.

Louis, A. K. [1984], "Orthogonal function series expansions and the null space of the Radon transform," *SIAM. J. Math. Anal.*, *15*, 621–633.

Louis, A. K. [1992], "Medical imaging: the state of the art and future development," *Inverse Problems*, *8*, 709–738.

Macowski, A. [1983], *Medical Imaging Systems*. Englewood Cliffs, New Jersey: Prentice Hall.

Maiman, T.H. [1960], "Stimulated optical radiation in Ruby," *Nature*, 187, 493–494.

Majorana, E. [1932], "Atomi orientati in campo magnetico variabile," *Nuovo Cimento*, *9*, 43–50.

Mallat, S. [1998], *A Wavelet Tour of Signal Processing*. New York: Academic Press.

Malus E. L. [1892], *L' agenda de Malus. Souvenirs de l' expédition d' Egypte, 1798–1801*, Paris.

Mansfield P. [1977], "Multi-planar image formation using NMR spin echoes," *Journal de Physique*, 10, C, pp. L55–58.

Mansfield, P., & Morris, G. P. [1982], *NMR Imaging in Biomedicine*. New York: Academic Press.

Marconi, S.G. [1922], "Radio Telegraphy," *proc. IRE,* vol.10 (4), p. 237.

Marr, D. [1982], *Vsion*. New York: Freeman.

Margaritondo, G. [1988], *Introduction to Synchrotron Radiation*. New York: Oxford University Press.

Mastrantonio, G. [2002], "Telesondaggio acustico del'atmosfera," article in *Tecniche e Strumenti per il Telerilevamento Ambientale,* (G. Galati and A. Gilardini Eds.) vol. 3, 655–882, Roma, Italy: Consiglio Nazionale delle Ricerche.

Mastrantonio, G., Fiocco, G. [1982], "Accuracy of wind velocity determination with Doppler sodar," *J. Appl. Meteorol.,* 21(6), 824–830.

Mathews, M. V. [1969], *The Technology of Computer Music*. Cambridge, Massachusetts: MIT Press.

Mathews, M. V., & Pierce, J. R. [1989], *Current Directions in Computer Music Research*. Cambridge, Massachusetts: MIT Press.

Mattson, J., & Simon, M. [1996], *The Pioneers of NMR and Magnetic Resonance in Medicine*. Jericho, New York: Dean Books.

Maxwell J. C. [1892], *A Treatise on Electricity and Magnetism*. London: Oxford University Press (Clarendon).

McAllister, L.G. [1968], "Acoustic sounding of the lower troposphere," *J. Atmos. Terrest. Phys.*, 30, 1439–1440.

McAllister, L.G. [1971], "Wind velocity measurements in the lower atmosphere using acoustic sounding techniques," *Tech. Note-A204 (AP),* 20 pp., Salisbury, Australia: Weapons Res. Estab.

McAllister, L.G., et al. [1969], "Acoustic sounding – A new approach to the study of atmospheric structure," *Proc. IEEE*, 57, 579–587.

McClung, F.J., Hellarth, R.W. [1962], "Giant optical pulsations from Ruby," *J. Appl. Phys.*, 33, 828–829.

McDonough, J.M. [2007], *Introductory Lectures on Turbulence – Physics, Mathematics, and Modeling*. (http://www.engr.uky.edu/acfd/lctr-notes634.pfd).

McLachlan, D. [1957], *X-Ray Crystal Structure*. New York: McGraw-Hill.
McPherson A. [1989], "Molecular crystal," *Scientific American*, (March), pp. 42–49.
Melfi, S.H., Lawrence, J.D., McCormick, M.P. [1969], "Observations of Raman scattering by water vapor in the atmosphere," *Appl. Phys. Letters*, 15, 295–297.
Meyer, Y. [1993], *Wavelets*. Philadelphia: SIAM.
Meyer-Arendt, J. R. [1972], *Introduction to Classical and Modern Optics*. Englewood Cliffs, New Jersey: Prentice-Hall.
Michelson A. A. [1898], *Nature* 58, p. 544, London.
Michelson, A. A., & Pease, F. G. [1921], "Measurement of the diameter of α Orionis with the interferometer," *Astrophys. J.*, *53*, 249–259.
Middleton, W.E.K., & Spilhaus, A.F. [1953], *Meteorological Instruments.* Toronto: University of Toronto Press.
Mie, G. [1908], "Beiträge zur optik trüber medien, spetiell kolloidaler metallösungen," *Annalen der Physik,* Vierte Folge, 25, 377–445.
Milankovitch, M. [1920], *Théorie Mathématique des Phénomènes Thermiques Produits par la Radiation Solaire*. Paris: Gauthier-Villars.
Milankovitch, M. [1941], *Kanon der Erdbestrahlung und seine Anwendung auf das Eiszeitenproblem,* Roy. Serb. Acad. Spec. Publ. 133, 1–633. (*Canon of Insolation and the Ice-age Problem,* published for the US Department of Commerce and the National Science Foundation, Washington D.C., by the Israel Program for Scientific Translations, Jerusalem 1969).
Miller, D. C. [1916], *The Science of Musical Sounds*. New York: McMillan.
Moore J. A. [1977], "Signal Processing Aspects of Computer Music –A Survey," *Computer Music* J., (July), pp. 4–37.
Monin, A.S. [1962], "Characteristics of the scattering of sound in a turbulent atmosphere," *Sov. Phys. Acoust., Engl. Transl.*, 7, 370–373.
Morgan, J. [1953], *Geometrical and Physical Optics*. New York: McGraw-Hill.
Morrison, P. [1967], "Extrasolar x-ray sources," *Ann. Rev. Astron. Astroph.*, *5*, 325–350.
Morse, P. M. [1948], *Vbration and Sound*. Melville, New York: The Acoustical Society of America.
Moseley, M. E., & Glover, G. H. [1995], "Functional MR Imaging," *Functional Neuroimaging*, *5*(2), 161–191.
Muzzolini, A. [1993], "The emergence of a food producing economy in the Sahara," 227–239, in Shaw et al. (Eds.), *The Archeology of Africa: Food, Metals and Towns*. London: Routledge.
Natterer, F. [1986], *The Mathematics of Computerized Tomography*. Chichester, England: Wiley.
Neugebauer, O. E. [1952], *The Exact Sciences in Antiquity*. Princeton, New Jersey: Princeton University Press.
de Niels, H., & Heathcote, V. [1953], *Nobel Prize Winners in Physics 1901–1950*. New York: Schuman.
Nyquist H. [1928], "Certain topics in telegraph transmission theory," *Trans. A. I. E.* E., Vol. 47, pp. 617–644.
Obukov, A.M. [1941], "On the distribution of energy in the spectrum of a turbulent flow," *Doklady Acad. Nauk. SSSR* 32, 22–24; "Spectral energy distribution in a turbulent flow," *Izvestiya Acad. Nauk. SSSR,* Ser. Geogr. Geofiz., No. 4-5, 453–466.
Oldendorf W. H. [1961], "Isolated flying spot detection of radiodensity discontinuities displaying the internal structural pattern of a complex object," *IRE Trans. Biomedical Electronics* 8, pp. 68–72.
Olson A. J., Goodsell D. S. [1992], "Visualizing biological molecules," *Scientific American*, (November), pp. 44–51.
Ottersten, H., Hardy, K.R., Little, C.G. [1973], "Radar and sodar probing of waves and turbulence in statically stable clear-air layers," *Boundary-Layer Meteorology*, 4, 47–89.
Ozaktas, H. M., et al. [2000], *The Fractional Fourier Transform*. New York: Wiley.
Papoulis, A. [1962], *The Fourier Integral and Its Applications*. New York: McGraw-Hill.
Papoulis, A. [1985], *Signal Analysis*. New York: McGraw-Hill.

Paratt, L. G. [1959], "Use of synchrotron orbit-radius in x-ray physics," *Rev. Sci. Instrum.*, *30*, 297–299.

Perley, R. A., Dreher, J. W., & Conwan, J. J. [1984], "The jet and filaments in Cygnus," *Astrophys. J.*, *285*, L35–L38.

Perutz M. F. [1964], "The hemoglobin molecule," *Scientific American*, (November), pp. 64–76.

Pierce, J. R. [1974], *Almost All About Waves*. Cambridge: MIT Press.

Pierce, J. R. [1983], *The Science of Musical Sound*. Freeman, San Francisco: Scientific American Books.

Poli G. De et al. (Eds.) [1991], *Representations of Musical Signals*, MIT Press.

Pope, S.B. [2009], *Turbulent Flows,* 6th ed., Cambridge University Press.

Prestini, E. [1979], "The phase problem for space curves," *Boll. Un. Mat. It.*, *5*, 316–321.

Prestini, E. [2016], "On the convergence of parabolically scaled two dimensional Fourier series in the linear phase setting," *Studia Math.* (to appear).

Price, E. W. [1992], "Solid rocket combustion instability –an American historical account," *Progress in Astronautics and Aeronautics*, Vol. 143, pp. 1–16, New York: American Institute of Aeronautics and Astronautics.

Probert-Jones, J.R. [1984], "Resonance component of backscattering by large dielectric sphere," *J. Opt. Soc. Amer.* A1, 822–883.

Pruppacher, H.R., & Beard, K.V. [1970], "A wind tunnel investigation of the internal circulations and shape of water drops falling at terminal velocity in air," *Quart. J. Roy. Meteorol. Soc.*, 96, 247–256.

Prusiner, S. B. [1982], "Novel proteinaceous infectious particles cause scrapie," *Science*, *216*, 136–144.

Prusiner, S. B. [1998], "Prions," *Proc. Natl. Acad. Sci.* U.S.A., 95, 13363–13383.

Pueschel, F. [1986], "Aerosol measurements in the boundary layer," article in *Probing the Atmospheric Boundary Layer,* (D.H. Lenschow Ed.) 57–86, Amer. Meteor. Soc., Boston.

Purcell, E. M., Torrey, H. C., & Pound, R. V. [1946], "Resonance absorption by nuclear magnetic moments in a solid," *Phys. Rev.*, *69*, 37–38.

Purcell, E. M., et al. [1946], "Resonance absorption by nuclear magnetic moments in a solid," *Phys. Rev.*, *69*, 37–38.

Purcell E. M. [1988], "Foreword," in *Magnetic Resonance Imaging*, 2nd ed., Partain et al. (Eds.), Philadelphia: Saunders.

Pykett I. L. [1982], "NMR imaging in medicine," *Scientific American*, (May), pp. 54–64.

Rabi, I. I. [1927], "On the principal magnetic susceptibilities of crystals," *Phys. Rev.*, *29*, 174–185.

Rabi, I. I. [1937], "Space quantization in a gyrating magnetic field," *Phys. Rev.*, *51*, 652–654.

Rabi, I. I., Zacharias, J. R., Millman, S., & Kusch, P. [1939], "The molecular beam resonance method for measuring nuclear magnetic moments: the magnetic moments of ${}_3Li^6, {}_3Li^7$ and ${}_9F^{19}$," *Phys. Rev.*, *55*, 526–535.

Radon, J. [1917], "Über die Bestimmung von Functionen durch ihr Integralwerte langs gewisser Mannigfaltigkeiten," *Berichte Saechsiche Akademie der Wissenschaften*, *69*, 262–277.

Ramachandran, G. N., & Srinivasan, R. [1970]. *Fourier Methods in Crystallography*. New York: Wiley.

Ramsey, N. F. [1950], "Magnetic shielding of nuclei in molecules," *Phys. Rev.*, *78*, 699–703.

Ramsey, N. F. [1951], "Dependence of magnetic shielding of nuclei upon molecular orientation," *Phys. Rev.*, *83*, 540–541.

Reader, J. [1998], *Africa: A Biography of the Continent*. New York: Knopf.

Reynolds, O. [1876], "On the refraction of sound in the atmosphere," *Phil. Trans. Roy. Soc. London*, 166, 315.

Reynolds, O. [1895], "On the dynamical theory of turbulent incompressible viscous fluids and the determination of the criterion," *Phil. Trans. Roy. Soc. London,* Ser. A 186, 123–164.

Rhodes, R. [1998], *The Making of the Atomic Bomb*. New York: Touchstone Books.

Richardson, L.F. [1922], *Weather Predictions by Numerical Process*. Cambridge (UK): Cambridge Univ. Press.

Riek, R., Wider, G., Billeter, M., Hornemann, S., & Glockshuber, R. [1998], "Prion protein NMR structure and familial human spongiform encephalopathies," *Proc. Natl. Acad. Sci. U.S.A., 95*, 11667–11672.

Riemann B., Weber E. W. [1927], *Differentialgleichungen der Physik*, 7th ed., Vol. II, Vieweg.

Riemann B. [1892], "Ueber die Darstellbarkeit einer Function durch eine trigonometrische Reihe," *Gesammelte Werke*, pp. 227–231, Leipzig.

Riemann B. [1898], "La possibilité de représenter une fonction par une série trigonométrique," *Oevres Mathématiques de Riemann*, pp. 225–279, Paris: Gauthier-Villars.

Risset J. C. [1991], "Timbre analysis by synthesis: representations, imitations and variants for musical composition," article in *Representation of Musical Signals*, G. de Poli, A. Piccialli and C. Road (Eds.), Massachusetts, Cambridge: MIT Press.

Ritman E. L. et al. [1997a], "Local reconstruction applied to X-ray microtomography," IMA Volumes in Mathematics and Its Applications, *Inverse Problems in Wave Propagation* 90, pp. 443–452, New York: Springer-Verlag.

Ritman, E. L., et al. [1997b], "Synchrotron-based micro-CT of in situ biological Basic Functional Units and their integration," *Proceedings of SPIE, Developments in X-ray Tomography, 3149*, 13–24.

Roads C. [1997], "Sound transformation by convolution," article in *Musical Signal Processing*, Roads C. et al. (Eds.), Lisse: Swets & Zeitlinger.

Roads, C., et al. (Eds.). [1997], *Musical Signal Processing*. Lisse: Swets & Zeitlinger.

Roberto, N., Adirosi, E., Baldini, L. et al. [2016], "Multi-sensor analysis of convective activity in central Italy during the HyMeX SOP 1.1," *Atmos. Meas. Tech.*, 9, 535–532.

Roederer, J. C. [1979], *Introduction to the Physics and Psychophysics of Music*. Berlin: Springer-Verlag.

Rogosinski, W. [1959], *Fourier Series*. New York: Chelsea.

Rogers, R.R. [1990], "The early years of Doppler radar in meteorology," in *Radar Meteorology* (D. Atlas ed.), Amer. Meteorol. Soc. Boston.

Romé de l'Isle J. B. L. [1783], *Crystallographie, ou description des formes propres a tous les corps du règne minéral*, 4 vols., Paris.

Rudin, W. [1966], *Real and Complex Analysis*. New York: Wiley.

Rudnick P. [1966], "Note on the Calculation of Fourier Series," *Math. Comput.*, Vol. 20 (3) (July), pp. 429–430.

Ruddiman, W.F. [2013], "The Anthropocene," *Annu. Rev. Earth Planet Sci.*, 41, 45–68.

Ruddiman, W.F. [2013], *Earth Transformed*. San Francisco: Freeman.

Ruelle, D., [1989], *Chaotic Evolution and Strange Attractors*. Cambridge (UK): Cambridge Univ. Press.

Ruffato, C., et al. [1986], *RMN in medicina*. Padova: Piccin.

Runge, C. [1903], "Über die Zerlegung empirisch gegebener Periodischer Funktionen in Sinusweellen," *Z. Math. Phys., 48*, 443–456.

Runge, C. [1905], "Über die Zerlegung einer empirischen Funktion in Sinusweellen," *Z. Math. Phys., 52*, 117–123.

Rybicki, G. B., & Lightman, A. P. [1979], *Radiative Processes in Astrophysics*. New York: Wiley.

Ryde, J.W. [1947], "The attenuation end radar echoes produced at centimeter wave-lengths by various meteorological phenomena," in *Meteorological Factors in Radio-Wave Propagation*. Report of a conference held 8 April 1946 by the Phys. Soc. and the Roy. Meteorol. Soc., published by the Phys. Soc., London, S.W.7, England.

Ryle, M., & Hewish, A. [1955], "The Cambridge radio telescope," *Mem. Roy. Astron. Soc., 67*, 97–154.

Ryle, M., & Hewish, A. [1960], "The synthesis of large radiotelescopes," *Mon. Not. Roy. Astron. Soc., 120*, 220–230.

Sagan, C., & Drake, F. [1975], "The search for extraterrestrial intelligence," *Scientific American, 232*(May), 80–89.

Schmidt, M., Ehrfeld, W., Feiertag, G., Lehr, H., & Schmidt, A. [1996], "Deep X-ray Lithography for Microfabrication," *Synchrotron Radiation News* 9, *No.3*, 36–41.

Schönflies, A. [1891], *Kristallsysteme und Kristallstruktur*. Leipzig: Teubner.

Schopper, H. (Ed.). [1993], *Advances of Accelerator Physics and Technologies*. River Edge, New Jersey: World Scientific.

Schott G. H. [1912], *Electromagnetic Radiation and the Mechanical Reactions Arising from It*, Cambridge University Press.

Schwartz, M., & Shaw, L. [1975], *Signal Processing*. New York: McGraw-Hill.

Schwinger, J. [1937], "On nonadiabatic processes in inhomogeneous fields," *Phys. Rev.*, *51*, 648–651.

Schwinger, J. [1946], "Electron radiation in high energy accelerators," *Phys. Rev.*, *70*, 798–799.

Scudder, H. J. [1978], "An introduction to computer aided tomography," *Proc. IEEE*, *68*, 628–637.

Shannon C. E. [1948], "A mathematical theory of communications," *Bell Syst. Tech. J.* 27, pp. 379–423.

Shepp, L. A., & Kruskal, J. B. [1978], "Computerized tomography: the new medical X-ray technology," *Amer. Math. Monthly*, *85*, 420–439.

Shepp L. A., Logan B. F. [1974], "The Fourier reconstruction of a head section," *Trans. Nucl. Sci.* NS-21, pp. 21–43.

Shirley, J.A. [1986], "Fiber Optic Raman Thermometer for Space Shuttle Main Engine Preburner Profiling," article in: *Advanced Earth-to-Orbit Propulsion Technology,* vol. I (R.J., Richmond and S.T., Wu Eds.), NASA Conference Publication CP, 2436, 107–123.

Shklovsky I. S. [1960], *Cosmic Radio Waves*, Harvard University Press.

Sica G. [1997], "Notations and interfaces for musical signal processing," article in *Musical Signal Processing*, Roads C. et al. (Eds.), Lisse: Swets & Zeitlinger

Sivers, E. A., Halloway, D. L., & Ellingson, W. A. [1993], "Obtaining high-resolution images of ceramics from 3-D x-ray microtomography by region-of-interest reconstruction," *Ceramic Eng. Sci. Proc.*, *14*(7–8), 463–472.

Skolnik, M.I. [1962], *Introduction to Radar Systems*. New York: McGraw-Hill.

Smith, K. T., & Keinert, F. [1985], "Mathematical foundations of computed tomography," Appl. *Optics*, *24*, 3950–3957.

Smullin, L.D., Fiocco, G. [1962], "Optical echoes from the Moon," *Nature*, 194, p.1267.

Sommerfeld, A. [1967], *Partial Differential Equations in Physics*. New York: Academic Press.

Srajer, V., et al. [1996], "Photolysis of the carbon monoxide complex of myoglobin: nanosecond time-resolved crystallography," *Science*, *274*, 1726–1729.

Stein, E. M., & Weiss, G. [1971], *Introduction to Fourier Analysis on Euclidean Spaces*. Princeton, New Jersey: Princeton University Press.

Stenonis N. [1669], *De Solido intra Solidum Naturaliter Contento Dissertationis Prodromus*, Florentiae.

Stent, G. S., & Calendar, R. [1978], *Molecular Genetics*. San Francisco: Freeman.

Stokes, G.G. [1857], *On the effect of wind on the intensity of sound,* Report, p. 22, Brit. Ass., Dublin; (reprinted in *Mathematical and Physical Papers of G.G. Stokes,* vol. 4, 110–111, Cambridge University Press, New York, 1904).

Stout, G. H., & Jensen, L. H. [1968], *X-ray Structure Determination*. New York: Macmillan.

Stroud, W.G., Nordberg, W., Walsh, J. [1956], "Atmospheric temperatures and winds between 30 and 80 km," *J. Geophys. Res.,* 61, 45–56.

Stryer, L. [1988], *Biochemistry*. New York: Freeman.

Stuhlinger, E., & Ordway, F. I. [1994], *Wernher von Braun*. Malabar, Florida: Krieger.

Stull, R.B. [1988], *An Introduction to Boundary Layer Meteorology*. Dordrecht, Holland: Kluver

Sullivan, W. T. [1984], *The Early Years of Radio Astronomy*. Cambridge: Cambridge University Press.

Sutton, G. P. [1992], *Rocket Propulsion Elements*. New York: Wiley.

Swords, S.S. [1986], *Technical history of the beginnings of RADAR*. London, England: Peter Peregrinus.

Tatarski, V.I. [1961], *Wave Propagation in a Turbulent Medium*. New York: McGraw-Hill.
Taylor, G.I. [1915], "Eddy motion in the atmosphere," *Phil. Trans. Roy. Soc. London,* Ser. A 215, 1–26.
Taylor, G.I. [1935], "Statistical theory of turbulence," *Proc. Roy. Soc. London,* Ser. A 151, 421–478.
Taylor, C. A., & Lipson, H. [1964], *Optical Transforms*. London: Bell.
Terras, A. [1985], *Harmonic Analysis on Symmetric Spaces and Applications*. New York: Springer-Verlag.
Titchmarsh E. [1937], *Introduction to the Theory of Fourier Integrals*. London: Oxford University Press (Clarendon).
Todd-Pokropek, A. E., et al. (Eds.). [1992], *Medical Images: Formation, Handling and Evaluation, NATO-ASI Series F (Vol. 98)*. Berlin: Springer-Verlag.
Thompson, A. R., Moran, I. M., & Sweson, W. J. [1986], *Interferometry and Synthesis in Radioastronomy*. New York: Wiley.
Tolstov, G. P. [1962], *Fourier Series*. Englewood Cliffs, New Jersey: Prentice Hall.
Tomboulian, D. H., & Hartman, P. L. [1956], "Spectral and angular distribution of ultraviolet radiation from the 300-Mev Cornell Synchrotron," *Phys. Rev.*, *102*, 1423–1447.
Treves, F. [1980], *Introduction to Pseudodifferential and Fourier Integral Operators*. New York: Plenum.
Tyndall, J. [1874], "On the atmosphere as a vehicle of sound," *Phil. Trans. Roy. Soc.* 164, 183–244.
Tyndall, J. [1875], *Sound,* 3rd ed., New York: Appleton.
Vainberg, E. I., Kazak, I. A., & Kurozaev, V. P. [1981], "Reconstruction of the internal three-dimensional structure of objects based on real time internal projections," *Soviet. J. Nondestructive Testing*, *17*, 415–423.
Vainberg, E. I., Kazak, I. A., & Faingoiz, M. L. [1985], "X-ray computerized back projection tomography with filtration by double differentiation. Procedure and information features," *Soviet J. Nondestructive Testing*, *21*, 106–113.
Vainshtein, B. K. [1981], *Modern Crystallography*. Berlin: Springer-Verlag.
Valley, S.L. (Ed.) [1965], *Handbook of Geophysics and Space Environments*. New York: McGraw-Hill.
van Bergeijk, W. A., Pierce, J. R., & Jr, David E. E. [1959], *Waves and the Ear*. New York: Doubleday.
Van der Hulst, H.C. [1957], *Light Scattering by Small Particles*. New York: Wiley.
Verschuur, G. L., & Kellermann, K. I. [1988], *Galactic and Extragalactic Radioastronomy*. Berlin: Springer-Verlag.
Vitagliano, L., Berisio, R., Mazzarella, L., & Zagari, A. [2001], "Structural bases of collagen stabilization induced by proline hydroxylation," *Biopolymers*, *58*(5), 459–464.
Vogl, T. J., et al. [1995], "Malignant liver tumors treated with MR imagingguided laser-induced thermotherapy: technique and prospective results," *Radiology*, *196*, 257–265.
Walker, J. S. [1988], *Fourier Analysis*. Oxford: Oxford University Press.
Walls, D. F., & Milburn, G. J. [1994], *Quantum Optics*. Berlin: Springer-Verlag.
Watson, G. N. [1944], *Theory of Bessel Functions* (2nd ed.). Cambridge: Cambridge University Press.
Watson, J. D. [1968], *The Double Helix*. New York: Atheneum.
Watson-Watt, R.A. [1957], "Three steps to victory," Long Acre London, England: Odhams Press.
Watson, J. D., & Crick, F. H. C. [1953a], "A structure for deoxyribosenucleic acid," Nature, *171*, 737.
Watson, J. D., & Crick, F. H. C. [1953b], "Genetical implications of the structure of deoxyribonucleic acid," *Nature*, *171*, 964.
Watson, J. D., et al. [1993], *Molecular Biology of the Gene*. Menlo Park, California: Benjamin/Cummings.
Weitkamp, C. (Ed.) [2005], *Lidar: Range-Resolved Optical Remote Sensing of the Atmosphere,* Springer Series in Optical Sciences, vol. 102, New York: Springer.
Wells, A. F. [1977], *Three-Dimensional Nets and Polyhedra*. New York: Wiley.

Werner, C. [2005], "Doppler wind lidar," article in *Lidar: Range-Resolved Optical Remote Sensing of the Atmosphere,* Springer Series in Optical Sciences, vol. 102, 325–354.

Weyl, H. [1952], *Symmetry*. Princeton: Princeton University Press.

Wheatley, P. J. [1981], *The Determination of Molecular Structure*. New York: Dover.

Whittaker, E. T. [1915], "On the functions which are represented by expansions of the interpolation-theory," *Proc. Royal Soc. Edinburgh, 35*, 181–194.

Whittaker, E. T. [1951], *A History of the Theories of the Aether and Electricity*. London: Nelson and Sons.

Whittaker, E. T., & Watson, G. N. [1946], *A Course of Modern Analysis* (4th ed.). Cambridge: Cambridge University Press.

Wiener, N. [1959], *[1933], The Fourier Integral and Certain of Its Applications, Cambridge University Press*. New York: London. Reprinted by Dover.

Wigner, E. P. [1970], *Symmetries and Reflections*. Bloomington, Indiana: Indiana University Press.

Wilbraham H. [1848], "On a certain periodic function," *Cambridge and Dublin Math. J.* 3, pp. 198–201.

Wilkins, M. H., Stokes, A. R., & Wilson, H. R. [1953], "A structure for deoxyribose nucleic acid," *Nature, 171*, 738–740.

Winckel, F. [1967], *Music*. Dover, New York: Sound and Sensation.

Winick H. [1987], "Synchrotron radiation," *Scientific American*, (November), pp. 88–99.

Winick, H. (Ed.). [1994], *Synchrotron Radiation Sources*. River Edge, New Jersey: World Scientific.

Woolfson, M. M. [1970], *An Introduction to X-Ray Crystallography*. Cambridge: Cambridge University Press.

Woolfson, M. M. [1987], "Direct methods from birth to maturity," Acta Cryst. *A*, *43*, 593–612.

Young, T. [1801], "On the theory of light and colours," *Phil. Trans. Roy. Soc.*, *92*, 12–48.

Young, T. [1802], "An account of some cases of the production of colours," *Phil. Trans. Roy. Soc.*, *92*, 387–397.

Zahn, R., Liu, A., Luhrs, T., Riek, R., von Schrûtter, C., Lopez, Garcia F., et al. [2000], "NMR solution structure of the human prion protein," Proc. Natl. Acad. Sci. U.S. *A.*, *97*, 145–150.

Zanotti C. et al. [1992], "Self-sustained oscillatory burning of solid propellents: experimental results," *Progress in Astronautics and Aeronautics*, Vol. 143, pp. 399–439, New York: American Institute of Aeronautics and Astronautics.

Zanotti, C., & Giuliani, P. [1993], "Experimental and numerical approach to the study of frequency response of solid propellants," IV International Symposium on Flame Structure, August 1992. *Novosibirsk, Siberia, published in Fizika Goreniya I Vzryva*, *3*, 36–41.

Zuegg, J., & Gready, J. E. [1999], "Molecular dynamics simulations of human prion protein: importance of correct treatment of electrostatic interactions," *Biochemistry*, *38*, 13862–13876.

Zuegg, J., & Gready, J. E. [2000], "Molecular dynamics simulation of human prion protein including both N-linked oligosaccharides and the GPI-anchor," *Glycobiology*, *10*, 959–974.

Zygmund, A. [1968], *Trigonometric Series* (2nd ed.). Cambridge, Massachusetts: Cambridge University Press.

Index

E. Prestini, *The Evolution of Applied Harmonic Analysis*,
Applied and Numerical Harmonic Analysis,
DOI 10.1007/978-1-4899-7989-6

Applied and Numerical Harmonic Analysis (74 volumes)

A. Saichev and W.A. Woyczynski: *Distributions in the Physical and Engineering Sciences* (ISBN 978-0-8176-3924-2)
C.E. D'Attellis and E.M. Fernandez-Berdaguer: *Wavelet Theory and Harmonic Analysis in Applied Sciences* (ISBN 978-0-8176-3953-2)
H.G. Feichtinger and T. Strohmer: *Gabor Analysis and Algorithms* (ISBN 978-0-8176-3959-4)
R. Tolimieri and M. An: *Time-Frequency Representations* (ISBN 978-0-8176-3918-1)
T.M. Peters and J.C. Williams: *The Fourier Transform in Biomedical Engineering* (ISBN 978-0-8176-3941-9)
G.T. Herman: *Geometry of Digital Spaces* (ISBN 978-0-8176-3897-9)
A. Teolis: *Computational Signal Processing with Wavelets* (ISBN 978-0-8176-3909-9)
J. Ramanathan: *Methods of Applied Fourier Analysis* (ISBN 978-0-8176-3963-1)
J.M. Cooper: *Introduction to Partial Differential Equations with MATLAB* (ISBN 978-0-8176-3967-9)
A. Procházka, N.G. Kingsbury, P.J. Payner, and J. Uhlir: *Signal Analysis and Prediction* (ISBN 978-0-8176-4042-2)
W. Bray and C. Stanojevic: *Analysis of Divergence* (ISBN 978-1-4612-7467-4)
G.T. Herman and A. Kuba: *Discrete Tomography* (ISBN 978-0-8176-4101-6)
K. Gröchenig: *Foundations of Time-Frequency Analysis* (ISBN 978-0-8176-4022-4)
L. Debnath: *Wavelet Transforms and Time-Frequency Signal Analysis* (ISBN 978-0-8176-4104-7)
J.J. Benedetto and P.J.S.G. Ferreira: *Modern Sampling Theory* (ISBN 978-0-8176-4023-1)
D.F. Walnut: *An Introduction to Wavelet Analysis* (ISBN 978-0-8176-3962-4)
A. Abbate, C. DeCusatis, and P.K. Das: *Wavelets and Subbands* (ISBN 978-0-8176-4136-8)

E. Prestini, *The Evolution of Applied Harmonic Analysis*,
Applied and Numerical Harmonic Analysis,
DOI 10.1007/978-1-4899-7989-6

O. Bratteli, P. Jorgensen, and B. Treadway: *Wavelets Through a Looking Glass* (ISBN 978-0-8176-4280-80)
H.G. Feichtinger and T. Strohmer: *Advances in Gabor Analysis* (ISBN 978-0-8176-4239-6)
O. Christensen: *An Introduction to Frames and Riesz Bases* (ISBN 978-0-8176-4295-2)
L. Debnath: *Wavelets and Signal Processing* (ISBN 978-0-8176-4235-8)
G. Bi and Y. Zeng: *Transforms and Fast Algorithms for Signal Analysis and Representations* (ISBN 978-0-8176-4279-2)
J.H. Davis: *Methods of Applied Mathematics with a MATLAB Overview* (ISBN 978-0-8176-4331-7)
J.J. Benedetto and A.I. Zayed: *Modern Sampling Theory* (ISBN 978-0-8176-4023-1)
E. Prestini: *The Evolution of Applied Harmonic Analysis* (ISBN 978-0-8176-4125-2)
L. Brandolini, L. Colzani, A. Iosevich, and G. Travaglini: *Fourier Analysis and Convexity* (ISBN 978-0-8176-3263-2)
W. Freeden and V. Michel: *Multiscale Potential Theory* (ISBN 978-0-8176-4105-4)
O. Christensen and K.L. Christensen: *Approximation Theory* (ISBN 978-0-8176-3600-5)
O. Calin and D.-C. Chang: *Geometric Mechanics on Riemannian Manifolds* (ISBN 978-0-8176-4354-6)
J.A. Hogan: *Time-Frequency and Time-Scale Methods* (ISBN 978-0-8176-4276-1)
C. Heil: *Harmonic Analysis and Applications* (ISBN 978-0-8176-3778-1)
K. Borre, D.M. Akos, N. Bertelsen, P. Rinder, and S.H. Jensen: *A Software-Defined GPS and Galileo Receiver* (ISBN 978-0-8176-4390-4)
T. Qian, M.I. Vai, and Y. Xu: *Wavelet Analysis and Applications* (ISBN 978-3-7643-7777-9)
G.T. Herman and A. Kuba: *Advances in Discrete Tomography and Its Applications* (ISBN 978-0-8176-3614-2)
M.C. Fu, R.A. Jarrow, J.-Y. Yen, and R.J. Elliott: *Advances in Mathematical Finance* (ISBN 978-0-8176-4544-1)
O. Christensen: *Frames and Bases* (ISBN 978-0-8176-4677-6)
P.E.T. Jorgensen, J.D. Merrill, and J.A. Packer: *Representations, Wavelets, and Frames* (ISBN 978-0-8176-4682-0)
M. An, A.K. Brodzik, and R. Tolimieri: *Ideal Sequence Design in Time-Frequency Space* (ISBN 978-0-8176-4737-7)
S.G. Krantz: *Explorations in Harmonic Analysis* (ISBN 978-0-8176-4668-4)
B. Luong: *Fourier Analysis on Finite Abelian Groups* (ISBN 978-0-8176-4915-9)
G.S. Chirikjian: *Stochastic Models, Information Theory, and Lie Groups, Volume 1* (ISBN 978-0-8176-4802-2)
C. Cabrelli and J.L. Torrea: *Recent Developments in Real and Harmonic Analysis* (ISBN 978-0-8176-4531-1)

M.V. Wickerhauser: *Mathematics for Multimedia* (ISBN 978-0-8176-4879-4)
B. Forster, P. Massopust, O. Christensen, K. Gröchenig, D. Labate, P. Vandergheynst, G. Weiss, and Y. Wiaux: *Four Short Courses on Harmonic Analysis* (ISBN 978-0-8176-4890-9)
O. Christensen: *Functions, Spaces, and Expansions* (ISBN 978-0-8176-4979-1)
J. Barral and S. Seuret: *Recent Developments in Fractals and Related Fields* (ISBN 978-0-8176-4887-9)
O. Calin, D.-C. Chang, and K. Furutani, and C. Iwasaki: *Heat Kernels for Elliptic and Sub-elliptic Operators* (ISBN 978-0-8176-4994-4)
C. Heil: *A Basis Theory Primer* (ISBN 978-0-8176-4686-8)
J.R. Klauder: *A Modern Approach to Functional Integration* (ISBN 978-0-8176-4790-2)
J. Cohen and A.I. Zayed: *Wavelets and Multiscale Analysis* (ISBN 978-0-8176-8094-7)
D. Joyner and J.-L. Kim: *Selected Unsolved Problems in Coding Theory* (ISBN 978-0-8176-8255-2)
G.S. Chirikjian: *Stochastic Models, Information Theory, and Lie Groups, Volume 2* (ISBN 978-0-8176-4943-2)
J.A. Hogan and J.D. Lakey: *Duration and Bandwidth Limiting* (ISBN 978-0-8176-8306-1)
G. Kutyniok and D. Labate: *Shearlets* (ISBN 978-0-8176-8315-3)
P.G. Casazza and P. Kutyniok: *Finite Frames* (ISBN 978-0-8176-8372-6)
V. Michel: *Lectures on Constructive Approximation* (ISBN 978-0-8176-8402-0)
D. Mitrea, I. Mitrea, M. Mitrea, and S. Monniaux: *Groupoid Metrization Theory* (ISBN 978-0-8176-8396-2)
T.D. Andrews, R. Balan, J.J. Benedetto, W. Czaja, and K.A. Okoudjou: *Excursions in Harmonic Analysis, Volume 1* (ISBN 978-0-8176-8375-7)
T.D. Andrews, R. Balan, J.J. Benedetto, W. Czaja, and K.A. Okoudjou: *Excursions in Harmonic Analysis, Volume 2* (ISBN 978-0-8176-8378-8)
D.V. Cruz-Uribe and A. Fiorenza: *Variable Lebesgue Spaces* (ISBN 978-3-0348-0547-6)
W. Freeden and M. Gutting: *Special Functions of Mathematical (Geo-)Physics* (ISBN 978-3-0348-0562-9)
A. Saichev and W.A. Woyczynski: *Distributions in the Physical and Engineering Sciences, Volume 2: Linear and Nonlinear Dynamics of Continuous Media* (ISBN 978-0-8176-3942-6)
S. Foucart and H. Rauhut: *A Mathematical Introduction to Compressive Sensing* (ISBN 978-0-8176-4947-0)
G. Herman and J. Frank: *Computational Methods for Three-Dimensional Microscopy Reconstruction* (ISBN 978-1-4614-9520-8)
A. Paprotny and M. Thess: *Realtime Data Mining: Self-Learning Techniques for Recommendation Engines* (ISBN 978-3-319-01320-6)
A. Zayed and G. Schmeisser: *New Perspectives on Approximation and Sampling Theory: Festschrift in Honor of Paul Butzer's 85th Birthday* (978-3-319-08800-6)

R. Balan, M. Begue, J. Benedetto, W. Czaja, and K.A Okoudjou: *Excursions in Harmonic Analysis, Volume 3* (ISBN 978-3-319-13229-7)
H. Boche, R. Calderbank, G. Kutyniok, J. Vybiral: *Compressed Sensing and its Applications* (ISBN 978-3-319-16041-2)
S. Dahlke, F. De Mari, P. Grohs, and D. Labate: *Harmonic and Applied Analysis: From Groups to Signals* (ISBN 978-3-319-18862-1)
G. Pfander: *Sampling Theory, a Renaissance* (ISBN 978-3-319-19748-7)
R. Balan, M. Begue, J. Benedetto, W. Czaja, and K.A Okoudjou: *Excursions in Harmonic Analysis, Volume 4* (ISBN 978-3-319-20187-0)
O. Christensen: *An Introduction to Frames and Riesz Bases, Second Edition* (ISBN 978-3-319-25611-5)
J.H. Davis: *Methods of Applied Mathematics with a Software Overview, Second Edition* (ISBN 978-3-319-43369-1)
E. Prestini: *The Evolution of Applied Harmonic Analysis: Models of the Real World, Second Edition* (ISBN 978-1-4899-7987-2)

For an up-to-date list of ANHA titles, please visit http://www.springer.com/series/4968

Printed in the United States
By Bookmasters